EXERCISES IN CHEMICAL PHYSICS

Exercises in Chemical Physics

J. R RITER Jr.

Chemistry Department, University of Denver

[redacted]

GORDON AND BREACH SCIENCE PUBLISHERS
New York London Paris

[redacted]

Gordon and Breach, Science Publishers, Inc.
440 Park Avenue South
New York, N.Y. 10016

Editorial office for the United Kingdom

Gordon and Breach, Science Publishers Ltd.
42 William IV Street
London W.C.2

Editorial office for France

Gordon & Breach
7 - 9 rue Emile Dubois
Paris 14^e

[redacted]

Library of Congress Catalog Card Number 72 - 92329
ISBN 0 677 02350 2. Printed in Great Britain.

[redacted]

PREFACE

It has been my observation that graduate students in the areas of physical and inorganic chemistry tend to become acquainted with the literature rather later than their counterparts in organic chemistry. In an effort to give those in my upper-division and graduate courses at least a slight introduction to this part of our scientific heritage I had from time to time constructed exercises based upon articles which had been published in The Journal of Chemical Physics. This book is a consequence of that activity, containing some 165 exercises drawn from articles appearing between 1952 and 1970; the greater number of them since 1963. My restriction to this source alone is solely a decision of convenience; it is obvious that there are other excellent journals reporting results of research in chemical physics.

The book is designed for graduate students wishing to see in what ways the concepts that they have studied are applied and extended in modern research. It may be that the undergraduate who is finishing a course in physical chemistry will find something of interest as well. While teachers soon learn that often students will tell them what they think their mentors might like to hear, I do believe that on occasion students do enjoy wrestling with the original literature. In the majority of the exercises most of the points raised can be dealt with without consulting that issue of the Journal; it is still hoped that curiosity or some other motive may bring the student to the library shelves.

It is hoped that the articles and exercises will help to acquaint the student with that part of research that can be transmitted through print. It is certainly recognized that some aspects of research cannot be experienced vicariously; walking up- or downstairs to obtain liquid nitrogen, rebuilding a vacuum line after one of the departmental theorists has enthusiastically described his new wavefunction with outstretched arms, and finally observing early one morning the long-sought spectral transition. These things must all be learned by doing.

A science that studies the materials of chemistry with the methods of physics presents us with a few problems in taxonomy. The first four chapters touch into what I have chosen as basic areas. The next pair, on solids and fluids, are generally concerned more with properties than with the nonspectroscopic methods that were usually employed. Chapter VII gives examples from the realm of impact phenomena including mass spectrometry.

Five types of spectroscopy form individual bases for successive chapters, concluded with one sampling other types of spectroscopy and structural determination methods. While several of the exercises could be slipped into two or more chapters, it is hoped that the overall scheme will meet the needs of the user.

Within the chapters I have attempted to introduce some variety in terms of length, difficulty, and subject matter.

There is an author index; the authors did the work. There is no subject index as it is felt that the Table of Contents by article title suffices. You might be reminded of the ruse used by T.E. Lawrence in his magnum opus: "Who would insult his 'Decline and Fall' by consulting it just upon a specific point?"

I am grateful to Academic Press for permission to reproduce Fig. 1 on p. 51, and to the American Institute of Physics, publisher of The Journal of Chemical Physics.

Most of all I am indebted to the some 350 authors upon whose articles this reader is based, and whose original research serves as an excellent example of what it is all about.

J.R.R., Jr.

Denver, Colorado

Thanks,

to C.M. Appleberry, who shared his enthusiasm for the subject with me some time ago, and

to WOCP, who have helped maintain my own since

TABLE OF CONTENTS

CHAPTER II QUANTUM MECHANICS

CHAPTER III STATISTICAL MECHANICS

CHAPTER VII MASS SPECTROMETRY AND IMPACT PHENOMENA

CHAPTER VIII INFRARED SPECTROSCOPY

NOTES

p. 16 — Fig. 5 as shown is a more correct version than that from the original article, and is included here by the courtesy of Dr. J. Berkowitz.

p. 21 — Fig. 1 and 2 are reproduced by permission of Los Alamos Scientific Laboratory.

pp. 46-48 — The CO wavefunction of Sahni that is given is now obsolete, though illustrative for our purposes.

p. 150 — Fig. 1 is due to B. Kamb, J.Chem.Phys. 43, 3918 (1965).

p. 200 — More accurate absorption contours than given in Fig. 1 are found in S.A. Clough, B. Schurin, and F.X. Kneizys, J.Chem.Phys. 43, 3410 (1965).

p.221 — Table I has been superseded in part by the Raman spectrum of ReF_7 vapor; H.H. Claassen, E.L. Gasner, and H. Selig, J.Chem.Phys. 49, 1803 (1968), Table IV.

pp. 275-7 — Higher-resolution NMR spectra have subsequently shown the belief that $|J_A - J_E| < 1$ cps to be in error: L.H. Sutcliffe and S.M. Walker, J.Phys.Chem. 71, 1555 (1967).

CHAPTER I

THERMODYNAMICS

1. "Measurement of the Heat of Dissociation of Fluorine by the Effusion Method" H. Wise 20, 927 (1952).

The mass flow $\dot{m}$ of a partially dissociated diatomic gas through an orifice of cross-sectional area A is given by

$$\dot{m} = PA[M/2\pi RT]^{\frac{1}{2}}[(1+0.414\alpha)/(1+\alpha)]$$

with P the total pressure, M the molecular weight of the diatomic, T the temperature, and α the degree of dissociation.

Can you derive this result from the effusion equation?

By determining α and P, at various values of $\dot{m}$, one has the equilibrium constant as a function of temperature as shown in Fig. 1.

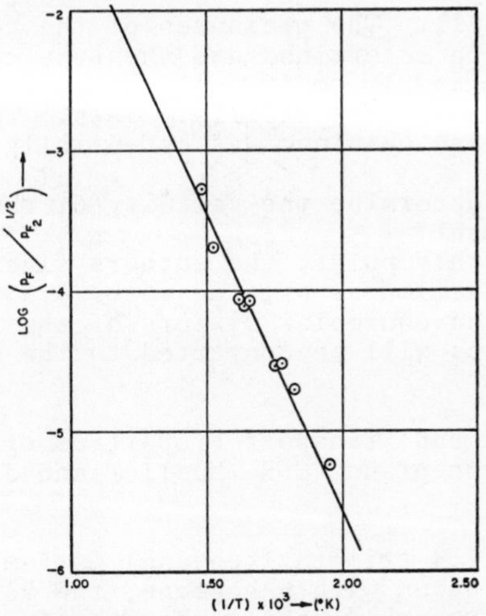

FIG. 1. Equilibrium constant for $\frac{1}{2}F_2(g) = F(g)$ as a function of temperature.

From the slope of this curve, using the Gibbs-Helmholtz equation, you should be able to match his value of 39.9±0.8 kcal/mole for ΔH°_{298} for the reaction $F_2(g) \rightarrow 2F(g)$.

Are these results precise enough to determine the temperature dependence of the dissociation energy?

Compute $\int \Delta C_p dT$ between the highest and lowest temperatures; is it within the overall experimental error?

2. "Determination of the Equilibrium Constant of the Reaction between BF_3 and BCl_3" T.H.S. Higgins, E.C. Leisegang, C.J.G. Raw, and A.J. Rossouw 23, 1544 (1955).

The extent of the reaction $BF_3 + BCl_3 \rightarrow BFCl_2 + BF_2Cl$ in the gas phase was followed by monitoring the intensity of the infrared absorption band at 955 cm^{-1}, due to BCl_3. The apparatus used is shown in Fig. 1.

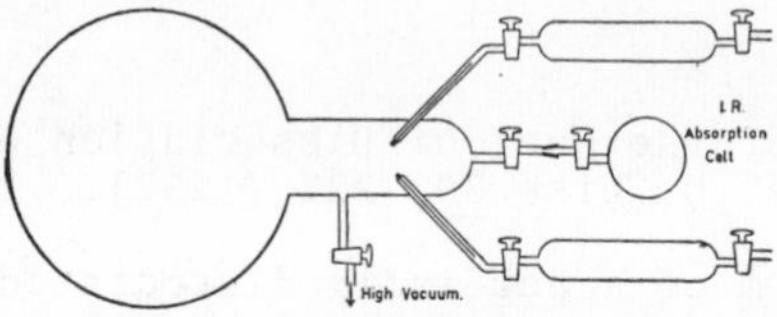

FIG. 1. Diagram of reaction vessel.

The Beer-Lambert law in the form

$$\log(I_0/I) = E\ell c$$

was assumed, with I_0 and I the incident and transmitted intensities respectively, E the extinction coefficient, ℓ the path length, and c the concentration. The volume of the small bulbs was known as were the volumes of the large reaction vessel and the IR absorption cell. The pressures of the reactants in their individual bulbs were determined and then the total pressure in the reaction vessel and 10-cm IR cell was known after expansion of the reactants through the capillary jets. The temperature at which the spectra were obtained was essentially constant at 27-29°C.

How would one determine the validity of the Beer-Lambert law in this experiment?

Upon checking this point, the authors find the equilibrium constant for the reaction as written to be 0.53±0.04.

Starting with an equimolar mixture of the reactants, what fraction of the gases will be converted to the mixed halides?

3. "Thermodynamic and Transport Properties of the System Iso-octane and Perfluoroheptane" C.R. Mueller and J.E. Lewis 26, 286 (1957).

Iso-octane (2,2,4-trimethylpentane) is familiar as the component, in a solution with n-heptane, the volume % of which gives the "octane number" directly when the fuel of this composition matches the knock characteristics of a test gasoline when both are run consecutively in a standard test engine.

Perfluoro-n-heptane was chosen as the companion liquid in this study of solution properties because the solution then does not exhibit hydrogen bonding, is an irregular solution, and has a critical solution temperature of 23.7°C. This last property permitted measurements to be carried out at or near room temperature and still be in the vicinity of the critical solution temperature.

The authors chose Guggenheim's form of the virial equation for calculating the fugacity f_A or f_B:

$$f_A = (1-x)P \exp\{[P/RT][B_{AA}+x^2(2B_{AB}-B_{AA}-B_{BB})]\}$$

$$f_B = xP \exp\{[P/RT][B_{BB}+(1-x)^2(2B_{AB}-B_{AA}-B_{BB})]\},$$

where the subscripts A and B denote iso-octane and perfluoroheptane respectively, P is the total vapor pressure, and the B's are the virial coefficients with B_{AB} that for the mixture. The vapor pressure data at 50°C are given in Table III.

TABLE III. Vapor pressure data at 50.000°C.

X_A	X_A'	P	f_A	f_B	ΔF^M
0	0	236.72	0	231.79	0
0.0560	0.1455	272.40	39.12	227.26	−59.04
0.1680	0.2940	303.91	87.74	209.11	−109.08
0.2555	0.3495	313.26	107.33	198.56	−123.17
0.2850	0.3615	313.66	111.12	195.16	−127.47
0.3760	0.3910	315.36	120.76	187.19	−129.52
0.4270	0.3997	315.05	123.31	184.36	−128.35
0.6450	0.4240	311.00	129.10	174.74	−112.03
0.7990	0.4540	301.88	134.20	160.96	−86.18
0.8258	0.4755	293.62	136.73	150.53	−84.49
0.8718	0.5015	284.23	139.64	138.65	−62.70
0.9085	0.5435	270.92	144.31	121.21	−40.15
0.9870	0.8790	174.81	151.42	20.97	−8.22
1	1	146.63	144.82	0	0

By making use of a square-well potential and a modified hole theory of liquids for the A-A, B-B, and A-B interactions they have obtained the virial coefficients as given in Table V(units of volume, here cm³/mole).

TABLE V. Virial coefficients.

	30°C	50°C	70°C
B_{AA}	−1970.47	−1705.23	−1494.47
B_{BB}	−2057.94	−1792.41	−1577.05
B_{AB}	−1429.11	−1247.64	−1100.52

By setting the mole fraction of perfluoroheptane x equal to one in the second equation above one may solve for the fugacity of pure B, f_B. Similarly with x=0 the first equation gives f_A°. Recall the activity of component i is given as $a_i \equiv f_i/f_i^\circ$ and γ_i, corresponding activity coefficient on a mole fraction basis, is then $\gamma_i = a_i/x_i$.

Making use of Tables III and V, can you compute activity coefficients for both components at 50°C?

The last column of Table III contains the free energy of mixing

$$\Delta F^M = \sum_i x_i RT \ln a_i$$

as can be verified from this equation. It can be rewritten in the form

$$\Delta F^M = \sum_i RT(x_i \ln x_i + x_i \ln \gamma_i)$$

where the first term is the free energy of mixing for ideal solutions (those that obey Raoult's law) and the second term corrects for the deviations from ideality; this is called the excess free energy of mixing

$$F^E = \sum_i x_i RT \ln \gamma_i,$$

usually a basic concept in solution theory. The Gibbs-Helmholtz equation in the form

$$H^E = -T^2 \frac{\partial (F^E/T)}{\partial T},$$

with pressure held constant, gives the excess enthalpy of mixing which is identical with the heat of mixing. This is true because the heat of mixing, as well as the volume change upon mixing, of an ideal solution are both zero. The restriction to constant pressure is not a very significant one in most cases as activity coefficients would be expected to change very little with the relatively small changes in vapor pressures with both composition and temperature (30-70°C) changes. The excess entropy is

$$S^E = -\frac{\partial F^E}{\partial T}$$

so that all three excess functions can be determined from a knowledge of the activity coefficients and their temperature derivatives. These functions for three temperatures are contained in Tables VI-VIII.

TABLE VI. Thermodynamic data at 30.000°C.

Mole fraction iso-octane	H^E	F^E	TS^E
0	0	0	0
0.10	231	128	103
0.20	384	215	169
0.30	412	277	135
0.40	416	315	101
0.50	402	329	73
0.60	367	320	47
0.70	314	290	24
0.80	246	234	12
0.90	164	140	24
0.95	93	76	17
1.00	0	0	0

TABLE VII. Thermodynamic data at 50.000°C.

Mole fraction iso-octane	F^E	H^E	TS^E
0	0	0	0
0.10	125	209	84
0.20	207	341	134
0.30	262	403	141
0.40	301	449	148
0.50	319	468	149
0.60	312	439	128
0.70	284	405	121
0.80	236	356	120
0.90	160	275	115
1.00	0	0	0

TABLE VIII. Thermodynamic data at 70.000°C.

Mole fraction iso-octane	F^E	H^E	TS^E
0	0	0	0
0.10	125	221	96
0.20	194	375	181
0.30	250	512	262
0.40	293	587	294
0.50	306	616	310
0.60	300	567	267
0.70	280	518	238
0.80	236	450	214
0.90	168	294	126
1.00	0	0	0

The units of these entries are cal/mole. The unit of pressure in Table III is mm as was probably obvious.

Abbreviated portions of Tables II and IV are presented below. Now activity coefficients for both components can be determined at each of the three temperatures, and hence the required derivatives $\partial \ln \gamma_i/\partial T$ for evaluation of the excess enthalpy and entropy as a function of x_A at each temperature as well.

Do this with the end in mind of evaluating F^E, H^E, and S^E for one or two compositions at 50° and compare your results with those of Table VII and Fig. 2.

TABLE II. Vapor pressure data at 30.000°C.

X_A	X_A'	P	f_A	f_B	ΔF^M
0	0	98.11	0	97.07	0
1	1	62.51	62.10	0	0

TABLE IV. Vapor pressure data at 70.000°C.

X_A	X_A'	P	f_A	f_B	ΔF^M
0	0	503.68	0	485.33	0
1	1	306.16	299.69	0	0

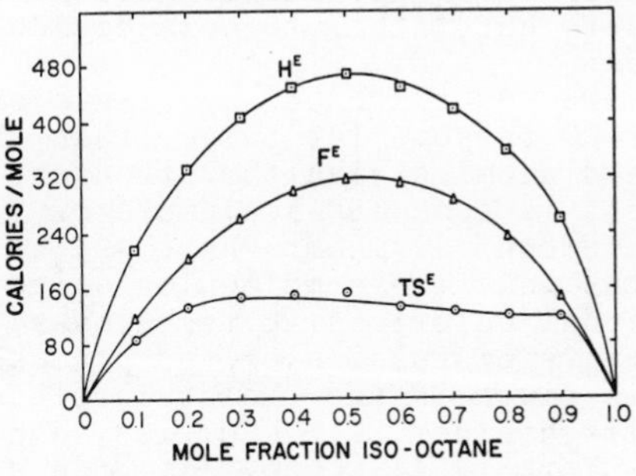

FIG. 2. Free energy, entropy, heat of mixing, 50°C.

At 30° for x_A=0.507 the heat of mixing was determined directly in a calorimeter as 421 cal/mole. Compute the error for the calculated value. Would you consider this good agreement?

The traditional plot of pressures vs x_A is shown in Fig. 5, with the large positive deviation from Raoult's law. Is the sign of the heat of mixing correct? What should the sign of the volume change upon mixing be?

Which of the two upper curves represents the total vapor pressure above the solution? A complete Table IV would answer this of course; still one should be able to determine this from the plot. Can you puzzle out what the other upper curve represents?

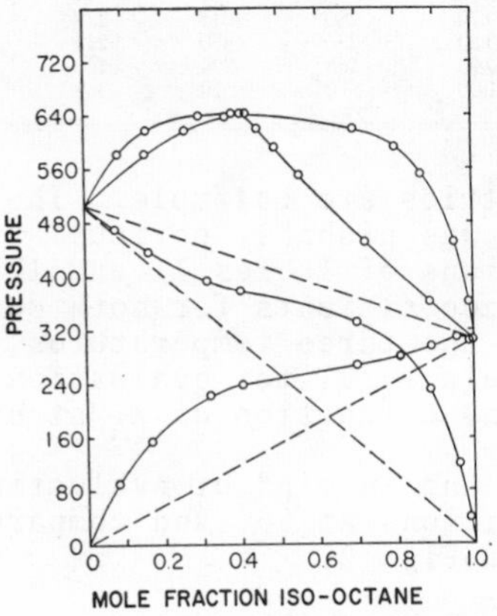

FIG. 5. Vapor pressure *vs* composition, 70°C.

4. "Dissociation Energies of Carbon Monoxide and Nitrogen from Reflected Shock Wave Studies" J.P. Toennies and E.F. Greene 26, 655 (1957).

The dissociation energies of CO and N_2 are now well-established as 256 and 225 kcal/mole respectively, but at the time this article was written there still remained some controversy over the spectroscopic and thermochemical data, particularly for CO. These shock tube experiments were not designed to give quantitative results but rather to help decide between a pair of proposed values.

In passing, are you aware of any simple explanation as to why the value for CO is some 14% larger than that of isoelectronic N_2? These diatomics have the strongest bonds of any; the third strongest is some 180 kcal/mole or less; see D.W. Setser and D.H. Stedman, J. Chem. Phys. 49, 467 (1968). The high bond strengths make these molecules quite difficult to study by conventional calorimetric techniques.

An authoritative discussion of the processes taking place in a shock tube is given in the "Molecular Theory of Gases and Liquids", J.O. Hirschfelder, C.F. Curtiss, and R.B. Bird (John Wiley & Sons, New York 1954), Chap. 11. See also the present authors' ref. 20 and 21.

They write the three conservation equations (mass, momentum, and energy) and the equation of state for a partially dissociated ideal diatomic gas for one-dimensional steady flow neglecting heat conduction and viscosity losses, in the form

$$\rho_1 u_1 = \rho_2 u_2$$

$$P_1+\rho_1 u_1^2 = P_2+\rho_2 u_2^2$$

$$\tfrac{1}{2}u_1^2+E_1+P_1/\rho_1 = \tfrac{1}{2}u_2^2+E_2+P_2/\rho_2$$

$$P_i/\rho_i = (1+\alpha_i)RT_i/M$$

in a coordinate system moving with the shock. ρ is the density, u the flow velocity, P the pressure, E the energy per unit mass, and α the degree of dissociation. The subscripts 1 and 2 refer, respectively, to the gas before and after its passage through the shock front. The molecular weight M was omitted in the article.

It might be useful to verify that the equations are all dimensionally correct; in the cgs system the gas constant R has the value 8.314×10^7 erg/mole.

This simplified treatment regards the shock wave as an infinitely thin mathematical discontinuity across which the conservation equations must apply.

Discuss the first equation in terms of the general result for mass flow $\dot{m}=\rho Au$ with A the cross-sectional area normal to the flow direction; can A vary for one-dimensional flow?

The second, or momentum conservation, equation is also called the equation of motion. Let the +x direction be that of flow and consider an element of compressible fluid of mass ρAdx. Neglecting friction and gravity the net force on the fluid element is $-AdP=\rho Adx \cdot du/dt=\rho Audu$. This would be as far as we could go were it not for the first equation which tells us that ρu is a constant and thus can be brought outside the integral sign. Complete the derivation.

The energy(per unit mass) equation is almost axiomatic; the p/ρ or flow work term is analyzed in discussions of the Joule-Thomson effect, for example.

Is the $1+\alpha$ term in the correct place in the equation of state? For fixed mass at constant temperature and pressure what happens to the density as dissociation takes place?

By knowing the incident shock Mach number (ratio of incident shock velocity to sound velocity in the undisturbed gas $=(\gamma RT_1/M)^{\frac{1}{2}}$ with $\gamma=C_P/C_V$), P_1 and T_1 we are left with four equations in as many unknowns; these are ρ_2, u_2, P_2 and T_2. The Mach number gives us u_1 in our coordinate system. E_2-E_1 is an explicit function of T_2-T_1. α is a function of T, P and the assumed dissociation energy.

Once the properties of the incident shock are known it is possible to calculate the properties of the reflected shock. The end plate of the shock tube requires the wave to "turn around" and pass through the shocked region again. As a result of this second passage the gas can be heated to incandescence. Several ionization probes located at various distances from the end plate triggered a drum camera which repeatedly photographed the reflected shock and also provided measuring intervals. In some of the tubes used these functions were performed by spaced photomultiplier tubes, responding to the luminous incident shock.

A time-distance diagram is given in Fig. 1. Region 5 is that of the gas heated by the reflected shock. It is this area that is heated to incandescence and provides the bright region in the drum camera photograph of Fig. 6.

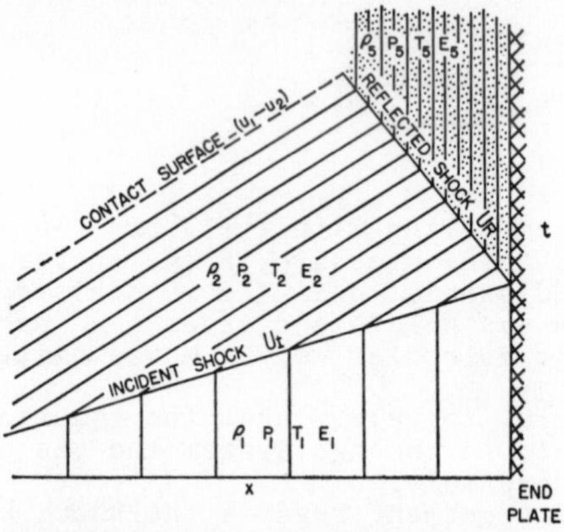

FIG. 1. Time-distance diagram for shock tube flow before and after reflection.

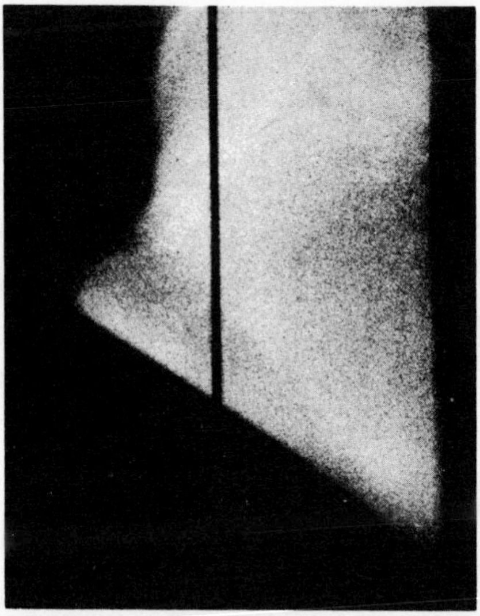

FIG. 6. Drum camera photograph of reflected shock in carbon monoxide. Incident Mach number 7.05, $P_1 = 5$ mm. The end plate is at the right and time increases upward. Reflected velocity $= 0.68$ mm/μsec.

The slope of the reflected shock front in Fig. 6 gives its velocity directly as the time derivative of distance. Verify from the curvature of the shock front that at least in this instance the reflected shock is accelerating away from the end plate. This is attributed to nonattainment of the steady state condition; the excited rotational, vibrational and electronic degrees of freedom relax and equilibrate with the translational degrees of freedom which manifest themselves as increased velocity imparted to the reflected shock. Soon it reaches a constant velocity some distance from the end plate and the steady state is attained. This relaxation-acceleration effect close to the end plate where the time interval between the incident and reflected shocks is too short to allow for equilibration of the degrees of freedom is an interesting one and can be studied also in Fig. 1 of the authors' ref. 15.

The reflected shock velocity U_R equals $-u_5$ in the coordinate system moving with the reflected shock. The conservation equations connecting region 2 and 5 are

$$\rho_5 u_5 = \rho_2(u_1 - u_2 + u_5)$$

$$P_5 + \rho_5 u_5^2 = P_2 + \rho_2(u_1 - u_2 + u_5)^2$$

$$\tfrac{1}{2}u_5^2 + E_5 + P_5/\rho_5 = \tfrac{1}{2}(u_1 - u_2 + u_5)^2 + E_2 + P_2/\rho_2.$$

The calculated properties in region 2 can now be used to calculate those in region 5 including U_R. The eight curves in Fig. 2 and 3 show this for N_2 and CO as functions of Mach number, P_1 and T_1 (room temperature), when dissociation is considered with two pairs of the dissociation energies.

Let's check one of the computed values of U_R. To make the calculation as easy as possible choose the extreme lower left point of the CO plot, i.e. Mach number of 4.0 and U_R of 0.45 mm/μsec. This will insure that dissociation and excitation of the electronic degrees of freedom for the heat capacity are both negligible. Let P_1 be 0.5 mm and T_1 be 300°K. To convert ΔE to ΔT you will need to approximate the mean value of C_V. It may then be useful to know that the two vibrational degrees of freedom, each contributing ½R to C_V when fully excited, are "dead" or completely unexcited at 300°, 40% excited at 1000°K, and 90% excited at 2000 °K.

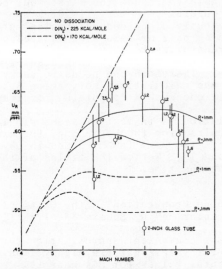

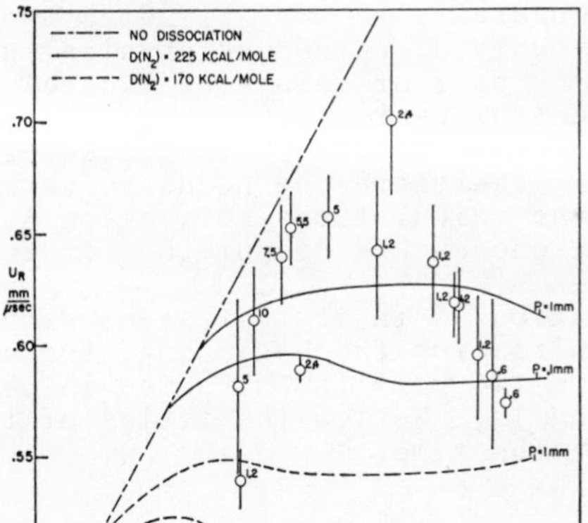

FIG. 2. Reflected shock velocity as a function of incident shock Mach number for nitrogen. The numbers beside the experimental points give P_1 in mm and the curves show the calculated values for P_1=0.1 and 1 mm.

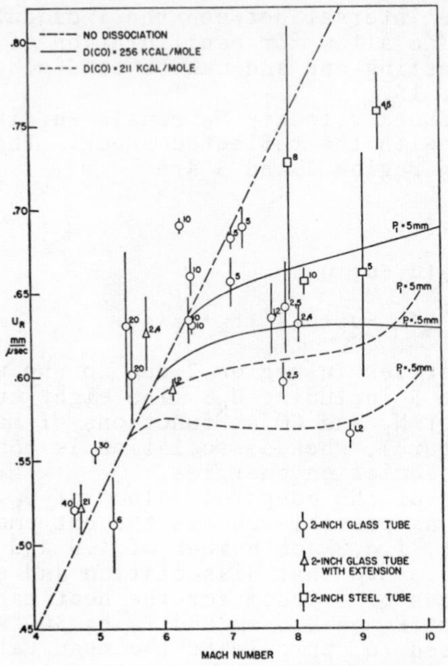

FIG. 3. Reflected shock velocity as a function of incident shock Mach number for carbon monoxide. The numbers beside the experimental points give P_1 in mm and the curves show the calculated values for $P_1 = 0.5$ and 5 mm.

You should be able to reduce the first four equations in four unknowns to a quadratic in $1/\rho_2$ yielding two roots neither of which can be obviously discarded on physical grounds. Further the corresponding pair of values of P_2 are both greater than P_1 as the correct pressure should be; the same is the case for the corresponding temperatures both exceeding T_1 as will be the case physically. Therefore one needs to carry a double set of values into the calculations for region 5 before making a decision. Finally one should compute U_R for comparison with 0.45 mm/μsec.

In this calculation as in most it seems simpler to go to a consistent set of units from the beginning, for example the cgs.

A final comment- two very famous names in physics are listed in ref. 16 and 17. Notice the titles of their work and their affiliation at the time.

5. "Comment on Sound Propagation in the Critical-Temperature Region of Hydrogen Chloride" W. Squire 38, 1785 (1963).

Squire derives a convenient approximation for computing the speed of sound for gases that are highly imperfect and compares with experimental values for HCl near its critical point where the pressure is about 82 atmospheres. He computes 190 m/sec to be compared with the experimental result of 180, both at the critical point.

The result for the velocity of vanishingly small pressure

waves, or the sonic velocity a is derived in Chapter 11 of the book by HCB referenced early in the last exercise. The result is

$$a^2 = (\partial P/\partial \rho)_s.$$

The needed relation between pressure and density for an isentropic path is readily available for an ideal gas with constant heat capacity in the form Pv^γ=const or $P=\rho^\gamma M\cdot\text{const}=\rho^\gamma\cdot\text{const}'$.

For this case you should be able to readily derive the equation for the speed of sound used in the previous exercise.

The author uses the compressibility form of the equation of state, $Z=PM/\rho RT$ with the compressibility factor Z a function of reduced temperatures and pressures. To compute the density derivative of pressure would then require temperature and pressure derivatives of Z, a procedure of inherently low accuracy.

The approach here is based upon the thermodynamic identity

$$(\partial P/\partial \rho)_s = (P/\rho)(\partial H/\partial E)_s.$$

Can you derive this result? The starting point would be to write the definition of enthalpy $H=E+Pv$ and differentiate wrt E holding s constant. From the combined form of the first and second law $dE=Tds+Pdv$ one divides through by dE, chooses s as the other independent variable, then keeps it constant. Substituting the partial derivative into the first equation, multiplying the rhs by $(\partial\rho/\partial\rho)_s$, using the relation between ρ and v and finally identifying the derivative $(\partial E/\partial v)_s=-P$ from the combined first and second laws completes the derivation.

Next he introduces the approximation

$$(\partial H/\partial E)_s \simeq H/E$$

to obtain his final result

$$a^2 \simeq ZRT[H/(H-ZRT)]/M.$$

Rationalize the approximation by showing that it is exact for an ideal gas and that departure due to nonideality is small because the correction must be applied only for a minor additive term.

Substitute through now for the final result.

6. "Heat of Sublimation of CaO" T.P.J.H. Babeliowsky <u>38</u>, 2035 (1963).

The heat of sublimation of CaO for $CaO(s) \rightarrow CaO(g)$ was determined mass-spectrometrically to be 139(±15?) kcal/mole at 2210°K. This was converted to ΔH°_{298} by means of $\int \Delta C_p dT$ giving the value 147±15 kcal/mole.

Evaluate this integral between the two temperatures. The vibrational frequency for the diatomic is 732.1 cm^{-1}, much lower than in CO so that a much larger fractional excitation of vibration will occur. An empirical result for the molar heat capacity of the solid is

$$C_p = 11.67+1.08\text{x}10^{-3}T-1.56\text{x}10^{5}/T^2 \text{ cal/mol}\cdot°\text{K}.$$

Construct a thermodynamic cycle to determine which of the several values quoted for the dissociation energy of the CaO molecule was used, viz 80, 120 or 136 kcal/mole, in his treatment of literature thermodynamic data to yield the result for the heat of sublimation of CaO of 133 kcal/mole. This is just inside his quoted experimental error. Other values you will need to construct this cycle at 298°K are the heat of sublimation of calcium, the dissociation energy of O_2, and the heat of formation of solid CaO; these are 42.2, 118.4 and -151.8 kcal/mole respectively.

As a sidelight notice the author's affiliation is given at the first of the article in Dutch and in a footnote in English.

7. "Heat Content of Uranium Dicarbide from 1484° to 2581°K" L.S. Levinson 38, 2105 (1963).

The enthalpy of uranium dicarbide(actual stoichiometry $UC_{1.93}$) was measured using a drop calorimeter. The results were referred to 310°K and are shown in Table I.

TABLE I. Heat content of $UC_{1.93}$.

Run No.	Temp. (°K)	Enthalpy experimental $H_T-H_{310°K}$ (cal/g)	Specific heat (cal/g/°K)
14	1484	87.80	0.091
1	1490	89.02	0.091
2	1587	97.07	0.097
3	1689	108.0	0.103
4	1794	119.1	0.109
5	1878	128.4	0.114
6	2016	145.0	0.123
13	2021	145.0	0.123
	2043 Transition temp. Tetragonal to cubic		
12	2081	162.0	0.100
7	2086	164.3	0.101
8	2213	176.2	0.107
15	2277	182.9	0.111
9	2393	197.5	0.117
10	2483	207.1	0.121
11	2581	219.4	0.127

X-ray diffraction analyses were made after runs No. 6, 11 and before run No. 1.

Below the solid-solid transition temperature shown in the Table the tetragonal form is the stable one while above it is the cubic as was confirmed by x-ray analyses. The heat content data are represented by the empirical equations

$$H(T)-H(310°K) = 20.21+7.978\text{x}10^{-4}T+3.024\text{x}10^{-5}T^2 \text{ cal/g (tetragonal)}$$

$$H(T)-H(310°K) = 67.91-9.113\text{x}10^{-3}T+2.629\text{x}10^{-5}T^2 \text{ cal/g (cubic)}$$

Check the experimental values at the extremes of the two stability regions by means of these equations; in particular does run no. 6 or no. 13 appear to be slightly in error? From the tabulated heat capacities the two enthalpies should differ by 0.6 cal/g.

What are the molar enthalpy and entropy for the tetragonal → cubic transition? Strictly speaking for the solid these quantities should be reported as per gram formula weight or gfw though probably most people still use per mole.

8. "Heat of Formation and Entropy of $(BOCl)_3$" J. Blauer and M. Farber 39, 158 (1963).

The equilibria

$$B_2O_3(\ell)+BCl_3(g) \rightleftarrows (BOCl)_3(g) \quad (1)$$

$$\tfrac{1}{3}B_2O_3(\ell)+\tfrac{1}{3}BCl_3(g) \rightleftarrows BOCl(g) \quad (2)$$

have been studied by the method of transpiration in the temperature range 536-825°K. At temperatures below its melting point of 723°K boric oxide is in the supercooled liquid or glassy state. The apparatus used is shown in Fig.1.

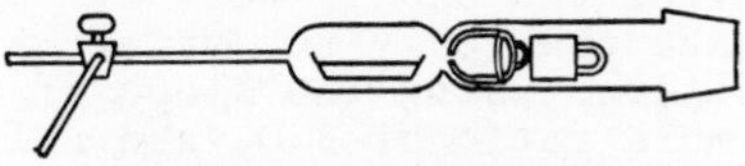

FIG. 1. Schematic drawing of transpiration cell.

A Vycor combustion boat is placed inside a Vycor tube which is sealed from the right by a hollow Vycor plug held in place by a solid quartz weight and connected on the left to a three-way stopcock by a capillary inlet. A weighed amount of boric oxide is placed in the transpiration cell which is then flushed with argon while the cell is held at a constant high temperature. BCl_3 was then introduced at varying partial pressures which were regulated by bleeding in a stream of argon as the BCl_3 passed the three-way stopcock. A given run at a fixed partial pressure of BCl_3 lasted up to two hours during which the transpired gases were passed into water. The transpired argon was collected over the water and the partial pressure of BCl_3 was calculated from the amount of argon collected and the acid titer(NaOH+phenolphthalein indicator). If both BOCl monomer and trimer are present in the product gases then the following equation describes conditions at equilibrium:

$$\frac{n_{B_2O_3}}{n_{BCl_3}} = K_1+\frac{K_2}{3}(V/RT)^{2/3}(1/n_{BCl_3})^{2/3}$$

with $n_{B_2O_3}$ the number of moles reacted, i.e. the weight loss of the Vycor boat, and n_{BCl_3} the number of moles present at equilibrium. K_1 and K_2 are equilibrium constant for reactions (1) and (2) and V is the total volume of transpired gases at temperature T.

Check the validity of this equation. Should the ideal gas law apply for the conditions of the experiment?

Fig. 2 shows a study of the influence of the partial pressure of BCl_3 in order to detect any appreciable concentration of monomer in the product gases. The volume flow rate for each run was held constant at 2.4±0.1 cc/min; t is the time in hours.

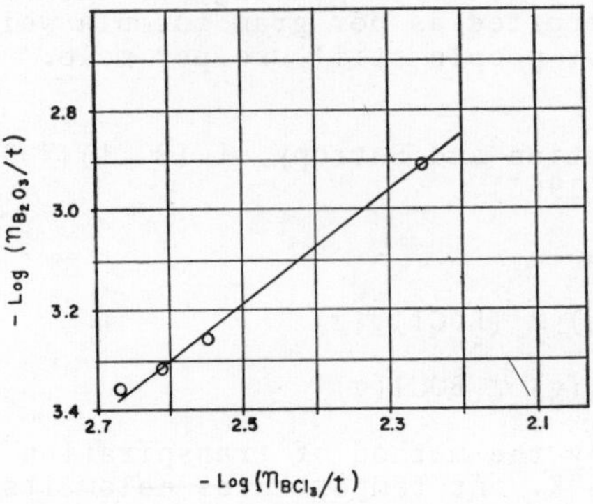

FIG. 2. Partial pressure study at 794°K.

The authors expect that the curve obtained under equilibrium conditions would parallel the one in Fig. 2.

Can you demonstrate that only the trimer is present under the experimental conditions using the slope of this curve and the equation for the mole ratio of B_2O_3 and BCl_3 which escape from the transpiration cell during a run?

Since there is no change in the number of moles of gas in reaction (1), and all of the B_2O_3 which escapes from the cell does so in the form $(BOCl)_3$, then K_1 is given by this mole ratio directly at equilibrium. It was found that the mole ratio decreased linearly with the BCl_3 flow rate so that by extrapolating the latter to zero, K_1 was determined. The results are shown for each temperature in Fig. 4.

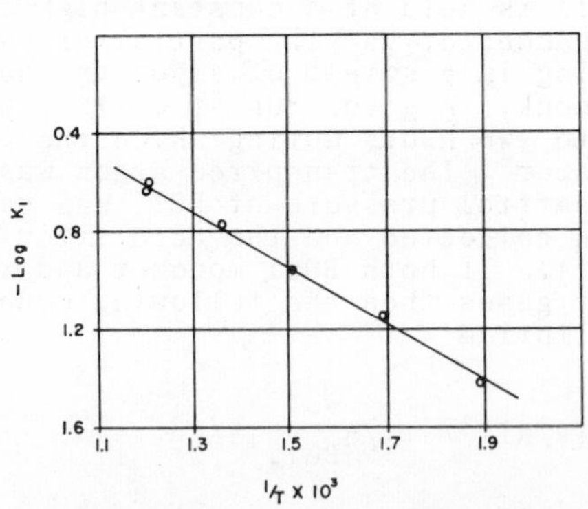

FIG. 4. Temperature dependence of the equilibrium constant.

Apply the van't Hoff equation now to determine the enthalpy change and a thermodynamic identity for the entropy change for reaction (1). From the entropies and heats of formation of the reactants, their ref. 1, these quantities are shown in Table III for $(BOCl)_3$ at two temperatures.

TABLE III. Thermodynamic properties of $(BOCl)_3$.

$H_{f695°K}$	−391.5±0.3 kcal/mole
$S_{695°K}$	124.2±0.5 cal/deg/mole
$H_{f298°K}$	−396.7±2 kcal/mole
$S_{298°K}$	92.5±2 cal/deg/mole

Working backwards, calculate the entropies and heats of formation at 695°K for $B_2O_3(\ell)$ and $BCl_3(g)$.

9. "Equilibrium Composition of Sulfur Vapor" J. Berkowitz and J.R. Marquart 39, 275 (1963).

The authors find, by mass spectrometer analyses of ionized effusing vapors from Knudsen cells containing pure sulfur and various binary metal sulfides, that the saturated vapor contains all possible S_n molecules between S_2 and S_8 in measureable concentration between room temperature and the boiling point of sulfur. Negligible concentrations of S_9 and S_{10} were also observed.

Their Introduction could serve as a model of a summary of the work, difficulties, and conclusions of earlier investigators. The straightforward statement "The present study was undertaken in order to correlate the mass of experimental information, with a full appreciation of the hazards encountered by previous workers" perhaps indicates the ideal relationships among observations and descriptions of nature, the scientific literature, and individual scientists themselves.

Sketch the S_8 molecule from their description in the third paragraph. For a polygon with n sides the sum of the interior angles is given by (n-2)x180°. Is this molecule planar? Does the adverb "regular" help to completely fix all of the bond angles?

For the reactions $(n/8)S_8 \rightarrow S_n (n=2,3,4,5,6,7)$ the enthalpy changes at ∿400°K are $\Delta H_2=23.5$, $\Delta H_3=22.5$, $\Delta H_4=20.5$, $\Delta H_5=14.3$, $\Delta H_6=6.2$, $\Delta H_7=5.7$ kcal/mole.

Check the value for ΔH_5 from Fig. 3. Data on the lhs of the plot are from vaporization and resultant electron bombardment of HgS while the rhs represents points for pure sulfur. K_6 and K_7 are both in good agreement over the entire range of 1/T while extrapolation of the HgS K_5 to lower temperatures falls well below the sulfur K_5. The explanation for this apparent discrepancy is that the S_5^+ ions detected by the mass spectrometer in the sulfur experiments are not due to ionization of parent molecules

$$S_5+e^- \rightarrow S_5^++2e^-$$

but rather to the fragmentation process

$$S_8+e^- \rightarrow S_5^++S_3+2e^-.$$

Using a number of metal sulfides with a range of dissociation pressures of sulfur to help understand the various mass spectra is a new technique that is introduced.

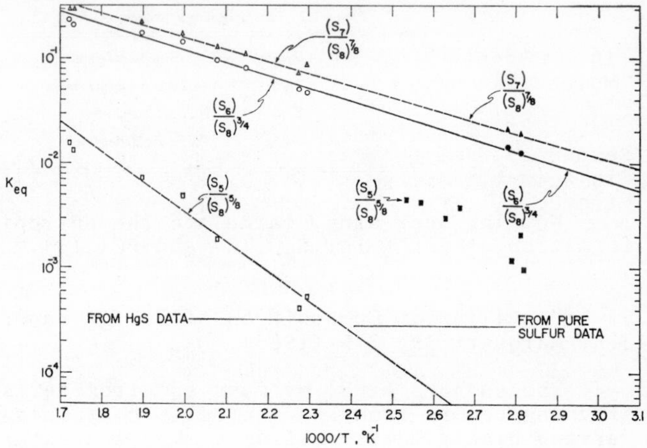

FIG. 3. The logarithms of the equilibrium constants $K_n = P(S_n)/P^{n/8}(S_8)$ $(n=5, 6, 7)$ vs $1/T$, using data from the HgS and pure sulfur experiments.

From the ΔH values above, make a plot of energy per bond for the ring structures $S_n (n=2,3,...10)$, considering the molecule S_2 to be a two-membered ring. The values should range between 51.0 and 63.0 kcal/mole. Assume enthalpy changes of zero for the reactions

$2S_8 \rightarrow S_7 + S_9$

$2S_8 \rightarrow S_6 + S_{10}$.

Fig. 5 shows mole fraction of the various sulfur species as a function of temperature. What is the most basic thermodynamic reason for the trend toward smaller rings at higher temperatures?

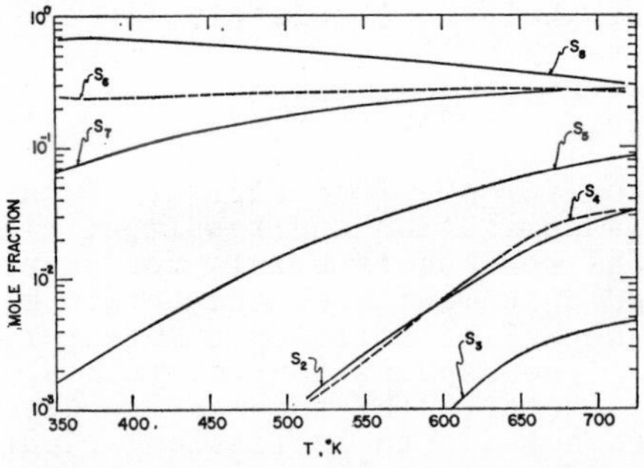

FIG. 5. Mole fraction of the sulfur species $S_n (n=2, 3, \cdots 8)$ in the saturated vapor from room temperature to the boiling point of sulfur.

A question which more properly would come under the subject matter of Chapters II and III; if S_8 belongs to point group D_{4d} and thus has a symmetry number σ of 8, while σ for both S_7 and S_9 is one, can you calculate 8.4 eu as the entropy change for the next to last reaction above? This change in σ should be the biggest contributor to the entropy change. The S dependence on σ is given by the additive term R ln (1/σ).

10. "Mass Spectrometric Detection of Polymers in Supersonic Molecular Beams" F.T. Greene and T.A. Milne 39, 3150 (1963).

Polymeric ions were observed for several common gases when expanded from room temperature and a few atmospheres pressure in a supersonic molecular beam and then analyzed in a mass spectrometer. Results for $(CO_2)_x^+$ and Ar_x^+ are shown in Fig. 1.

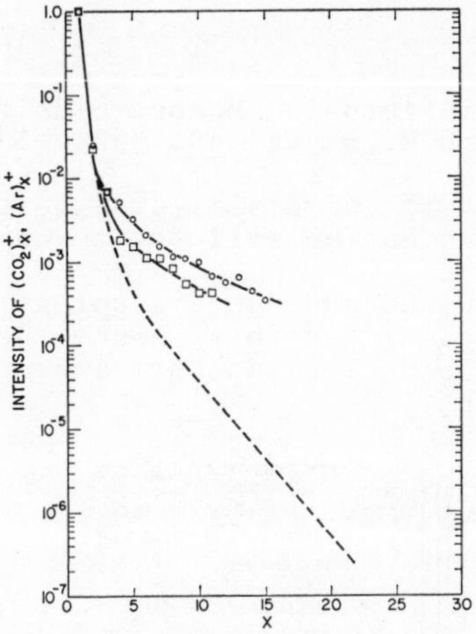

FIG. 1. Relative polymer ion intensities for CO_2 and Ar, sampled at room temperature and 5 atm, versus degree of polymerization, *X*. ○, CO_2 (this work); □, Ar (this work); – – –, CO_2 (Bentley).

The dashed curve represents the data of Bentley(Nature 190, 432 [1961]) who first reported an observation of the effect.

Here it is claimed that $(CO_2)_x^+$ ions were detected qualitatively beyond mass 1200; to what value of x would this correspond?

According to the measurements of their ref. 5, Mach numbers of 10 or greater are probable in this type of beam system. The beam achieves these very high velocities at the expense of the internal and translational energies of the molecules, so that the gas becomes supersaturated with respect to the liquid or solid phase and condensation or polymerization begins to take place. This technique may then provide a unique method of looking at the kinetics and mechanisms of nucleation phenomena. This explanation is quite plausible and is reinforced by some approximate calculations of the very low temperatures reached after expansion.

For an isentropic expansion of an ideal gas with a constant ratio of specific heats γ the Mach number after expansion N_2 is related to the temperature ratio T_1/T_2 by

$$N_2^2 = 2/(\gamma-1)[(T_1/T_2)-1]$$

This result can be derived by combining the differential form of the equation of motion(p. 7) $-dP=\rho u du$, the definition of Mach number using local rather than undisturbed sonic velocity as done earlier(p. 7), $u_2=(\gamma RT_2/M)^{\frac{1}{2}}$, and the relation for an isentropic process with constant γ, viz $P=\rho^{\gamma}\cdot const'$ (p. 11). Integrating and setting $u_1=0$ gives the above equation.

Having verified this, check their claim that expansion from room temperature with atoms leads to a final temperature of 8.8°K and that with diatomics the final temperature is 14.5°K. What will actually happen to γ for diatomics at these very low temperatures? How will this affect the calculated value of the final temperature?

11. "Heat Capacity of Plutonium Monocarbide from 400° to 1300°K" O.L. Kruger and H. Savage 40, 3324 (1964).

This exercise covers quite similar experimental work to that of exercise no. 7, but we will try to emphasize some rather different aspects here.

Solid solutions of UC and PuC are potentially valuable as fuel materials for nuclear reactors; hence the interest in their thermodynamic properties. The isothermal drop calorimeter used is shown in Fig. 1.

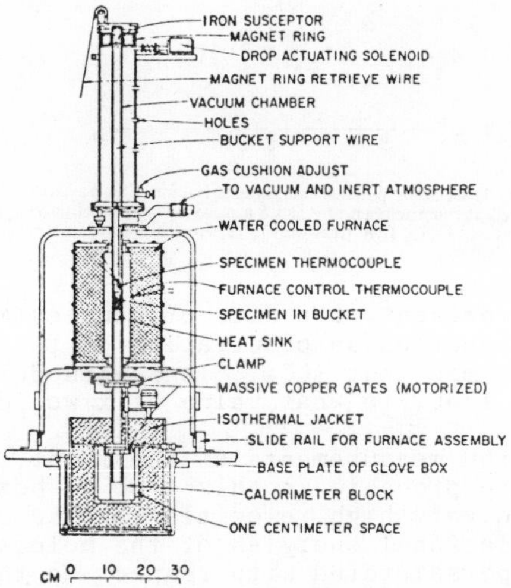

FIG. 1. Isothermal drop calorimeter.

The entire apparatus was installed inside a glove box. Why? What are the functions of the two copper gates? If just one is water-cooled which one would it be and why? After the PuC specimen was wrapped in tantalum foil and placed in the platinum bucket before heating the furnace, the system was evacuated and then purged several times with argon to remove traces of oxygen. Why would a residual pressure of 10 mm of argon be left in the sys-

tem? Two reasons. The enthalpy of the foil and bucket were subtracted from the total enthalpies much as the fuse wire correction is made in combustion calorimetry, though larger now. The carbon content was 46.5 atomic %(wt. %?) as determined by chemical analysis. Would it be accurate to represent the formula of this somewhat nonstoichiometric solid by $PuC_{0.87}$? If so this would represent a "mole" of material. We will discuss the departure from a strict 1:1 ratio a bit later in connection with the heat capacity data.

Metallographic examination showed that the specimen was single-phase PuC, with its microstructure as shown in Fig. 2. No Pu_2C_3 phase was detected on examination of the sample in the as-polished or etched condition.

FIG. 2. Single-phase PuC (as polished; 150×).

The self-heating rate of plutonium, due to radioactive alpha decay, is quoted from their ref. 10 as $(1.923 \pm 0.019)10^{-3}$ watt/gram. See if you can approximate this value given that the preponderant isotope ^{239}Pu decays with α emission in the range 5.10-5.15 mev and a half-life of 24,360 years. Neglect the accompanying γ emission. The self-heating effect on the time-temperature curve is shown in Fig. 3.

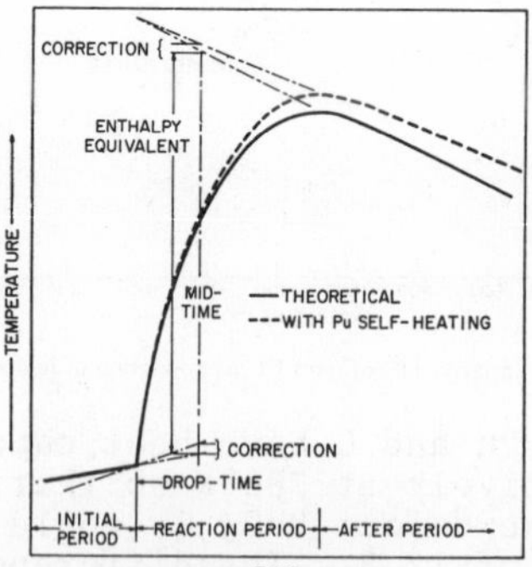

FIG. 3. Curve showing change in temperature of calorimeter.

The empirical heat capacity equation is cast into the fo

$$C_p = 13.08 + 11.44 \times 10^{-4} T - 3.232 \times 10^5 T^{-2} \text{ cal/mole} \cdot {}^\circ K,$$

the same as that for CaO(s) in exercise no. 6 but slightly different from that for $UC_{1.93}$ in no. 7. Integration of Cp and Cp/T will give equations for enthalpy and entropy referred to the reference temperature of 298°K. The smoothed data is shown in Table II.

TABLE II. Molar thermodynamic properties of plutonium monocarbide.[a]

Temperature (°K)	H_T-H_{298} (cal mole^{-1})	C_P (cal mole^{-1} °K^{-1})	S_T-S_{298} (cal mole^{-1} °K^{-1})
400	1097	11.52	3.16
500	2296	12.36	5.83
600	3559	12.87	8.13
700	4865	13.22	10.15
800	6201	13.49	11.94
900	7562	13.71	13.52
1000	8943	13.90	15.01
1100	10 342	14.07	16.34
1200	11 757	14.23	17.57
1300	13 188	14.38	18.68
1400	14 633	14.51	19.77
1500	16 091	14.66	20.77
1600	17 564	14.78	21.72
1700	19 049	14.92	22.61
1800	20 547	15.04	23.47
1900	22 057	15.16	24.30

[a] Last digit retained for comparison purposes only.

Integrate these two equations and check your results by means of one or two lines in the Table.

The heat capacity curve itself is discussed further and is compared with that for UC in Fig. 4. The difference in heat capacities between PuC and UC, e.g. 0.8 cal/mol·°K at 700°K, was attributed to the nonstoichiometric or defect structure of PuC. State the general form of the Neumann-Kopp rule from their

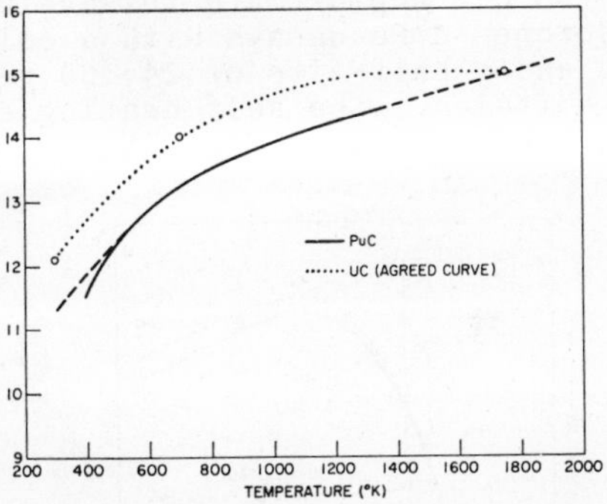

FIG. 4. Heat capacity of PuC and UC as a function of temperature.

specific application: Pu and C have heat capacities of 9.0 and 4.4 cal/mol·°K respectively at 700°K so that of stoichiometric PuC would be 13.4. The defect $PuC_{0.87}$ would then have a calculated heat capacity of 12.8. The difference 0.6 is close to the difference of 0.8 between the observed PuC and UC values.

Does this argument assume that the UC heat capacity curve is actually for the 1:1 composition?

12. "Vapor Pressure of Neptunium" H.A. Eick and R.N.R. Mulford <u>41</u>, 1475 (1964).

In this work vapor pressure of metallic neptunium was determined with the Knudsen effusion cell shown in Fig.1. The experiment was possible only because of the availability of neptunium in more than milligram quantities for the first time.

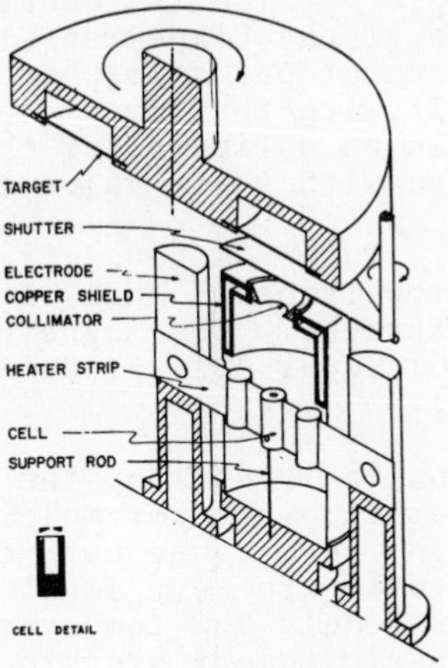

FIG. 1. Vapor-pressure apparatus.

The water-cooled copper shield condensed all of the effusate except that which passed through the collimator and was itself then caught by the water-cooled copper annular ring target. The target could be rotated through a seal so that ten or more exposures could be made without breaking the vacuum. After a series of exposures the target was removed, cut into segments of one exposure each, and weighed(why?).

Fig. 2 presents the vapor pressure results. The four letters keyed represent four different targets.

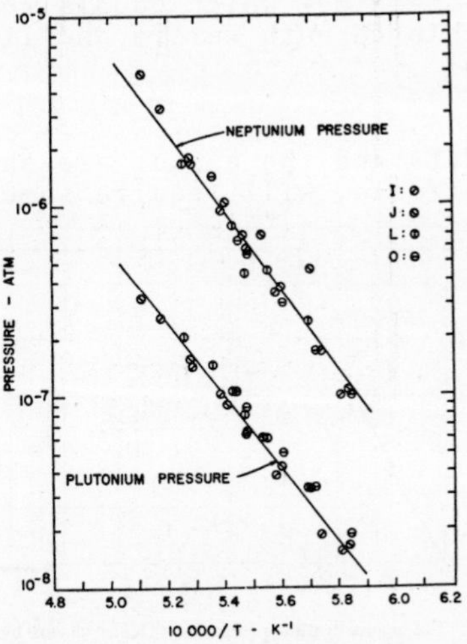

FIG. 2. Plot of the observed neptunium and plutonium pressures.

The sample actually contained 0.56 wt % plutonium. The amount of each metal deposited on the target as effusate could be readily determined(how?). This is an advantage of working with radioactive materials, but there are disadvantages!

The equation describing the vapor pressure of neptunium is

$$\log_{10}p(\text{atm}) = -(20{,}610\pm1280)/T + (5.099\pm0.702).$$

Check a point or two on the plot. From the equation find the heat of vaporization by means of the Clapeyron-Clausius relation and the entropy of vaporization at one atmosphere vapor pressure. Compare this last result with that predicted by Trouton's rule. By extrapolation obtain an estimate for the normal boiling point of neptunium and compare with their value of 4175°K.

13. "Magnetothermodynamic Properties of $MnCl_2$ from 1.3° to 4.4°K at 90 kG. A Zero Entropy Reference. The Magnetomechanical Process at Absolute Zero" W.F. Giauque, G.E. Brodale, R.A. Fisher, and E.W. Hornung <u>42</u>, 1 (1965).

The magnetic work done on the system by the magnetic field $\underline{H}$ is $\underline{H}d\underline{M}$ where $\underline{M}$ is the magnetic moment per mole. The ordinary expansion work term PdV is ignored because both the expansion of the solid and the pressure itself are very small at the near-vacuum conditions of these extremely low temperatures. The magnetic field and molal magnetic moment are brought into the definition of the heat content H through $H=E-\underline{H}\underline{M}$.

What property does dH have at constant magnetic field $\underline{H}$? What is the analogous property of dH in the absence of a magnetic field but including the PdV term at constant pressure?

A number of interesting relationships ensue from their measurements and the accompanying discussion. We have chosen just one, the energy change at 0°K for the magnetization of a mole of $MnCl_2$. This is well within the scope of classical thermodynamics. Read the sections on "structure" and "pernt".

For reversible processes the combined form of the first and second law becomes $TdS=dE-\underline{H}d\underline{M}$ which equals zero at 0°K. Therefore you should be able to both derive and illustrate the result

$$\Delta E = \int_0^{M_{sat}} \underline{H}d\underline{M}.$$

Both the necessary data and the answer are shown in Fig. 2. The units of $\underline{M}$, gauss·cm³/mole, will require some conversion.

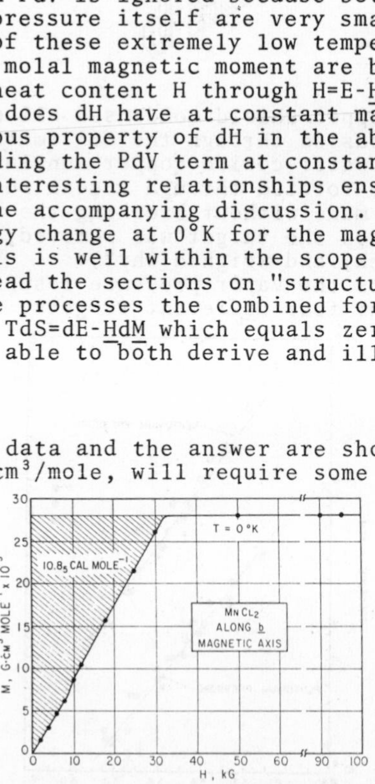

FIG. 2. The magnetization process at 0°K as shown by the extrapolation of isoerstedic measurements of magnetic moment.

14. "Heat Capacity and Other Thermodynamic Properties of MoF_6 between 4° and 350°K" D.W. Osborne, F. Schreiner, J.G. Malm, H. Selig, and L. Rochester 44, 2802 (1966).

Enthalpies for the solid-solid(orthorhombic(?)→bcc) transition at 263.50°K, fusion at 290.76°K, and vaporization as a function of temperature through the vapor pressure and liquid density together with heat capacities for the condensed phases were measured. Our application here will be to use the heat capacity data of Fig. 1, the density data of Table VI, and the vapor pressure data of Table VII to follow the third law calculation of the entropy of MoF_6 in the ideal gas(standard) state at 298.15°K which is summarized in Table VIII.

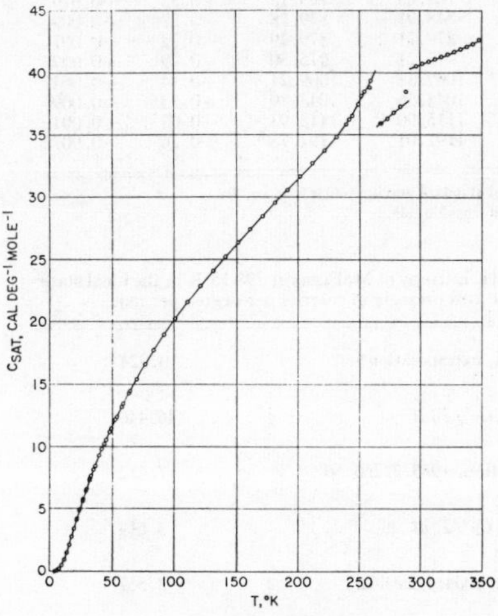

FIG. 1. Heat capacity of MoF_6.

TABLE VI. Density of MoF_6.

T (°K)	Method	Observed density (g/cm³)	Calculated density (g/cm³)
	Low-temperature solid[a]		
77.16	He displacement	3.51₉	3.519
173.83	He displacement	3.39₃	3.393
237	x-ray[b]	3.27±0.03	3.311
	High-temperature solid[c]		
278	x-ray[b]	2.88±0.04	2.88
	Liquid[d]		
294.33	Pycnometer	2.544	2.544
307.08	Pycnometer	2.491	2.492
344.63	Pycnometer	2.341	2.341

[a] Calculated density=3.619−0.00130T g/cm³.
[b] D. Northrup and S. Siegel (private communication).
[c] Calculated density=3.464−0.00210T g/cm³.
[d] Calculated density=3.733−0.00404T g/cm³.

TABLE VII. Vapor pressure of liquid MoF_6.

T (°K)	Observed P (mm Hg)[a]	Calculated[b] P (mm Hg)[a]	ΔP (mm Hg)[a]	ΔT (°K)
291.15	412.85	412.83	+0.02	+0.001
293.15	447.51	447.73	−0.22	−0.012
294.15	466.32	466.04	+0.28	+0.015
295.15	484.64	484.95	−0.31	−0.016
296.15	504.85	504.47	+0.38	+0.019
297.15	524.47	524.60	−0.13	−0.006
298.15	545.39	545.37	+0.02	+0.001
299.15	566.66	566.78	−0.12	−0.006
301.15	611.57	611.62	−0.05	−0.002
302.15	635.03	635.06	−0.03	−0.001
303.15	659.39	659.21	+0.18	+0.007
305.15	709.71	709.68	+0.03	+0.001
306.15	735.75	736.02	−0.27	−0.010
307.15	763.65	763.13	+0.52	+0.019
310.15	848.91	849.18	−0.27	−0.009
311.15	879.71	879.49	+0.22	+0.007
314.15	975.21	975.50	−0.29	−0.009
315.15	1009.38	1009.24	+0.14	+0.004
316.15	1043.56	1043.89	−0.33	−0.009
318.15	1115.90	1115.93	−0.03	−0.001
320.15	1192.01	1191.75	+0.26	+0.007

[a] At 0°C and standard gravity (980.665 cm sec^{-2}).
[b] Calculated from Eq. (3).

TABLE VIII. Entropy of MoF_6 gas at 298.15°K in the ideal state at 1 atm pressure in calories per degree per mole.

0°–5°K, extrapolation[a]	0.024
$\int_5^{263.50} (C_P^\circ/T)dt$	46.440
Transition, 1953.2/263.50	7.412
$\int_{263.50}^{290.76} (C_P^\circ/T)dt$	3.614
Fusion, 1034.1/290.76	3.557
$\int_{290.76}^{298.15} (C_P^\circ/T)dT$	1.014
S° of liquid at 298.15°K	62.061±0.06
$(\partial S_{liq}/\partial P)_T(P-P^\circ)$[b]	0.001
Vaporization, 6630/298.15	22.237±0.08
Correction to ideal state	0.108
Compression to 1 atm	−0.659
S° of gas from thermal data	83.75±0.10
S° of gas from molecular data[c]	83.77

[a] Assumed heat capacity varies as T^3.
[b] P is the vapor pressure at 298.15°K (545.37 mm Hg), $P^\circ = 1$ atm.
[c] Assumed the frequency assignment of Weinstock and Goodman,[21] O_h symmetry, and an Mo–F distance of 1.840 Å.[10]

The heat capacity curve of Fig. 1 is actually very slightly different from that at one atmosphere total pressure, being obtained under the vapor pressure of the solid and liquid phases. This is the only correction we will allow ourselves the luxury of ignoring, though the authors themselves have determined it in the article. Some of the terms in Table VIII are quite small and it may be one of the phenomena often attributed to Prof. Murphy arises here; the smaller an effect is the more trouble it requires to evaluate it. At any rate let's proceed term by term through Table VIII.

The extrapolation formula $C_p=5.81\text{x}10^{-4}T^3$ has the correct theoretical temperature dependence.

The integrals or "sensible" entropies you should be able to approximate graphically by plotting Cp(Csat)/T vs T. The entropies of transition and fusion are simple arithmetic with the enthalpies and temperatures given. Notice how the C_p curve breaks at these temperatures; the discontinuities define first-order transitions, whereas a continuous curve with a discontinuous slope would indicate a second-order transition.

The vapor pressure results given by the equation

$$\log_{10}P_{mm} = -2047.15/T - 4.28004 \log_{10}T + 20.19354 \tag{3}$$

are shown in Table VII. The liquid density is given in a footnote to Table VI. The equation of state in virial form

$$PV = RT+B(T)P \tag{4}$$

was used with

$$B(T) = -8.20\text{x}10^7T^{-2} \text{ cm}^3/\text{mole} \tag{5}$$

in the Clausius-Clapeyron equation to obtain the heat of vaporization shown. Can you follow this route?

The barely significant pressure dependence of the liquid at 298.15°K can be evaluated using the Maxwell relation $(\partial S/\partial P)_T=-(\partial V/\partial T)_P$ for the liquid. The rhs can be evaluated using the liquid density equation.

Can you derive the result needed for the next term?

$$S_{ideal}-S_{real} = (\partial B/\partial T)_P P. \tag{7}$$

The compression term should now be easy. The calorimetric entropy is very close to the spectroscopic entropy. We will examine this in detail in an exercise in Chapter III.

Would you say the third law is valid for MoF_6?

One final say on small effects. "A difference is not a difference unless it makes a difference".

15. "Phase Diagrams of Sodium Nitrite and Potassium Nitrite to 40 kbar" E. Rapoport 45, 2721 (1966).

The phase diagrams were determined by differential thermal analysis(DTA) and volume-discontinuity methods. The diagram for $NaNO_2$ is shown in Fig. 2. Five solid phases and the liquid are identified. Phase II has a very narrow range of stability at pressures of zero to one atmosphere, 163.4°-164.7°C. What can you say about the stability range of phase V at atmospheric pressure? What phases are present at each of the triple points i) 9.8 kbar, 343°C and ii) 24.2 kbar, 252 °C? What about -1.7 kbar, 155°C?

Observe the striking similarity between the III-II and II-I boundaries on the one hand and the I-V and V-liquid boundaries on the other. The author's ref. 13 suggested that $NaNO_2$ retained a solid-like structure in the liquid state up to 15°C

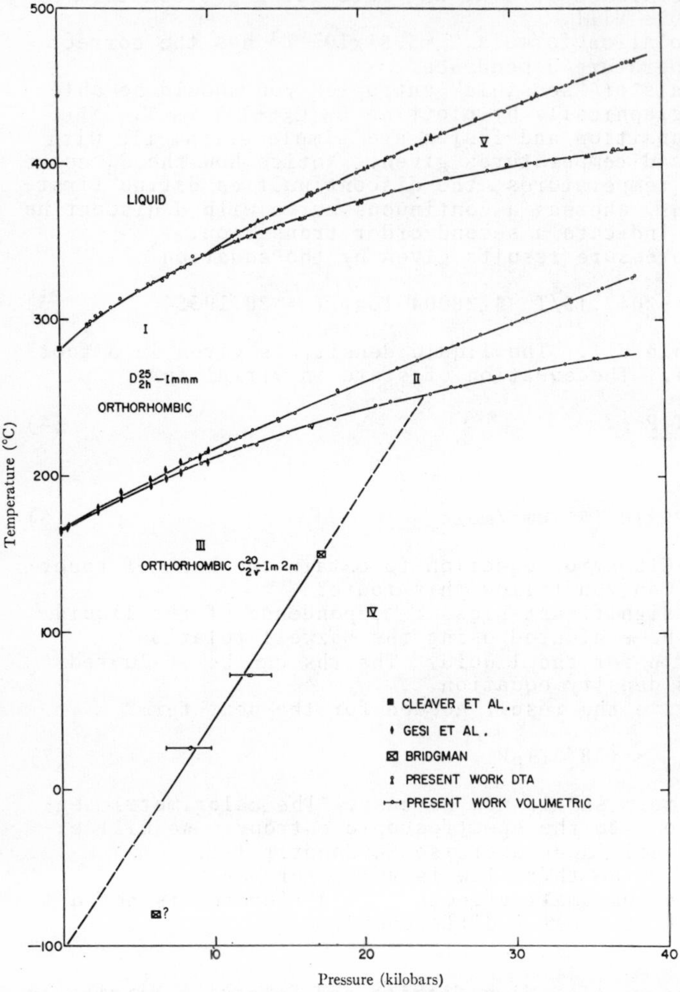

FIG. 2. The phase diagram of $NaNO_2$ to 40 kbar.

above the melting point; this was before the author's discovery of the distinct phase V. What type of evidence regarding the latent heats V→liquid, I→liquid and I→V would support the existence of this phase V? Such evidence is presented in the article.

Compute ΔH for I→liquid from dP/dT at zero pressure and $\Delta V = \pm 4.3\ cm^3/mole$. The slope will let you determine the sign. Compare with his value of 6±1 kcal/mole.

On the argument right under Table IV in the article can you rationalize an increase in the entropy of fusion from the formation of association complexes in the melts? If this was true it would lead to a lower than normal melting temperature assuming the heat of fusion was not greatly increased. Are the melting points of these nitrites lower than those of the halides?

16. "Spectroscopic Evidence for Binary O_2 Clusters in the Gas Phase" R.P. Blickensderfer and G.E. Ewing 47, 331 (1967).

Fig.1 presents the absorption spectrum of oxygen gas near 12 600 Å. The sharp features are transitions of the monomer O_2 while the broad feature is attributed chiefly to the binary cluster $(O_2)_2$.

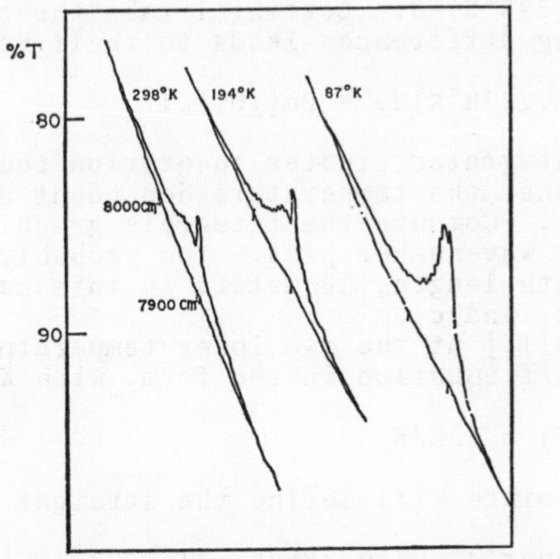

FIG. 1. Absorption spectrum of oxygen gas near 12 600 Å at a density of 0.1 mole/liter with a 30-m path length. Gas pressure is 2.4 atm at 298°K, 1.6 atm at 194°K, and 0.7 atm at 87°K. The vertical marks on the background curves indicate 8000 and 7900 cm^{-1} on each spectrum.

For the binary clusters to be present in any significant concentration it is necessary that the rather weak intermolecular potential energy not be overwhelmed by the translational kinetic energy. Is this going to happen at lower or higher temperatures? Does the spectrum bear out your answer?

The authors estimate the energy of dimerization to be about -380 cal/mole which they feel could be in error by ±100 cal/mole because of the assumptions made and the difficulty in determining the integrated absorption of the weak features.

We will try and obtain this value for ΔE, even if with rather poor precision.

First it is necessary to replot the data in Fig. 1 using optical density $\log_{10}(T_0/T)$, with T transmission, rather than % T, as the ordinate. You may assume that the abscissa in Fig. 1 is already in wavenumbers ν; even if it is linear in wavelength the error should be minor over this relatively small range.

They introduce the reasonable approximation that the concentration of the cluster is negligible at room temperature which allows the data to be handled rather easily. Since the fraction of oxygen molecules that dimerize is expected to be small even at low temperatures it is assumed that the monomer concentration $[O_2]$ remains constant at 0.1 mole/liter. They then subtract the room temperature absorption, due only to the monomer, from the low temperature absorption due to both monomer and dimer to obtain $[(O_2)_2]$ as a function of temperature and pave the way for an application of thermodynamics.

For a system ot two components, each of which obey's Beer's law, the optical density $\alpha=\alpha(\nu,T)$ is

$$\alpha = \alpha(\nu,T) = \log_{10}(T_0/T) = \varepsilon_1(\nu)c_1(T) + \varepsilon_2(\nu)c_2(T)$$

with $\varepsilon_{1,2}(\nu)$ the extinction or absorption coefficient at frequency or wavenumber ν and $c_{1,2}(T)$ the temperature-dependent concentration for monomer and dimer. Here $c_1(T)=c_1(298°K)=0.1$ mole/liter and $c_2(298°K)=0$. Letting T take the two lower values in turn and forming differences leads to their result

$$\int\alpha(\nu,T)d\nu-\int\alpha(\nu,298°K)d\nu = Bc[(O_2)_2].$$

Show that their integrated cluster absorption coefficient Bc is our $\int\varepsilon_2(\nu)d\nu$ and that the temperature-dependent dimer concentration $[(O_2)_2]$ is c_2. Compute the integrals graphically from your optical density vs wavenumber plot. You probably noticed that we absorbed the path length, 30 meters in this case, right into the coefficients c_1 and c_2.

Knowing $Bc[(O_2)_2]$ at the two lower temperatures lets us apply the van't Hoff equation in the form, with $K_C=[(O_2)_2]/[O_2]^2$

$$d \ln K_C/d(1/T) = -\Delta E/R$$

to find ΔE. Two points will define the straight line though not allow us a check.

In a later paper(J. Chem. Phys. 51, 873 (1969)) these same authors show that the assumption of negligible dimer concentration at room temperature is in error. See the last paragraph of this paper titled "Clustering of Oxygen".

Notice however that the paper under discussion here is a Communication, a special type of Letter to the Editor, the definition of which appears in the same issue right after p. 348. Read this definition and see if this particular article doesn't meet the spirit of the requirements.

17. "Thermodynamic Properties of the Potassium-Graphite Lamellar Compounds from Solid-State emf Measurements" S. Aronson, F.J. Salzano, and D. Bellafiore 49, 434 (1968).

Graphite reacts with the heavier alkali metals to form lamellar compounds. Fig. 1 of ref. 5 shows the arrangements of the K^+, Rb^+, and Cs^+ ions in the graphite lattice. The compounds that have been reported in each system have the composition(M= alkali metal) C_8M, $C_{10}M$, $C_{24}M$, $C_{36}M$, $C_{48}M$, and $C_{60}M$.

In this work thermodynamic properties of the potassium-graphite compounds were determined in the temperature range 200°-350°C using a solid-state emf technique. The combination of Knudsen effusion and tracer techniques that had been used earlier with the rubidium and cesium compounds could not be used with those of potassium. Why not?

The apparatus used is shown in Fig. 1. The emf cells employed were of the form

K(ℓ) | K-glass | K-graphite

where K(ℓ) represents molten potassium, K-glass is a glass electrolyte reversible to potassium ion, and K-graphite represents a mixture of two potassium-graphite compounds in thermodynamic equilibrium with each other(a two-phase region).

Given the cell as written above, with the two-phase region containing $C_{24}K$ and $C_{36}K$ for illustrative purposes, write the half-reactions for the left-hand electrode(anode) and the right-hand electrode(cathode). The overall reaction is

$$2C_{36}K + K(\ell) \rightarrow 3C_{24}K$$

for one equivalent of potassium transferred.

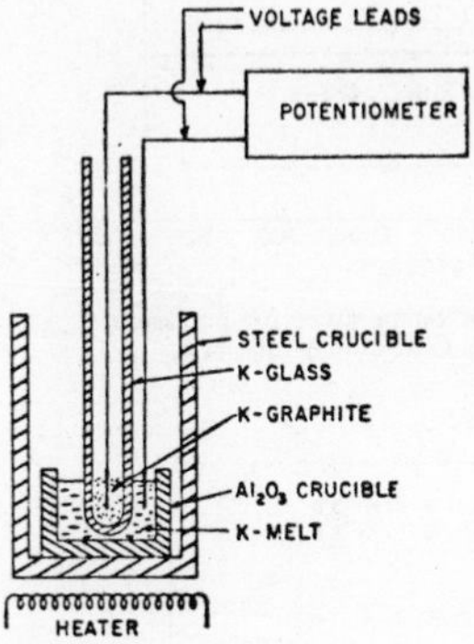

FIG. 1. Apparatus for emf measurements on cells of the type K (liquid) | K-glass | K-graphite.

Below 200°C the cell sensitivity was low and the response of the emf to temperature change was sluggish. The use of an open container for the potassium above 350°C was not practical; why not?

Fig. 2-4 present emf data as a function of cell temperature over this region. The C/K ratios shown are the overall composi-

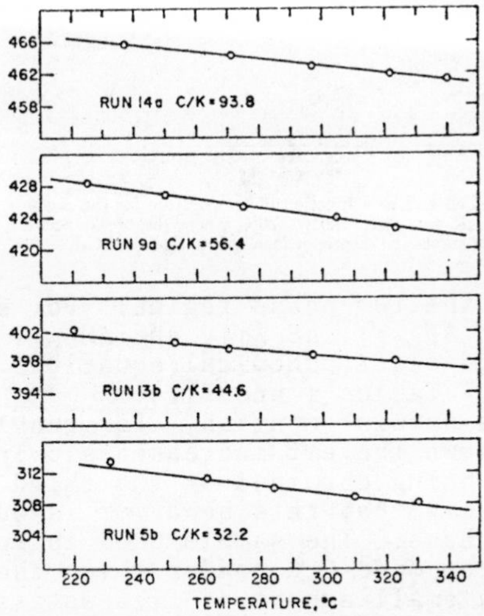

FIG. 2. Cell emf as a function of temperature for dilute potassium-graphite compounds.

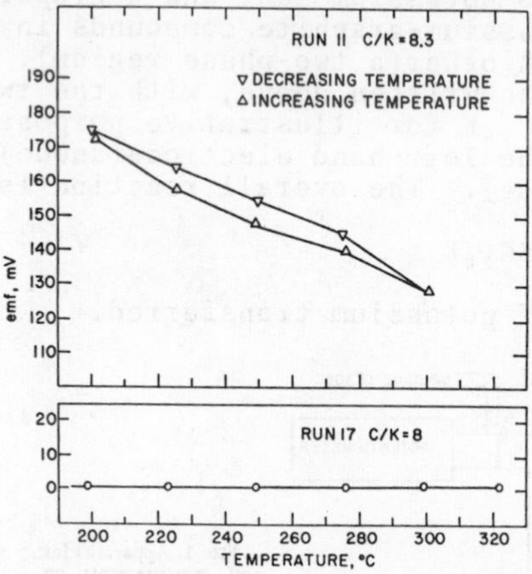

FIG. 3. Cell emf as a function of temperature for potassium-graphite compositions near a C/K atomic ratio of 8.

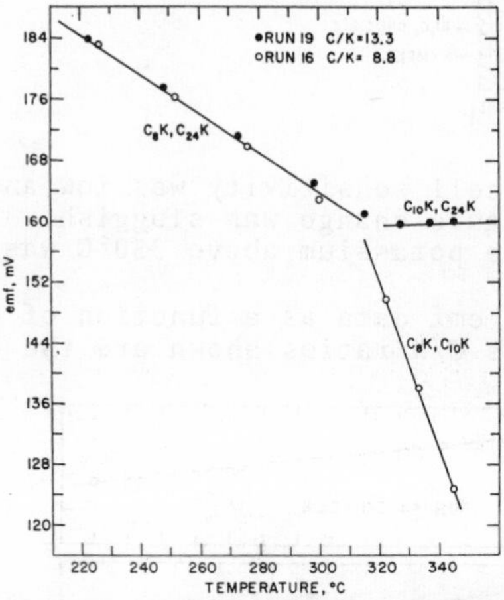

FIG. 4. Cell emf as a function of temperature for the region between C_8K and $C_{24}K$. Below 320°C where the breaks occur, $C_{10}K$ is not stable and disproportionates into $C_{24}K$ and C_8K.

tion of K-graphite, the two-phase region. For each of the eight linear curves obtain ΔF_{cell}, ΔS_{cell}, and ΔH_{cell} at 277°C by means of the standard electrochemical equations. Compare your results with those of Tables I and II.

Fig. 3 deserves special mention. For run 11 in addition to the hysteresis shown the emf increased slowly from run to run, the data shown being obtained on the third cycle. The authors think it likely that this behavior is due to nonstoichiometry in the C_8K phase. The sample used in run 17 was either a commercial or custom-made C_8K powder which the authors feel probably contained a small amount of free potassium. What would you expect the emf to be of a liquid-potassium vs liquid-potassium cell? Would it be temperature dependent?

Fig. 2 yields the interesting trend that the emf increases with decreasing potassium content. What would you expect to be the emf when graphite powder was substituted for K-graphite in the cell? The authors tried this with the results discussed under equation (4) in the paper.

In Fig. 4 at temperatures above about 320°C two different curves are obtained depending upon the sample composition. The large negative entropy change of Stage 1b of Table III partially reflects the increase in concentration of potassium ions between each graphite layer as $C_{10}K$ is converted to C_8K. How is entropy affected in general by concentration increases? Another contribution to the entropy change may be the increase in three-dimensional order, i.e. interaction among layers of potassium as $C_{10}K$ is converted to C_8K.

Would each of these factors favor the appearance at higher temperatures of the intermediate phase $C_{10}K$, that is increase its stability relative to the C_8K phase?

With the values for the heat and entropy of vaporization for liquid potassium of 20.2 kcal/mole and 19.4 cal/mole·°K you should be able to produce all of the thermodynamic values in the Potassium-graphite columns of Table III from Tables I and II.

Table IV, not given here, presents heats and entropies of formation for the known alkali metal-graphite compounds.

TABLE I. Thermodynamic data in the region C_8K to $C_{24}K$ obtained from the electrochemical measurements.

C/K atom ratio	Run No.	Phases present	emf at 277°C (mV)	$-\Delta F_{cell}$ at 277°C (kcal/mole K)	$-\Delta H_{cell}$ (kcal/mole K)	$-\Delta S_{cell}$ (cal/°K·mole K)
8.0	17	C_8K, K	0.5	0.01	...	...
8.3	18	C_8K	146	3.4	...	...
8.3	11	C_8K	132–145	3.0–3.3	...	...
8.8	16a	C_8K, $C_{24}K$	169.2	3.90	7.5	6.5
	(314°C)[a]	C_8K, $C_{10}K$	144.4[b]	3.33	18.7	25.6
9.0	10	C_8K, $C_{24}K$	169.8	3.92	7.4	6.3
9.1	20	C_8K, $C_{14}K$	169.8	3.92	7.2	3.3
	(316)[a]	C_8K, $C_{10}K$	148.2[b]	3.42	17.8	23.9
10.0	6	C_8K, $C_{10}K$	170.4	3.92	7.1	5.7
13.3	19	C_8K, $C_{24}K$	170.6	3.93	7.1	5.7
	(323°C)[a]	$C_{10}K$, $C_{24}K$	169.6[b]	3.68	3.2	−0.5
16.2	4	C_8K, $C_{24}K$	171.6	3.96	6.9	5.4
17.8	21	C_8K, $C_{24}K$	171.4	3.95	7.3	6.0
	(318°C)[a]	$C_{10}K$, $C_{24}K$	160.7[b]	3.71	3.7	0

[a] The values in parentheses are the temperatures at which breaks in the emf-vs-temperature plots occur. See text for further discussion.

[b] These emf values were measured at 327°C. The thermodynamic data correspond to temperatures above the break points.

TABLE II. Thermodynamic data for the dilute potassium–graphite compounds obtained from the electrochemical measurements.

C/K atom ratio	Run No.	Phases present	emf at 277°C (mV)	$-\Delta F_{cell}$ at 277°C (kcal/mole K)	$-\Delta H_{cell}$ (kcal/mole K)	$-\Delta S_{cell}$ (cal/°K·mole K)
32.2	5	$C_{24}K$, $C_{36}K$	307.5	7.09	7.5	0.7
32.2	5a	$C_{24}K$, $C_{36}K$	306.8	7.07	7.5	0.7
32.2	5b	$C_{24}K$, $C_{36}K$	310.6	7.16	7.9	1.4
44.6	13a	$C_{36}K$, $C_{48}K$	398.6	9.19	9.7	0.9
44.6	13b	$C_{36}K$, $C_{48}K$	399.8	9.22	9.8	1.0
56.4	9a	$C_{48}K$, $C_{60}K$	421.7	9.7	10.1	0.6
56.4	9b	$C_{48}K$, $C_{60}K$	425.3	9.8	10.6	1.3
93.8	14a	$C_{84}K$, $C_{96}K$[a]	464.1	10.70	11.3	1.1
93.8	14b	$C_{84}K$, $C_{96}K$[a]	465.7	10.73	11.3	1.1
93.8	14c	$C_{84}K$, $C_{96}K$[a]	462.2	10.66	11.7	1.6

[a] This two-phase region is hypothetical since compounds more dilute in K than $C_{60}K$ have not been identified.

TABLE III. Thermodynamic properties of potassium–graphite, rubidium–graphite, and cesium–graphite compounds.

		Potassium–graphite		Rubidium–graphite		Cesium–graphite	
Stage	Equilibrium reaction	$-\Delta H°$ (cal/mole k)	$-\Delta S°$ (cal/mole K·°K)	$-\Delta H°$ (cal/mole Rb	$-\Delta S°$ (cal/mole Rb·°K)	$-\Delta H°$ (cal/mole Cs)	$-\Delta S°$ (cal/mole Cs·°K)
1a	$1/3C_{24}M(s)+M(s)\rightleftarrows C_8M(s)$	27 400	25.7				
1b	$4C_{10}M(s)+M(g)\rightleftarrows 5C_8M(s)$	38 000	44	33 900	42 2	43 800	43 4
2	$5/7C_{24}M(s)+M(g)\rightleftarrows 12/7C_{10}M(s)$	24 000	24	25 300	19 2	29 600	19.6
3	$2C_{36}M(s)+M(g)\rightleftarrows 3C_{24}M(s)$	27 800	20.6	27 200	17.0	32 700	18 7
4	$3C_{48}M(s)+M(g)\rightleftarrows 4C_{36}M(s)$	30 000	20.7	29 500	17.6	34 200	18.6
5	$4C_{60}M(s)+M(g)\rightleftarrows 5C_{48}M(s)$	30 600	20.8	31 100	18.3	34 900	18.6
6	$5C_{72}M(s)+M(g)\rightleftarrows 6C_{60}M(s)$			31 800	18 1	35 800	18.8
7	$6C_{84}M(s)+M(g)\rightleftarrows 7C_{72}M(s)$						
8	$7C_{96}M(s)+M(g)\rightleftarrows 8C_{84}M(s)$	31 700	20.9				

18. "Thermodynamics of the Solubility and Permeation of Hydrogen in Metals at High Temperature and Low Pressure" D.S. Shupe and R.E. Stickney 51, 1620 (1969).

"The permeation, diffusion, and solution of hydrogen in metals have been widely studied because of the fact that a very small amount of absorbed hydrogen can lead to drastic changes in the mechanical properties of some metals." Examples of extreme temperature and pressure conditions are nuclear reactors and ultrahigh vacuum systems. The authors reference studies that show that both the density of absorbed atoms and permeation rate appear to pass through a maximum as T is increased at constant total pressure $p_t=p_H+p_{H_2}$, when operating in the high-temperature-low-pressure regime. Both the equilibrium and kinetic effects are discussed; we shall follow their analysis of the equilibrium situation by means of classical thermodynamics.

This maximum in the number of H atoms absorbed per atom of solid, called $x(x<<1)$, as a function of T is shown as the solid line in Fig. 1.

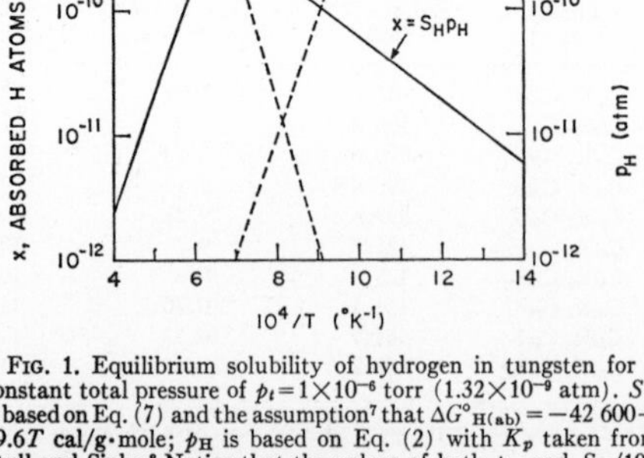

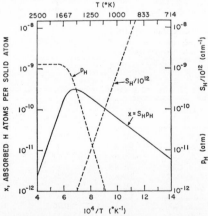

FIG. 1. Equilibrium solubility of hydrogen in tungsten for a constant total pressure of $p_t=1\times10^{-6}$ torr (1.32×10^{-9} atm). S_H is based on Eq. (7) and the assumption[7] that $\Delta G°_{H(ab)}=-42\ 600+29.6T$ cal/g·mole; p_H is based on Eq. (2) with K_p taken from Stull and Sinke.[8] Notice that the values of both p_H and $S_H/10^{12}$ are given along the right-hand ordinate.

$S_H \equiv x/p_H$. For the reaction

$$\tfrac{1}{2}H_2(g) \rightleftarrows H(g)$$

the equilibrium constant

$$K_p = p_H/p_{H_2}^{\frac{1}{2}} \tag{2a}$$

is related to the standard free energy change for the above reaction by

$$K_p = \exp(-\Delta G^\circ_{H(g)}/RT). \tag{2b}$$

What are the standard states for the atom and the diatomic? By the use of pressures(what unit? necessarily!) rather than fugacities have we implicitly assumed the ideal gas law to be valid? Is this legitimate in the regime under discussion?

The authors next restrict consideration to solids which, rather than react with hydrogen to form a new solid phase, absorb hydrogen atoms as expressed by $H(g) \rightleftarrows H(ab)$. Equilibrium between gaseous and absorbed atoms requires that the chemical potentials of the atoms in the two phases be equal

$$\mu_{H(ab)} = \mu_{H(g)} = \mu^\circ_{H(g)} + RT \ln p_H.$$

Is the choice of standard state for the gaseous atom the same as it was in the discussion of the dissociation reaction?

For the absorbed atoms it is surely an assumption of convenience that they form an ideal solution with the solid, that is that the mole fraction of absorbed H atoms in the solid is equal to the partial pressure due to gaseous H atoms in equilibrium with the absorbed atoms divided by the vapor pressure that the solid would exert if it were 100% H atoms, with all substances at the same temperature T. Of course the solid composed of pure hydrogen atoms is fictional, but the fiction is useful here.

Does it seem realistic to assume the solid solution is ideal? Do you think activity coefficients have been measured as they had been in exercise no. 3? Show that the mole fraction of H atoms absorbed in the solid reduces to x as defined earlier when $x \ll 1$ which was one of our initial premises and is actually the case experimentally.

In one phrase the authors seem to equate "dilute" and "ideal" solutions; this isn't correct as a solution may be dilute and still be quite nonideal. It is of little matter as their equation for the chemical potential of the absorbed H atoms

$$\mu_{H(ab)} = \mu^\circ_{H(ab)} + RT \ln x$$

fixes the choice of the standard state for the absorbed atoms.

The last two equations together let us define the stability constant

$$S_H \equiv x/p_H = \exp(-\Delta G^\circ_{H(ab)}/RT). \tag{7}$$

How is it related to the equilibrium constant for the reaction $H(g) \rightleftarrows H(ab)$? The free energy change on the rhs is the difference of standard chemical potentials for this reaction.

Combine these results to get their result for x

$$x = S_H p_H = \tfrac{1}{2} S_H K_p^2 [-1 + (1 + 4p_t/K_p^2)^{\frac{1}{2}}]. \quad (10)$$

This is plotted in Fig. 1 with numerical values in the caption. Notice that x is indeed very small. It is the product of the two dotted curves; do they each have the correct temperature-dependence? The equation for p_H as a function of p_t and T is given by dividing equation (10) by S_H. We do not know the exact form of the result, but we can ask if the leveling off of the p_H curve at higher temperatures, $\simeq 1700°K$, is what we would expect physically. Why? With the assumption of the indicated form for $\Delta G^\circ_{H(ab)}$ does the linear curve follow? Put these observations about the two factors of x into words that describe its overall temperature dependence.

In the limit of high temperature when $p_H \to p_t$ show that

$$\lim_{T\to\infty} x = S_H p_t = p_t \exp(-\Delta G^\circ_{H(ab)}/RT).$$

Except at very low pressures, this limit is not often approached in experimental studies; why not?

The low-temperature limit($p_t \to ?$) is

$$\lim_{T\to 0} x = S_H K_p p_t^{\frac{1}{2}} = p_t^{\frac{1}{2}} \exp[-(\Delta G^\circ_{H(g)} + \Delta G^\circ_{H(ab)})/RT].$$

Write the reaction corresponding to the sum of the standard free energy changes in the above equation, called ΔG°_S. Since most experimental conditions correspond to the low-temperature limit the more common definition of solubility is

$$S_{H_2} \equiv x/p_{H_2}^{\frac{1}{2}} = S_H K_p = \exp(-\Delta G^\circ_S/RT);$$

verify the equalities. Under what condition may p_t be substituted for p_{H_2} in this equation?

From the two limit equations what is the dependence of x on p_t as temperature increases? Does the order of the dependence change?

19. "Thermodynamic Properties and the Cohesive Energy of Calcium Ammoniate" S. Dickman, N.M. Senozan, and R.L. Hunt <u>52</u>, 2657 (1970).

This article reports vapor pressure measurements and the thermodynamic properties associated with the dissociation process

$$Ca(NH_3)_6(s) \rightleftarrows Ca(s) + 6NH_3(g)$$

as well as some calculations and discussion of cohesive energies for calcium ammoniate and the alkali and alkaline earth metals.

The alkaline earth ammoniates were reported just before the turn of the century in their ref. 1. Upon evaporation of ammonia from solutions of calcium, strontium, and barium in liquid ammonia, lustrous golden-bronze colored metallic compounds first form. This is unlike the solutions of the alkali metals which yield the metals themselves. Though the alkaline earth compounds have been referred to in the literature as metallic, apparently no quantitative studies of their electrical or thermal conductivities, thermal expansion, reflectivity, magnetic susceptibility, or any such physical property have been made. This must be due in part to the difficulties in preparing sufficiently stable samples. The irreversible decomposition is

$$Ca(NH_3)_6(s) \rightarrow Ca(NH_2)_2(s) + 4NH_3(g) + H_2(g)$$

catalyzed by even minute amounts of impurities, particularly iron compounds and amides.

Fig. 1 gives the experimental vapor pressure measurements;

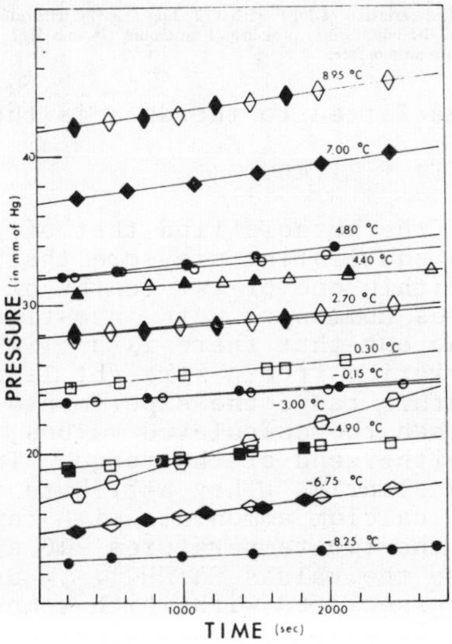

FIG. 1. The variation of pressure above calcium ammoniate with time. Different shapes correspond to different samples; two consecutive set of readings taken on the same sample at the same temperature are differentiated by shading the first set of points.

to what would you attribute the slight increase in pressure with time? If it is due to the evolution of hydrogen by the decomposition reaction what is the reaction order? Is this what you would expect from a surface reaction? Should the decomposition reaction be a surface one? What value(s) of time would you choose in order to obtain the equilibrium pressure of ammonia? You should know that the calcium metal and ammonia, twice distilled over sodium(why?), were originally brought together at dry ice temperature. After solution was complete the solution was warmed to -45°C and the ammonia pumped off; then a methanol bath regulated to one of the constant temperatures shown in Fig. 1 was placed around the sample tube at time zero. Notice that no pressure data was taken during the ensuing transient periods; would these be meaningful? Can you suggest a mechanism or effect to account for the larger slope of the sample marked by octagonal points?

Check your decision as to the value of time used from Fig. 1 to determine equilibrium pressures of ammonia by means of the points in Fig. 2. Notice that here "log" is $\log_e$ or ln.

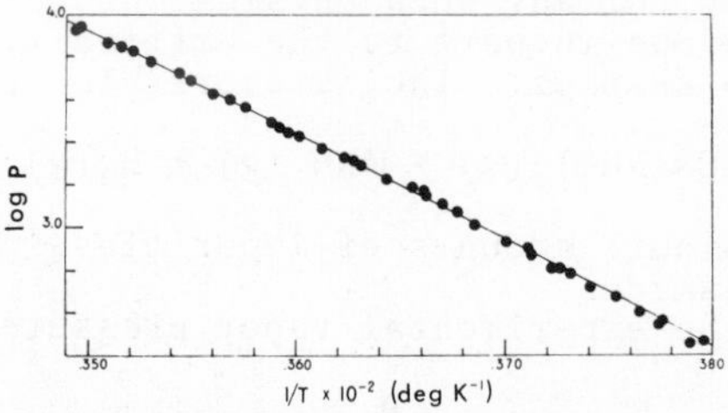

FIG. 2. The variation of log$P$ with $1/T$. Log P is the natural logarithm of the equilibrium pressure of ammonia (in mm Hg) above calcium ammoniate.

The straight line fitted to the data is the equation

$$\ln P_{mm} = (-5007/T) + 21.46.$$

Interestingly enough, the authors find that of three other investigations of the equilibrium pressure that of the earliest (1908) was closest, within one or two tenths of a millimeter at the two temperatures common to both investigations.

The authors point out that there is a slight but definite concave(downward) deviation from a straight line in Fig. 2. In about the mid-temperature range the experimental values are consistently higher than the calculated values, whereas the opposite is true at either end of the range. Table I, not given here, shows this more clearly. They attribute this to a change in the composition of calcium ammoniate with temperature. Their ref. 11 says that at the two temperatures -10 and 13°C the actual compositions have the values $Ca(NH_3)_{5.87}$ and $Ca(NH_3)_{5.89}$; which composition is associated with which temperature if the ammonia-rich compound is assumed more volatile?

Back to the equation for ln P. Recognition of the form of this equation and a routine operation lets us compute the enthalpy of dissociation to be compared with their value of $\Delta H°=9.95\pm0.10$ kcal/mole NH_3 formed. Thinking about the utter disdain of thermodynamics for our models of atomic and molecular structure and mechanism we see that the derivative of ln P wrt 1/T gives $-\Delta H/R$, if the vapor is an ideal gas, where ΔH is the enthalpy difference between condensed and vapor phases when the two are in equilibrium. You may think of the condensed phase to be solid, glass or liquid and the dissociation process to be a vaporization or better, sublimation; we still have "ΔH"! Two other nonprofound remarks: the superscript zero is added because the isothermal compression of the vapor to one atmosphere produces no enthalpy change for an ideal gas, and the units of ΔH are those of RT, cal/mole(vapor) - where did R come from anyway? - and the smart money bets that you had the answer as about 10 000 cal/mole or 10 kcal/mole before you thought about the coefficients in the dissociation equilibrium equation as written earlier.

Multiplying all of these coefficients by 1/6 allows us to now write the equilibrium constant for the dissociation reaction as $K=P_{NH_3}$ by assuming that activities of the solid phases are unity. In passing do you detect in their discussion of the equilibrium constant a place where the word fugacity should be replaced by activity? Now you compute values of ΔF°_{273} and ΔS°_{273} to be compared with theirs of 1.890±0.010 kcal and 29.5±0.4 eu. Why is greater accuracy reported for ΔF° than for ΔH°? Which of these two controls the accuracy of ΔS°?

In their discussion of the cohesive energy they define the modified cohesive energy of a solid complex as the energy difference between a gaseous molecule ion that has lost the valence electrons of the metal atom and the solid complex itself

$$Ca(NH_3)_6(s) \rightarrow Ca(NH_3)_6^{2+}(g) + 2e^-(g).$$

What is the difference between this definition and the more usual one of cohesive energy?

The energy change for the above reaction is evaluated by computing the enthalpy changes for the following sequence of four reactions, adding them, and converting from ΔH to ΔE by means of ΔnRT:

dissociation of calcium ammoniate	?→?	ΔH= 59.7 kcal
sublimation of calcium	?→?	ΔH= 42.3
double ionization of calcium	?→?	ΔH= 416.5
gaseous ammoniation of Ca^{2+}	?→?	ΔH=-282

They find the modified cohesive energy of calcium ammoniate to be 234 kcal/mole. Could this cycle be considered as a special case or extension of a general cycle developed to compute crystal lattice energies in ionic solids?

Table II presents cohesive energies computed for alkali and and alkaline earth metals in this way for comparision with the last result. You should be able to duplicate one of the metal entries in the Table from data given above.

TABLE II. Comparison of the cohesive energies. The cohesive energy of M is defined as the energy of the reaction $M(s) \rightarrow M^{+n}(g) + ne^-(g)$.

Metal	Modified Hartree cohesive energy (kcal/mole)	Actual cohesive energy (kcal/mole)
Li	108	162
Na	98	144
K	86	121
Rb	82	115
Cs	77	108
Mg	478	558
Ca	420	456
Sr	394	424
Ba	379	393
$Ca(NH_3)_6$	102	234

The lower value for calcium ammoniate, compared to calcium, is their reason for stating that the former is more metallic or has more valence electron delocalization.

20. "Properties of Fluorine along the Vapor-Liquid Coexistence Boundary" R. Prydz, G.C. Straty, and K.D. Timmerhaus 53, 2359 (1970).

Having opened this Chapter with an exercise on a chemical property of fluorine, we now close it with one on some physical properties. The need for knowledge of these properties comes not only from a desire to increase our understanding of fluids but also from engineers designing nonairbreathing engines. Note the full institutional affiliation of the first two authors and the sponsoring organization acknowledged at the end in this connection.

TABLE I. Derived fluorine saturation densities.

T (°K)	Dens exptl mol/liter	Dens. calc mol/liter	T (°K)	Dens. exptl mol/liter	Dens. calc mol/liter
53.4811	44.86	44.862	121.682	31.288	31.287
54.541	44.690	44.696	123.443	30.761	30.757
55.572	44.541	44.533	123.944	30.600	30.602
57.516	44.223	44.224	129.378	28.765	28.766
59.837	43.858	43.851	131.805	27.830	27.829
59.840	43.857	43.851	134.753	26.535	26.545
61.601	43.563	43.566	136.977	25.427	25.421
63.581	43.234	43.243	139.075	24.172	24.166
63.697	43.216	43.224	140.614	23.031	23.046
65.717	42.887	42.891	141.950	21.876	21.822
67.709	42.564	42.560	141.945	21.868	21.827
69.842	42.206	42.202	142.948	20.672	20.594
69.850	42.205	42.201	143.782	19.822	19.002
71.703	41.888	41.887	144.052	18.753	18.123
73.796	41.528	41.528	144.415	17.928	
73.824	41.525	41.523	144.978	15.773	
73.834	41.521	41.522	144.507	14.498	
75.381	41.252	41.254	144.414	13.423	
77.421	40.894	40.897	144.121	11.928	12.380
79.813	40.470	40.474	143.608	10.869	10.873
79.811	40.470	40.474	142.937	9.7727	9.8008
81.837	40.112	40.110	142.264	9.0213	9.0381
81.838	40.112	40.110	141.530	8.3853	8.3801
83.832	39.747	39.747	140.606	7.6858	7.7038
83.840	39.745	39.746	139.209	6.8755	6.8793
85.898	39.370	39.366	138.660	6.5942	6.6004
87.855	39.005	39.000	137.526	6.0828	6.0837
89.622	38.667	38.665	136.012	5.4888	5.4899
91.284	38.343	38.345	133.730	4.7445	4.7443
93.466	37.919	37.919	129.430	3.6640	3.6611
93.470	37.915	37.918	123.881	2.6486	2.6493
95.544	37.499	37.505	120.060	2.1167	2.1181
97.722	37.066	37.063	114.481	1.5138	1.5127
99.972	36.596	36.596	110.080	1.1438	1.1445
99.996	36.586	36.591	104.344	0.77679	0.77647
101.566	36.255	36.258	99.742	0.55397	0.55436
101.723	36.223	36.224	94.251	0.35765	0.35702
103.274	35.892	35.890	88.461	0.21181	0.21210
105.589	35.380	35.378	84.355	0.14035	0.14025
107.242	35.006	35.004	80.000	0.08598	0.08603
109.943	34.373	34.374	78.746	0.07396	0.07392
109.969	34.363	34.368	75.000	0.04538	0.04540
111.813	33.925	33.923	73.703	0.03787	0.03785
113.792	33.434	33.432	70.000	0.02158	0.02157
113.784	33.432	33.434	65.000	0.00897	0.00897
115.916	32.887	32.886	60.000	0.00315	0.00315
117.676	32.419	32.418	54.000	0.00067	0.00067
119.552	31.903	31.900	53.4811	0.00057	0.00057
119.587	31.890	31.890			

Table I presents saturated liquid and vapor densities at temperatures from the triple point to the critical point, close to 100 measurements in all. Table II lists the critical parameters of fluorine

TABLE II. Fluorine critical parameters.

T_c, IPTS 1968 (°K)	ρ_c (mol/liter)	P_c (MN/m²)	Ref.
144	...	5.57	17
	12.4		15
	15.0		16
143.9	15.1		4
144.31±0.05	15.10±0.04[a]	5.215	This research

[a] The estimates of the uncertainty in these critical parameters were based on the experience gained through the analysis of the data since no direct statistical measure is obtained from this iterative fitting procedure.

determined by fitting a complicated empirical equation to the data above 120°K in Table I for ρ_C and T_C, then determining P_C from a very accurate vapor pressure equation by knowing T_C. Separate equations were then used for the vapor-saturated liquid and the liquid-saturated vapor with each equation constrained to pass through the critical point. Notice in Table I that experimental densities were determined at several temperatures just higher than the critical temperature.

Beginning with the triple point temperature in Table I and continuing up to the critical point at 10 or 15 degree intervals compute ρ/ρ_C and T/T_C for both liquid and vapor, taking care to use the same value of T/T_C for each; this will necessitate some interpolation. Make a plot of these two quotients and compare for other fluids. Evaluate a and b for fluorine in one of the oldest results of physical chemistry, the law of rectilinear diameters(1886)

$$\rho_{avg} = a - bT,$$

where ρ_{avg} is the arithmetic mean of the liquid and vapor density.

The triple point marks the lower end of the vapor-liquid coexistence boundary. The temperature and pressure are 53.4811 °K and 2.52×10^{-4} MN/m². Given that 1 MN/m²=9.86923 atm, can you determine what an MN is if it is an unfamiliar unit?

The vapor pressure data in their ref. 1 were fitted to an older temperature scale there. In the present article they were fitted to the equation

$$\ln (P/P_t) = D_1X + D_2X^2 + D_3X^3 + D_4X(1-X)^{D_5}$$

where

$$X = (1-T_t/T)/(1-T_t/T_c),$$

and D_1 through D_5 taking the values 7.89592346, 3.38765063, -1.34590196, 2.73138936, and 1.4327 respectively. The subscripts t and c stand for triple point and critical.

We want to use this vapor pressure equation for four problems. In increasing order of difficulty they are i) determine the triple point pressure and compare with that given above,

ii) setting D_2 through D_4 equal to zero evaluate an approximate value for the heat of vaporization, iii) check and see if the value given for the normal boiling point of fluorine, $84.95_0 \pm 0.003$°K fits the equation, and iv) determine the exact value for the heat of vaporization at the nbp using saturated liquid and vapor molar volumes calculated from data in Table I and the exact form of the Clapeyron equation

$$dP/dT = \Delta S/\Delta V$$

for two-phase equilibrium. What is the error in the heat of vaporization as determined in problem ii)?

CHAPTER II

QUANTUM MECHANICS

21. "The Spectra and Electronic Structure of the Tetrahedral Ions MnO_4^-, CrO_4^{--}, and ClO_4^-" M. Wolfsberg and L. Helmholz 20, 837 (1952).

This paper is a pioneering one in theoretical inorganic chemistry in that the authors "have applied to inorganic complex ions a semiempirical method similar to that which has yielded significant results in the case of organic molecules". The ions chosen were permanganate, chromate, and perchlorate.

The last ion has no visible or near ultraviolet electronic transitions, in common with the isoelectronic sulfate and phosphate ions, though the energy of the first transition that would be encountered in the uv will be predicted later. The ions of the fourth-row transition elements with the tetrahedron of oxygen atoms as shown in Fig. 1 and with the same number of electrons show characteristic visible and near uv absorption, two strong maxima with the corresponding peaks displaced toward shorter wavelengths with decreasing atomic number of the central metal atom. This trend is also observed for the corresponding isoelectronic compounds of the fifth- and sixth-row elements. As the authors point out, a satisfactory theory explains these differences and regularities.

In Fig. 1 the x,y and z axes of the orbitals centered on the metal atom(called X orbitals, for XO_4, in Table I) are those perpendicular to the cube faces. The z axes of the oxygen ligands point toward the metal atom. They are labeled σ_i while the x and y axes of the oxygens are labeled π_{xi} and π_{yi}; is this in accord with the usual convention for 2p orbitals?

From the symmetry orbitals listed in Table I one sees that only the valence orbitals are used in the construction of the former. Is this quite a reasonable approximation? It has only been within the last several years that all-electron calculations have been performed for ions and molecules of this size. Are the atomic orbitals for the metal already symmetry orbitals belonging to the given irreducible representations of the tetrahedral point group T_d? What assumption is made about the overlap integrals between atomic orbitals on different oxygen nuclei? The oxygen 2s orbitals were not considered to be valence orbitals in this work though this restriction was removed later. Could the σ_i atomic orbitals represent an optimized combination of $2s_i$ and $2p_i$ orbitals without changing the symmetry situation

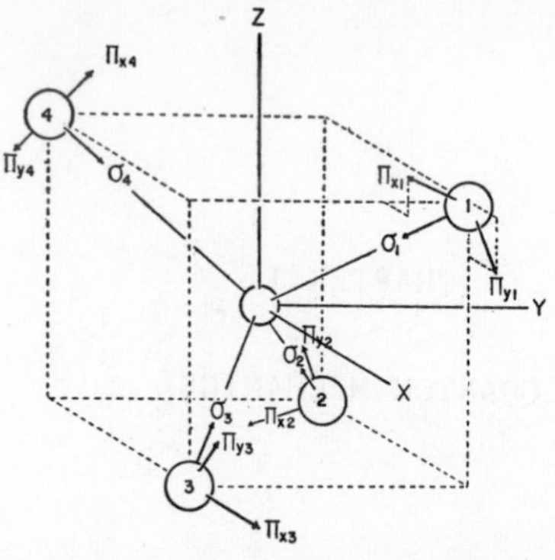

FIG. 1. Orientation of the orbitals. The orientation of the oxygen orbitals relative to the axes on the central atom is shown in this figure. The direction cosines of the O_1 orbitals are σ_1: $-1/3^{1/2}$, $-1/3^{1/2}$, $-1/3^{1/2}$; π_{x1}: $1/6^{1/2}$, $-2^{1/2}/3^{1/2}$, $1/6^{1/2}$; π_{y1}: $1/2^{1/2}$, 0, $-1/2^{1/2}$. The direction cosines of the other oxygen orbitals may be obtained from these by the action of the twofold axes of T_d, which coincide with the x, y, and z axes shown. The central atom orbitals are defined with respect to the axes on the central atom.

TABLE I. Molecular orbital combinations—XO_4, symmetry T_d.

X orbitals	O orbitals		Irreducible representations
s	$(\frac{1}{2})[\sigma_1+\sigma_2+\sigma_3+\sigma_4]$		A_1
d_{z^2}[a]		$(\frac{1}{4})[\pi_{x1}+\pi_{x2}+\pi_{x3}+\pi_{x4}-3^{1/2}(\pi_{y1}+\pi_{y2}+\pi_{y3}+\pi_{y4})]$	E
$d_{x^2-y^2}$		$(\frac{1}{4})[\pi_{y1}+\pi_{y2}+\pi_{y3}+\pi_{y4}+3^{1/2}(\pi_{x1}+\pi_{x2}+\pi_{x3}+\pi_{x4})]$	
p_x, d_{yz}	$(\frac{1}{2})[\sigma_1+\sigma_3-\sigma_2-\sigma_4]$	$(\frac{1}{4})[\pi_{x4}+\pi_{x2}-\pi_{x1}-\pi_{x3}+3^{1/2}(\pi_{y4}+\pi_{y2}-\pi_{y1}-\pi_{y3})]$	T_2
p_y, d_{xz}	$(\frac{1}{2})[\sigma_1+\sigma_2-\sigma_3-\sigma_4]$	$(\frac{1}{2})[\pi_{x1}+\pi_{x2}-\pi_{x3}-\pi_{x4}]$	
p_z, d_{xy}	$(\frac{1}{2})[\sigma_1+\sigma_4-\sigma_2-\sigma_3]$	$(\frac{1}{4})[\pi_{x3}+\pi_{x2}-\pi_{x1}-\pi_{x4}+3^{1/2}(\pi_{y4}+\pi_{y1}-\pi_{y2}-\pi_{y3})]$	
		$(\frac{1}{4})[\pi_{y2}+\pi_{y4}-\pi_{y3}-\pi_{y1}+3^{1/2}(\pi_{x1}+\pi_{x3}-\pi_{x2}-\pi_{x4})]$	T_1
		$\frac{1}{2}[\pi_{y1}+\pi_{y2}-\pi_{y3}-\pi_{y4}]$	
		$(\frac{1}{4})[\pi_{y2}+\pi_{y3}-\pi_{y1}-\pi_{y4}+3^{1/2}(\pi_{x2}+\pi_{x3}-\pi_{x1}-\pi_{x4})]$	

[a] The designation of the d-orbitals is that given by Eyring, Walter, and Kimball, *Quantum Chemistry* (John Wiley and Sons, Inc., New York, 1944). We have also used throughout the group theoretical symbolism which is essentially that of these same authors.

of the A_1 and T_2 symmetry species? What about the E and T_1 species?

The calculations of the molecular orbital energies ε("which we employ in the same manner as is usual in the semiempirical methods for organic molecules") and the coefficients of the symmetry orbitals that make up the molecular orbitals require solution of the secular equation

$$|H_{ij}-G_{ij}\varepsilon| = 0,$$

one for each symmetry species; 2x2 determinants for the A_1 and E species and a 4x4 for the T_2 species. The orbital of symmetry T_1 is nonbonding and is thus fully determined by symmetry. The H_{ii} are the valence-state ionization energies for electrons in each of the appropriate atomic orbitals. These are given in Fig. 2. It turns out that these values were somewhat parameterized though the distinction is not important for our purposes. The method used to approximate the off-diagonal elements H_{ij} has been used very much since this paper appeared and is known as the Wolfsberg-Helmholz approximation(F_σ=1.67 and F_π=2.00)

$$H_{ij} = F_x G_{ij}(H_{ii}+H_{jj})/2.$$

The G_{ij} are the group overlap integrals, that is the overlap integrals between symmetry orbitals within the same irreducible representations. These reduce ultimately to constants multiplied by ordinary two-center diatomic overlap integrals. Verify that the group overlap integral within the A_1 species is

$$G_{A_1}(s,\sigma) = 2S(s,p_\sigma)$$

with S the diatomic overlap. Fig. 2 shows the H_{ii} and the molecular orbital energies and Table III the G_{ij} of the three ions involved.

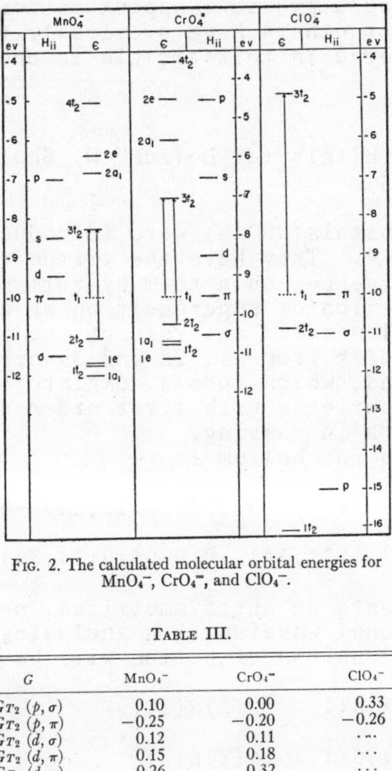

FIG. 2. The calculated molecular orbital energies for MnO_4^-, $CrO_4^=$, and ClO_4^-.

TABLE III.

G	MnO_4^-	$CrO_4^=$	ClO_4^-
$G_{T_2}(p,\sigma)$	0.10	0.00	0.33
$G_{T_2}(p,\pi)$	−0.25	−0.20	−0.26
$G_{T_2}(d,\sigma)$	0.12	0.11	...
$G_{T_2}(d,\pi)$	0.15	0.18	...
$G_E(d,\pi)$	0.26	0.32	...
$G_{A_1}(s,\sigma)$	0.28	0.27	0.56

What are the values of G_{ii} within the same symmetry species and of G_{ij} when symmetry orbitals of different symmetry species are involved?

Solve the quadratic secular equations for A_1 and E and compare the resultant values with those for the permanganate and chromate ions in Fig. 2. Why are the energies for the nonbonding T_1 orbitals placed equal to those of the $2p\pi$ oxygen orbitals?

Place the 24 valence electrons(again not counting the oxygen 2s electrons) into the appropriate molecular orbitals and thus construct the ground state configurations for the three ions. Also construct the excited configurations corresponding to the upper end of the transitions shown in Fig. 2. Measure the transition energies and compare with those given in Table IV. Can the differences and regularities now be explained?

TABLE IV. Calculated and observed excitation energies.

	MnO_4^- Obs	MnO_4^- Calc	CrO_4^- Obs	CrO_4^- Calc	ClO_4^- Obs	ClO_4^- Calc
1st transition	2.29 ev	1.68 ev	3.25 ev	2.42 ev	6 ev	5.23 ev
2nd transition	3.96	2.78	4.59	3.15	···	···

Advances in semiempirical methods are made, with a few exceptions, by improving a previous approximation or relaxing a single assumption that others have previously felt forced to make; the method covered in this article is one of the exceptions.

22. "Natural Spin Orbitals for Helium" H. Shull and P.-O. Löwdin 23, 1565 (1955).

Natural spin orbitals(NSO's) were introduced by the second author in their ref. 4. They have the virtue that they provide the most rapid convergence, on a term by term basis, of any configuration interaction(or superposition of configurations) scheme. We will follow their analysis of the best spatial wavefunction, for Z=2, built from 1s, 2s and 3s orbitals taken from the Laguerre functions, which form a complete orthogonal set. We will gain some experience with first order density matrices and occupation numbers in passing.

The wavefunction for helium is

$$\Psi(1,2) = 0.963175(1s^2) - 0.250366(1s,2s) - 0.032613(2s^2)$$

$$+0.092211(1s,3s) - 0.000310(2s,3s) - 0.006692(3s^2).$$

Here (ns,n's) represents an antisymmetrized, normalized two-electron configurational wavefunction including the spin. If n=n' this configurational wavefunction will be a single Slater determinant, e.g.

$$(1s^2) = \frac{1}{\sqrt{2}}\begin{vmatrix} 1s(1)\alpha(1) & 1s(2)\alpha(2) \\ 1s(1)\beta(1) & 1s(2)\beta(2) \end{vmatrix}.$$

If n≠n' it will be a normalized sum or difference of Slater determinants.

Write out the three (ns,n's) configurational wavefunctions and show that each is normalized.

Remembering that the different configurations are orthogonal show that Ψ(1,2) is normalized.

As covered by these authors elsewhere(Phys. Rev. 101, 1730 (1955) and J. Chem. Phys. 30, 617 (1959)) the first order density matrix takes a particularly simple form for a two-electron wavefunction expressed in terms of different configurations

$$\Psi(1,2) = \sum_k C^\circ_{kk}\psi_k(1)\psi_k(2) + \sum_{k<j} C^\circ_{kj}[\psi_k(1)\psi_j(2)+\psi_j(1)\psi_k(2)]/\sqrt{2},$$

where here $\psi_k(1)\psi_k(2)$ corresponds to our (ns^2) and $[\psi_k(1)\psi_j(2)+\psi_j(1)\psi_k(2)]/\sqrt{2}$ to (ns,n's). The first-order density matrix may be obtained in terms of the coefficients C_{kj}, where

$$C_{kk} = C^\circ_{kk}$$

$$C_{kj} = C^\circ_{kj}/\sqrt{2}.$$

Demonstrate that for the C°_{kj}

$$\sum_k |C^\circ_{kk}|^2 + \sum_{k<j} |C^\circ_{kj}|^2 = 1.$$

Thus the C_{kj} are themselves not normalized. The first-order density matrix $\underline{\Gamma}$ is given by the product of the $\underline{C}$ matrix and its transpose conjugate("tranjugate") $\underline{C}^\dagger$

$$\underline{\Gamma} = \underline{CC}^\dagger$$

or

$$\Gamma_{rs} = \sum_t C_{rt}C^*_{st}.$$

In our case the 3x3 matrix can be readily diagonalized; the eigenvalues are called occupation numbers and are given below. The coefficients obtained from the diagonalized density matrix are just the coefficients of (ns) in the normalized NSO's as shown below:

NSO	Occupation Number	Normalized NSO
X_1	0.995660	0.983545(1s) -0.168992(2s) +0.063880(3s)
X_2	0.004265	0.178369(1s) +0.964488(2s) -0.194800(3s)
X_3	0.000075	-0.028690(1s) +0.202991(2s) +0.978760(3s)

When these NSO's are introduced into the helium wavefunction we find

$$\Psi(1,2) = \sqrt{n_1}(X_1)^2 - \sqrt{n_2}(X_2)^2 - \sqrt{n_3}(X_3)^2$$

in which the coefficients are the square roots of the occupation numbers.

Carry through this process beginning with the numerical values given for the C°_{kk} and C°_{kj} used in constructing $\underline{\Gamma}$. You can avoid diagonalization by testing each of the occupation numbers in turn to see if it is an eigenvalue as alleged. Then determine the coefficients of (ns) in the NSO's and finally verify the last result given above.

The occupation numbers represent the average number of electrons in each NSO. What relation would you expect among the three of these in the present circumstances? Is it obeyed?

The first NSO $(X_1)^2$ in the limit $n\to\infty$ is closely related to the Hartree-Fock function; to what is the minus sign in the second term related?

23. "Electronic Population Analysis on LCAO-MO Molecular Wave Functions. I" R.S. Mulliken <u>23</u>, 1833 (1955).

Mulliken's population analysis is just one of his many contributions to molecular quantum mechanics and chemical physics.

We follow here his development and application to the SCF wavefunction obtained for the CO molecule by Sahni, ref. 12, and shown in Table I. The row heading 2σ is a misprint and

TABLE I. Computed SCF-LCAO MOs for CO (by R. C. Sahni, reference 12).

ϕ_i \ χ_r	$2s_O$	$2p\sigma_O$	$2s_C$	$2p\sigma_C$	$2p\pi_O$	$2p\pi_C$	calc $-\epsilon_i(ev)$	obs $I_i(v)$
2σ	0.675	0.231	0.270	0.227			43.37	
4σ	0.718	−0.607	−0.493	−0.168			20.01	19.70
1π					0.8145	0.4162	15.97	16.58
5σ	0.187	−0.189	0.615	−0.763			13.37	14.01

should be 3σ. The electron configuration of the ground state of CO is $(1\sigma)^2(2\sigma)^2(3\sigma)^2(4\sigma)^2(1\pi)^4(5\sigma)^2$. The MO's 1σ and 2σ were approximated by Sahni as $1s_O$ and $1s_C$; which AO is identified with which MO and why? In Table I each entry is the coefficient of the AO in its column heading that go together to make up the expansion of the MO in its row heading, that is c_{ir_k} in

$$\phi_i = \sum_{r_k} c_{ir_k}\chi_{r_k} \tag{1'}$$

for AO χ_{r_k} and MO ϕ_i. For the special case of a normalized MO for a simple diatomic written in approximate form for atoms k, ι

$$\phi = c_r\chi_r + c_s\chi_s \tag{1}$$

the MO population of N electrons (usually N=2) may be considered as divided into three subpopulations whose detailed distributions in space are given by the three terms in equation (2)

$$N\phi^2 = Nc_r^2(\chi_r)^2 + 2Nc_rc_sS_{rs}(\chi_r\chi_s/S_{rs}) + Nc_s^2(\chi_s)^2, \tag{2}$$

where S_{rs} is the overlap integral $\int\chi_r\chi_s dv$. Show that each term in parentheses is a normalized distribution and that upon integration of each term in (2) one obtains the breakdown of N into three parts

$$N = Nc_r^2 + 2Nc_rc_sS_{rs} + Nc_s^2. \tag{3}$$

Verify that (3') is the appropriate generalization of (3) corresponding to (1'):

$$N(i) = N(i)\sum_{r_k} c_{ir_k}^2 + 2N(i)\sum_{s_\iota > r_k} c_{ir_k}c_{is_\iota}S_{r_k s_\iota} \tag{3'}$$

Overlap integrals for CO and H_2O, also treated in this article, are given in Table III. Can you explain the zero entries for both molecules?

TABLE III. Values of overlap integrals $S_{rs} = \int \chi_r \chi_s dv$.

χ_r \ χ_s	$2s_O$	$2p\sigma_O$	$2p\pi_O$
$2s_C$	0.4063	0.3147	0
$2p\sigma_C$	0.4807	0.3018	0
$2p\pi_C$	0	0	0.2409

χ_r \ χ_s	$1s_O$	$2s_O$	$2pz_O$	$2py_O$	$1s_b(H)$
$1s_a(H)$	0.0610	0.4946	0.2118	0.2760	0.3479
$a_1(H_2)$	0.0736	0.5965	0.2554	0	
$b_2(H_2)$	0	0	0	0.4935	

From equations (3) and (3') one sees that the total population is a sum of net atomic populations and overlap populations which are positive or negative according to the sign of S_{rs}. Referring to the simplest two-center case, equation (3), it is seen that the overlap term is related completely symmetrically to the two centers, even if they are unlike and the coefficients c_r and c_s therefore unequal. Hence it appears necessary to assign exactly half of the overlap population, plus, of course, the appropriate net atomic population, to each center. From equation (3) one then obtains for the gross atomic populations N(k) and N(ι) on atoms k and ι the expressions

$$\begin{aligned} N(k) &= N(c_r^2 + c_r c_s S_{rs}) \\ N(\iota) &= N(c_r c_s S_{rs} + c_s^2). \end{aligned} \qquad (6)$$

The sum of N(k) and N(ι) should be N; is it?
In the general case corresponding to equation (3') one has

$$N(i;r_k) = N(i) c_{ir_k} (c_{ir_k} + \sum_{\iota \neq k} c_{is_\iota} S_{r_k s_\iota}) \qquad (6')$$

and

$$\begin{aligned} N(i;k) &= \sum_r N(i;r_k) \\ N(k) &= \sum_i N(i;k) = \sum_r N(r_k) \\ N(i) &= \sum_{r_k} N(i;r_k) = \sum_k N(i;k) \\ N(r_k) &= \sum_i N(i;r_k) \\ N &= \sum_i N(i) = \sum_{r_k} N(r_k). \end{aligned} \qquad (7)$$

$N(i;r_k)$ is the partial gross population in MO ϕ_i and AO χ_{r_k}; $N(i;k)$ is the subtotal in MO ϕ_i on atom k. N(k) is the total gross population on atom k. N(i) is the total population in MO ϕ_i; is this number necessarily an integer and if so why? $N(r_k)$ is the total gross population in AO χ_{r_k}. N is the overall total population of electrons in the molecule. It turns out that $N(r_k)$, N(k), and N are invariant wrt orthogonal transformations of the occupied LCAO MO's.

From the definitions (6') and (7) the author further defines the "gross charges"

$$Q(r_k) = N_o(r_k) - N(r_k) \qquad (8a)$$

in any AO, and

$$Q(k) = N_0(k) - N(k) \tag{8b}$$

on any atom, where $N_0(r_k)$ is the number of electrons in the AO χ_r and $N_0(k)$ the total number of electrons in the ground state of the free neutral atom k. The Q's in equations (8) are in units of +e.

Table IV contains the application of these ideas to the CO LCAO-MO-SCF wavefunction of Sahni's. You should be able

TABLE IV. Gross atomic populations and charges in CO (see Eqs. (6′), (7), (8).

ϕ_i \ χ_{r_k}	Partial populations $N(i; r_k)$: $2s_O$	$2p\sigma_O$	$2p\pi_O$	$2s_C$	$2p\sigma_C$	$2p\pi_C$	$N(i; O)$	$N(i; C)$	$N(i)$
3σ	1.207	0.178		0.333	0.282		1.385	0.615	2.000
4σ	0.627	0.985		0.386	0.002		1.612	0.388	2.000
1π			2.980			1.020	2.980	1.020	4.000
5σ	0.026	0.085		0.776	1.113		0.111	1.889	2.000
$N(r_k)$	1.860	1.248	2.980	1.495	1.397	1.020	$N(O)=$ 6.088	$N(C)=$ 3.912	$N=$ 10.000
$Q(r_k)$ in e units	+0.140	−0.248	+0.020	+0.505	−0.397	−0.020	$Q(O)=$ −0.088	$Q(C)=$ +0.088	0.000

to produce all of the entries from data given earlier.

From Table IV write the effective electron configurations for the C and O atoms <u>in the CO molecule</u>, to be compared with

$$1s_C^{2.00}2s_C^{2.00}2p_C^{2.00} \text{ and } 1s_O^{2.00}2s_O^{2.00}2p_O^{4.00}$$

for the ground states of the neutral atoms.

Is it evident that there has been a charge transfer accompanying molecule formation of -0.09e from C to O? Can we answer the questions how much of this transfer has been out of $2s_C$ and how much out of $2p_C$, and how much as been into $2s_O$ and how much into $2p_O$? If these questions have no definite answer do they indeed have any real meaning? Mulliken does suggest, however, that most of the charge transfer has been out of $2p_C$ and into $2p_O$, because the net 2s promotions which exist in CO, both out of $2s_C$ and into $2p\sigma_C$ and out of $2s_O$ and into $2p\sigma_O$, must have arisen mainly as a response to the possibility of gains in stability by hybridization; and if so, the extent of these promotions should be relatively independent of loss or gain of charge in, say, the 2p AO's.

If one defines the amounts of promotion in the atoms in CO on the basis that charge transfer involves only 2p AO's, what is the amount of $2s \rightarrow 2p\sigma$ in each atom?

Interpretations of the very small dipole moment and some of the chemical properties of CO, for example its ability to function as an electron donor

$$B_2H_6 + 2CO \rightarrow 2H_3BCO$$

are given in the article; can you follow the arguments given there in terms of population analyses?

These ideas have proven to be quite fruitful, particularly so when applied to larger molecules where the wavefunctions are necessarily still very approximate.

24. "Hartree-Fock Approximation of CH_4 and NH_4^+" M. Krauss 38, 564 (1963).

Gaussian orbitals, to be examined in exercise no. 25, have come to rival Slater-type orbitals as bases functions for molecular calculations. This paper by Krauss reports the calculation of the energies of methane and the isoelectronic ammonium ion with a gaussian basis set. It will give us an opportunity to discuss the partitioning of various energy terms in a molecular calculation.

Table III presents the results of his calculations.

TABLE III. Summary of results.

	CH_4	NH_4^+
Calculated Hartree–Fock approximation	−40.1668	−56.5038
Experimental energy	−40.522[a]	−56.857[b]
Sum of Hartree–Fock energies for separated atoms	−39.689[c]	−55.901[c]
Relativistic correction	−0.012[c]	−0.024[c]
Upper limit to molecular correlation energy	−0.343	−0.330
Hartree–Fock binding energy	−0.478	−0.603
Experimental binding energy	−0.625[d]	−0.769[b]
Upper limit to correlation contribution to binding energy	−0.147	−0.166

[a] See reference 6.
[b] See T. L. Cottrell, *The Strength of Chemical Bonds* (Butterworths Scientific Publications, Ltd., London, 1958); reference 9; E. L. Wagner and D. F. Hornig, J. Chem. Phys. **18**, 296 (1950).
[c] See reference 8.
[d] G. Glockler, J. Chem. Phys. **21**, 1242 (1953).

The Hartree-Fock limit is defined to be the lowest(best in the sense of the variation theorem?) energy obtained with the minimum number of Slater determinants necessary to be eigenfunctions of the various angular momentum operators. How many determinants will be required for closed shell species such as these? His calculated Hartree-Fock approximation for methane is a considerable improvement over all values published to that time. Is the gap between his SCF value and the CH_4 H-F limit easily estimated?

The experimental energy for species on the quantum mechanical scale can be computed from thermodynamic data and ionization potentials; do this for CH_4 and NH_4^+ using the value of ref. 9 for the proton affinity of ammonia.

Unlike molecules other than some diatomics the Hartree-Fock limits for the ground states of the lighter atoms have been reached. Can you duplicate his results for the Hartree-Fock energies of the separated atoms given the H-F energies of $C(^3P)$, $N^+(^3P)$, and $N(^4S)$ are -37.689, -53.888, and -54.401 atomic units respectively? By comparison of the ionization potentials of N and H, which of the systems $N^+ + 4H$ or $N + 3H + H^+$ is favored energetically? Your answer here will determine which energy values to use to compute the H-F energies of the separated atoms of NH_4^+.

The relativistic corrections, always negative, are fortunately small here as they are difficult to compute. For our species they reside almost completely in the K shell of the heavier atoms. A crude check is available from the one-electron problem where the leading spin-orbit and relativity energy term (called jointly the "relativity correction" in this and other applications) has a Z^4 dependence. Apply this check.

The correlation energy, also a negative quantity, is taken to be the difference between the exact nonrelativistic and the H-F limit energies. The experimental energy is then given by the sum of three negative quanties, the H-F limit, the correlation energy, and the relativistic correction.

Why are the molecular correlation energies upper limits?

Verify the entries for the H-F binding energies. This quantity is sometimes called the "rationalized" binding energy, especially when the H-F limit for the molecule has been reached.

You should be able to produce the experimental binding energies as byproducts of your calculation of the total energies done earlier.

Do you see why the correlation contributions to the total energies are labeled upper limit?

25. "Use of Gaussian Functions in the Calculation of Wavefunctions for Small Molecules. I. Preliminary Investigations" C.M. Reeves 39, 1 (1963).

The use of gaussian functions as a basis set for molecular calculations was proposed by S.F. Boys in Proc. Roy. Soc. A200, 542 (1950). In this exercise we will examine some of their properties.

First and foremost gaussian functions are useful in many-center(molecular) calculations because the product of two gaussians having different centers A and B is itself a gaussian (apart from a constant factor) with a center somewhere on the line segment AB. Thus the evaluation of multicenter integrals involving nuclear attraction terms $1/r_i$ or electron repulsion terms $1/r_{ij}$ can be made into an analytical rather than a numerical procedure. To sketch the proof of this theorem we will follow the method of I. Shavitt in Methods in Computational Physics, Chapter I "The Gaussian Function in Calculations of Statistical Mechanics and Quantum Mechanics", p.3 (B. Alder, S. Fernbach, and M. Rotenberg eds.) Academic Press New York (1963). The proposition is

$$G_i(r_A)G_j(r_B) = KG_k(r_C)$$

where

$$G_i(r_A) \equiv \exp(-\alpha_i r_A^2),$$

$$K = \exp[-\alpha_i\alpha_j\overline{AB}^2/(\alpha_i+\alpha_j)],$$

and

$$\alpha_k = \alpha_i+\alpha_j.$$

Shavitt's Fig. 1 shows the quantities involved.

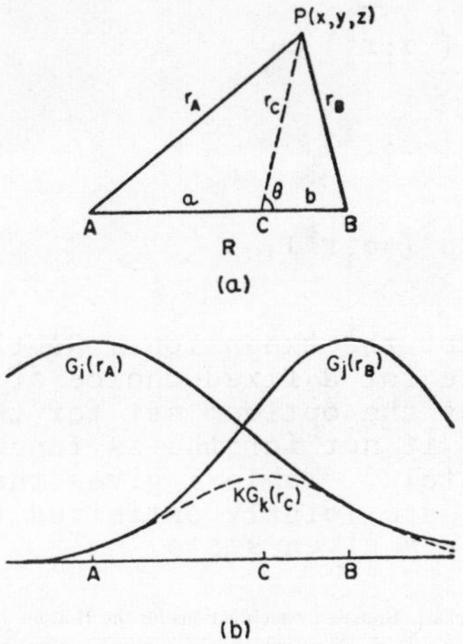

FIG. 1. The product of two Gaussians. (a) Definition of symbols. (b) Values of the Gaussians along *AB*.

If the point C is placed so that its coordinates wrt a specified origin are

$$C_x = (\alpha_i A_x + \alpha_j B_x)/(\alpha_i + \alpha_j)$$

$$C_y = (\alpha_i A_y + \alpha_j B_y)/(\alpha_i + \alpha_j)$$

$$C_z = (\alpha_i A_z + \alpha_j B_z)/(\alpha_i + \alpha_j)$$

then, if $R \equiv \overline{AB}$

$$a \equiv \overline{AC} = \alpha_j R/(\alpha_i + \alpha_j)$$

$$b \equiv \overline{CB} = \alpha_i R/(\alpha_i + \alpha_j).$$

In Fig. 1(a) apply the cosine law expressed in terms of cos θ to both of the smaller triangles with the common side r_C. Multiply one of these equations by a and the other by b and add them to obtain the result

$$br_A^2 + ar_B^2 = a^2b + ab^2 + (a+b)r_C^2 = R(ab+r_C^2).$$

Substituting for a and b from the above line segment equations leads to

$$\alpha_i r_A^2 + \alpha_j r_B^2 = \alpha_i\alpha_j R^2/(\alpha_i+\alpha_j) + (\alpha_i+\alpha_j)r_C^2,$$

the equality of the arguments in the proposition.

Following Reeves we will now do simple one-center calculations for the hydrogen atom; he continues on with H_2, He, Be, N, NH and the four-center NH_3 system in this and the following paper though we will be content with just one or two states of the H atom. What is the disadvantage of gaussians orbitals?

The known exact wavefunctions for hydrogen are represented approximately by the expansions

$$H_{n,1s} = \sum_{i=1}^{n} A_i \exp(-\alpha_i r^2)$$

$$H_{n,2s} = \sum_{i=1}^{n} A_i \exp(-\alpha_i r^2)$$

$$H_{n,2p_x} = \sum_{i=1}^{n} A_i x \exp(-\alpha_i r^2),$$

where n is the length of the expansion. Routine application of the variation principle for a fixed choice of the α_i to the 1s and $2p_x$ functions gives the optimum set for the respective coefficients A_i. Why will it not for the 2s function as well, which must be handled separately? Table I gives the results for each state for $n \leq 4$. The α_i are jointly optimized for each n value within a calculation on a given state.

TABLE I. Gaussian wavefunctions for the H atom.

State	α A				E
$H_{1,1s}$	0.28294 0.65147				−0.424413
$H_{2,1s}$	0.2015 0.1760	1.3320 0.2425			−0.485813
$H_{3,1s}$	0.1483 0.1084	0.6577 0.2164	4.2392 0.1575		−0.496967
$H_{4,1s}$	0.1233 0.0756	0.4552 0.1874	2.0258 0.1620	13.7098 0.0947	−0.499276
$H_{2,2s}$	0.0163 −0.03335	0.3211 0.14093			−0.116833
$H_{3,2s}$	0.01934 −0.03909	0.2586 0.10736	1.6844 0.08625		−0.123668
$H_{4,2s}$	0.02006 −0.04064	0.2276 0.09059	1.0334 0.08108	7.042 0.04646	−0.124631
$H_{1,2p}$	0.04527 0.02976				−0.113177
$H_{2,2p}$	0.03240 0.01539	0.1393 0.03952			−0.123289
$H_{3,2p}$	0.02465 0.00803	0.07970 0.02870	0.3363 0.03369		−0.124728
$H_{4,2p}$	0.02018 0.00452	0.05572 0.02042	0.1743 0.02944	0.7342 0.02544	−0.124952

Notice how the energy improves as we add basis functions. See also that it appears to approach the exact value. For a fixed n does the 2s or $2p_x$ energy lie lower? Which converges faster? Why is the minimum n for 2s shown as 2? Does this help to answer the earlier question about this state?

For n=1 analytical solutions are available for the optimum exponent α for both the 1s and $2p_x$ states. What is the relation then between A_1 and the normalization constant? Verify that it has the value for the 1s state of

$$[4/(3\pi)]^{3/2} = 0.27649.$$

This does not agree with the entry in the Table - see if the square root has been taken rather than the three-halves power. For this 1s state determine the variation energy $\xi=\xi(\alpha)$ and show that the optimum α is $8/(9\pi)$ leading to the minimum energy of $-4/(3\pi)$. Verify the numerical values.

Check and see if the virial theorem is satisfied with the optimum value of α; our optimization for the one-term expansion is the simplest illustration of satisfying the virial theorem by scaling.

Find the normalization constant, optimum α, and best energy for the $2p_x$ state with n=1, both analytical expressions and numerical values. Does the normalization constant for the $2p_{x,y,z}$ orbitals differ? Can you give a physical interpretation?

All-electron calculations computing all integrals have recently been performed on molecules of quite reasonable size using the gaussian bases. See the work of E. Clementi on the NH_3+HCl potential energy surface in J. Chem. Phys. 46, 3842 (1967), 47, 2323 (1967), and with J.N. Gayles 47, 3837 (1967), and, with H. Clementi and D.R. Davis, on pyrrole 46, 4725 (1967), on pyridine 46, 4731 (1967) and its positive ion 47, 4485 (1967), and on pyrazine 46, 4737 (1967).

These SCF calculations are not claimed to be near the Hartree-Fock limit which is even now just imminent for the H_2O molecule, but they yield information that was only a matter of conjecture before. For example the traditional assumption that sigma and pi electrons were approximately separable has been shown to be on very shaky grounds.

A feeling for the status of computational chemistry of large molecules today and a look at some authoritative projections into the future may be had by studying the article by E. Clementi "Study of the Electronic Structure of Molecules. XI. Comments on Some Present Aspects and Tentative Extrapolations" found in the International Journal of Quantum Chemistry III S, 179 (1969).

26. "Calculation of the Barrier to Internal Rotation in Ethane" R.M. Pitzer and W.N. Lipscomb 39, 1995 (1963).

The work reported here was important in that all electrons were considered and all integrals evaluated so that the question of whether or not the state of the art in SCF calculations could yield meaningful information about chemical questions could be answered without recourse to the nature of the semiempiricism(s) employed and the corresponding rationalizations. The answer to the question is a provisional yes, as you may judge for yourself.

The SCF energies were computed for both the eclipsed and staggered forms of C_2H_6. The basis set was composed of Slater orbitals, one each for 1s, 2s, $2p_x$, $2p_y$, and $2p_z$ for each carbon atom and a 1s for each hydrogen atom. The orbital exponents ζ, for the exponential part of the orbital exp $(-\zeta r)$, were chosen by the rules of J.C. Slater as may be verified by consulting any one of a number of textbooks or the original paper in Phys. Rev. 36, 57 (1930). The fixed nuclear geometries used are shown in Table I. 7H, 8H, and 9H are the nuclei of the second methyl group when it has been rotated into the staggered configuration.

TABLE I. Nuclear positions in cylindrical coordinates.

Nucleus	r(a.u.)	ϕ(rad)	z(a.u.)
$1C$	0.0		−1.45793806
$2C$	0.0		1.45793806
$1H$	1.96163013	0	−2.15708616
$2H$	1.96163013	$2\pi/3$	−2.15708616
$3H$	1.96163013	$4\pi/3$	−2.15708616
$4H$	1.96163013	0	2.15708616
$5H$	1.96163013	$2\pi/3$	2.15708616
$6H$	1.96163013	$4\pi/3$	2.15708616
$7H$	1.96163013	π	2.15708616
$8H$	1.96163013	$5\pi/3$	2.15708616
$9H$	1.96163013	$\pi/3$	2.15708616

The entries are given to as many figures as they are in order to maintain internal consistency. How many independent structural parameters are there for each form of ethane? Determine them from the entries in the Table. Also compute the total nuclear repulsion energies for both forms and compare with the results given in Table III. The crudest possible model of the potential.barrier must be the two bare nuclear frameworks. Show that this does give about 140% of the computed barrier height; what about the sign?

What are the appropriate point groups for the two forms? Table II, not given here, lists the appropriate symmetry orbitals for the eclipsed form. The molecular orbitals themselves are given in Table III along with the various energy quantities. The first two molecular orbitals for each conformation are shown as having the lowest orbital energies by far. Could these correspond to the K shells of the carbon atoms? Of course. Let's look more closely at the coefficients. From top to bottom the symmetry orbitals are the sums, for $1a_1'$ and $1a_{1g}$(differences for $1a_2''$ and $1a_{2u}$), of [(1H1s)+(2H1s)+(3H1s)]±[(4H1s)+(5H1s)+(6H1s)], (1C1s)±(2C1s), (1C2s)±(2C2s), and $(1C2p_z)\pm(2C2p_z)$. Each term within parentheses represents a single normalized Slater type orbital. Are the coefficients for these four molecular orbitals about what you'd expect?

Let's be a bit more quantitative. Notice that the symmetry orbitals above are unnormalized. Neglecting the overlap between carbon 1s orbitals does the normalization constant of these particular symmetry orbitals make their coefficients in the appropriate MO's very close to unity? Should this $1s_c$-$1s_c$ overlap be large? Can you pick out the difference in the correct normalization constants, viz $(2\pm 2S)^{-\frac{1}{2}}$, from the Table? What about the sign of the small shift?

From the total energies shown which form is more stable? What is the computed height of the potential barrier in kcal/mole? What is the experimental value? With a method similar to that used in exercise no. 24 compute the experimental total energy of (staggered)ethane. What fraction of the total error in the calculation, SCF minus experimental, is the barrier height? The authors performed experiments with integrals truncated to five decimal places rather than the six-eight carried throughout and feel the potential barrier is accurate to two significant figures. Charge density contours are presented.

This work did stimulate the hopes of chemists that there are somehow large energy contributions having their concomitant errors cancel when comparing different conformations.

TABLE III. Wavefunctions and energies.

Eclipsed conformation							
MO's	$1a_2''$	$1a_1'$	$2a_1'$	$2a_2''$	$1e'$	$3a_1'$	$1e''$
Coefficients	−0.004232	−0.003932	0.071369	0.148427	0.220326	−0.141869	0.296820
	0.703549	0.703981	−0.152646	−0.135350	0.399236	−0.013892	0.419375
	0.022998	0.016488	0.486501	0.479241		0.081629	
	0.002266	−0.002479	0.060215	−0.147537		0.543534	
Orbital energies	−11.343743	−11.343636	−1.039015	−0.857780	−0.627649	−0.534921	−0.512443

Kinetic energy		77.98356
Nuclear repulsion energy	41.93988	
H nuclear attraction energy	−39.00916	
C nuclear attraction energy	−226.02746	
Electron repulsion energy	66.12725	
Potential energy		−156.96949
Total energy		−78.98593

Staggered conformation							
MO's	$1a_{2u}$	$1a_{1g}$	$2a_{1g}$	$2a_{2u}$	$1e_u$	$3a_{1g}$	$1e_g$
Coefficients	−0.004177	−0.003971	0.071052	0.147841	0.220672	−0.141586	0.295821
	0.703561	0.703976	−0.152732	−0.135365	0.401163	−0.013791	0.415455
	0.022889	0.016551	0.487240	0.479746		0.080876	
	0.002281	−0.002498	0.059832	−0.148237		0.544000	
Orbital energies	−11.345968	−11.345881	−1.040014	−0.858843	−0.627326	−0.536073	−0.514999

Kinetic energy		77.96342
Nuclear repulsion energy	41.93239	
H nuclear attraction energy	−38.98580	
C nuclear attraction energy	−226.00184	
Electron repulsion energy	66.10068	
Potential energy		−156.95457
Total energy		−78.99115

27. "Study of Two-Center Integrals Useful in Calculations on Molecular Structure. V. General Methods for Diatomic Integrals Applicable to Digital Computers" A.C. Wahl, P.E. Cade, and C.C.J. Roothaan 41, 2578 (1964).

This exercise will serve to introduce you to a very useful two-center coordinate system, that of prolate spheroidal or elliptical coordinates, if you are not already familiar with it. We will express normalized complex Slater type orbitals (STO's) in spherical coordinates and then by means of transformation relations between the two systems we will express the STO's in prolate spheroidal coordinates. Finally we will have the simplest application of the two-center coordinate system and evaluate the overlap integral S between 1s STO's on different nuclei as a function of internuclear distance R and orbital exponents $\zeta_A = \zeta_B$.

Fig. 1 shows the prolate spheroidal coordinate system with

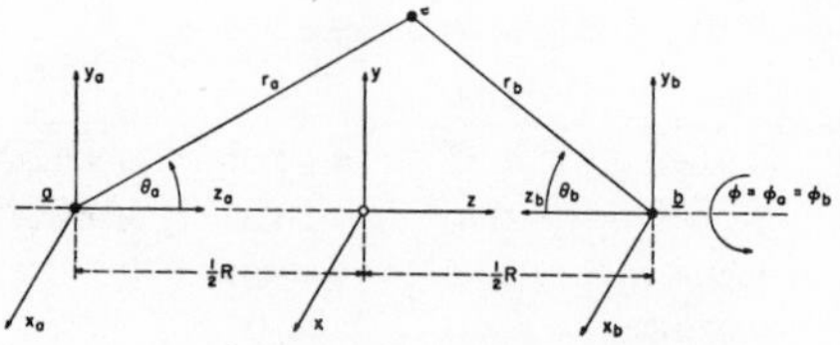

FIG. 1. Coordinate systems for diatomic molecules.

the foci a and b at the position of the two nuclei and separated by the distance R which is regarded as fixed for the coordinate system; it will appear in all integration formulas however so that our results will be functions of R in general. Superimposed in Fig. 1 are shown a cartesian system(right or left handed?) with origin at the midpoint of the foci and z axis pointing from a to b, and spherical systems with origins on each foci, z axes pointed toward one another, and x and y axes oriented such that the three y axes point in the same direction, as do the three x axes. Does this necessarily mean that one of the spherical systems will be based on a right handed frame and one on a left handed? Following the authors we will use subscripts a and b for the two spherical systems and we will introduce X,Y,Z for the cartesian systems.

The prolate spheroidal coordinates ξ, η, and ϕ are defined by

$$\xi \equiv (r_a+r_b)/R \qquad \eta \equiv (r_a-r_b)/R \qquad \phi = \phi_a = \phi_b.$$

Verify the reciprocal relations

$$X = \frac{R}{2}\sqrt{(\xi^2-1)(1-\eta^2)}\cos\phi$$

$$Y = \frac{R}{2}\sqrt{(\xi^2-1)(1-\eta^2)}\sin\phi$$

$$Z = \frac{R}{2}\xi\eta,$$

and

$$r_a = \frac{R}{2}(\xi+\eta) \qquad r_b = \frac{R}{2}(\xi-\eta)$$

$$\cos\theta_a = (1+\xi\eta)/(\xi+\eta)$$

$$\cos\theta_b = (1-\xi\eta)/(\xi-\eta)$$

$$\sin\theta_a = \sqrt{(\xi^2-1)(1-\eta^2)}/(\xi+\eta)$$

$$\sin\theta_b = \sqrt{(\xi^2-1)(1-\eta^2)}/(\xi-\eta).$$

The volume element in prolate spheroidal coordinates is

$$dV = \frac{R}{8}(\xi^2-\eta^2)d\xi d\eta d\phi;$$

verify the factor before the product of differentials by means of the Jacobian $\partial(\xi,\eta,\phi)/\partial(X,Y,Z)$. Rationalize the integration limits of $-1\leq\xi\leq 1$, $1\leq\eta\leq\infty$, and $0\leq\phi\leq 2\pi$ (easy!).

The normalized complex STO's are given in one of the spherical systems as, suppressing the subscripts,

$$(n,\ell,m) = (2\zeta)^{n+\frac{1}{2}}[(2n)!]^{-\frac{1}{2}}r^{n-1}e^{-\zeta r}Y_{\ell m}(\theta,\phi)$$

with $Y_{\ell m}$ the normalized complex spherical harmonic given by the product of the normalized associated Legendre function $P_{\ell m}(\cos\theta)$ and $(2\pi)^{-\frac{1}{2}}\exp(im\phi)$. Transform this general STO into the prolate spheroidal system to get (normalized?)

$$(n,\ell,m) = (2\zeta)^{n+\frac{1}{2}}[(2n)!]^{-\frac{1}{2}}(\tfrac{1}{2}R)^{n-1}(\xi+\eta)^{n-1}$$

$$\times \exp[-\tfrac{1}{2}R\zeta(\xi+\eta)]P_{\ell m}[(1+\xi\eta)/(\xi+\eta)](2\pi)^{-\frac{1}{2}}\exp(im\phi).$$

Show that the overlap integral for 1s STO's with the same ζ is

$$S = e^{-\zeta R}(1 + \zeta R + \zeta^2R^2/3).$$

What is the value of the overlap for hydrogen atoms with $\zeta=1$ at the separation of H_2? for the 1s carbon orbitals with the Slater value for ζ and the separation in ethane? Does this latter value shed any light on the discussion in the third paragraph of p. 54?

28. "Ethylene Molecule in a Gaussian Basis. I. A Self-Consistent Field Calculation" J.W. Moskowitz and M.C. Harrison <u>42</u>, 1726 (1965).

Fig. 1 shows the atom numberings and coordinate system used.

FIG. 1. Planar form of ethylene. CH=1.07Å, CC=1.35 Å, ∢HCH=120°.

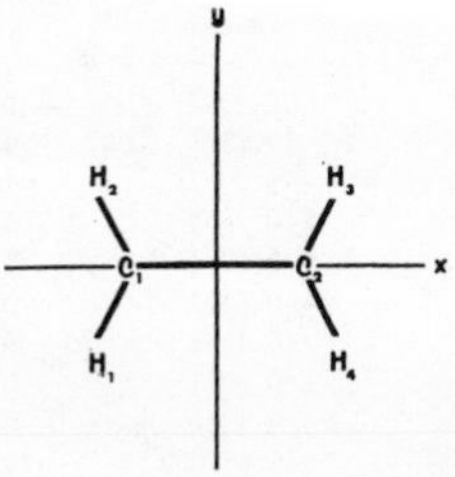

Table I presents the symmetry orbitals for the point groups corresponding to different twist angles θ; D_{2h} for the planar case $\theta=0$, D_2 for $0<\theta<\pi$, and D_{2d} for $\theta=\pi$. From character tables for these groups show that the symmetry orbitals shown do behave as the irreducible representations under which they are listed. The twisting is a symmetric one in that each methylene group is twisted, in opposite directions of course, by $\theta/2$. Why is only one set of s orbitals listed for carbon? Is the basis set less than minimal? Look at the expansions for 1s and 2s atomic orbitals in terms of gaussians back on p. 52. Actually the set is much larger than minimal which it must be to be competitive with STO bases, in terms of total energies.

TABLE I. The symmetry orbitals of ethylene and their correlation under different angles of twist.

Symmetry orbitals	D_{2h}	D_3	D_{2d}
s_1+s_2	a_g	a	a
x_1-x_2			
$h_1+h_2+h_3+h_4$			
s_1-s_2	b_{3u}	b_3	b_2
x_1+x_2			
$h_1+h_2-h_3-h_4$			
y_1+y_2	b_{2u}	b_2	e
$h_1-h_2-h_3+h_4$			
y_1-y_2	b_{1g}	b_1	e
$h_1-h_2+h_3-h_4$			
z_1+z_2	b_{1u}	b_1	e
z_1-z_2	b_{2g}	b_2	e

Table XI gives individual orbital and total energies as functions of the twist angle θ. By least squares the authors have fit the total energy points to a quadratic force constant for the twisting or torsional vibrational mode, such that the contribution to the vibrational potential energy is $\frac{1}{2}k\theta^2$ with k taking the value 4.35 eV. Show that the computed torsional frequency is 1168 cm^{-1}. An excellent discussion is given in "Molecular Spectra and Molecular Structure Vol.II. Infrared and Raman Spectra of Polyatomic Molecules", G. Herzberg (D. Van Nostrand Co. Inc. Princeton, New Jersey 1945), pp. 151, 183. Compute the total energy at each of the twist angles shown using this force constant; does the fit seem good even at the larger angles? Compare with the experimental k and torsional frequency quoted.

TABLE XI. Total electronic (including nuclear repulsion) and orbital energies of twisted ethylene for basis set 2-(3221). Energy in atomic units (2 Ry).

D_{2h}	$\theta=0°$	$\theta=5°$	$\theta=10°$	$\theta=60°$	$\theta=80°$	D_2
$1a_g$	−10.8351	−10.8347	−10.8351	−10.8324	−10.8309	$1a_1$
$1b_{3u}$	−10.8330	−10.8325	−10.8329	−10.8302	−10.8289	$1b_3$
$2a_g$	−1.0038	−1.0032	−1.0033	−1.0013	−1.0007	$2a_1$
$2b_{3u}$	−0.7656	−0.7653	−0.7654	−0.7648	−0.7641	$2b_3$
$1b_{2u}$	−0.6231	−0.6224	−0.6221	−0.6001	−0.5837	$1b_2$
$3a_g$	−0.5417	−0.5414	−0.5415	−0.5396	−0.5390	$3a_1$
$1b_{1g}$	−0.4693	−0.4698	−0.4719	−0.5251	−0.5482	$1b_1$
$1b_{1u}$	−0.3683	−0.3671	−0.3648	−0.2934	−0.2563	$2b_1$
$1b_{2g}$	0.1860	0.1856	0.1840	0.1274	0.0925	$2b_2$
Total energy	−74.640217	−74.639909	−74.638226	−74.559738	−74.498737	

29. "Molecular Scientists and Molecular Science: Some Reminiscences" R.S. Mulliken 43, S2(Supplementary Part 2 of 15 November Issue) (1965).

Read about the occasion on which this address was given. Read the entire article. Try and associate all of the surnames mentioned(with certain obvious exceptions such as Hitler's personal physician!) with specific areas of research. Chemical Abstracts for the years under discussion, and since in many cases, will be useful.

Footnote 1 uses the noun perturbations in perhaps a little different context than spectroscopist and theorists might usually.

Would a knowledge of the rules of punctuation of German be necessary to appreciate Mulliken's aside about Dr. Stich?

What did Marlene Dietrich have in common with Hitler?

Why was it necessary for the Leipzig champion table tennis player to have the artificial long gray beard affixed to his face? Who was the affixer?

30. "Two-Dimensional Chart of Quantum Chemistry" J.A. Pople 43, S229 (1965).

Find two or three articles by each of the persons shown on the curve and check their approximate position; can you substantiate the path shown for Roothaan for both values of the ordinate? Write the PPP entry in expanded form.

The article by C.A. Coulson, in ref. 1, is a synopsis of a 1959 conference on quantum mechanics and chemistry and introduces the distinction between Group I and Group II theorists. Coulson's article also deserves to be read in its entirety.

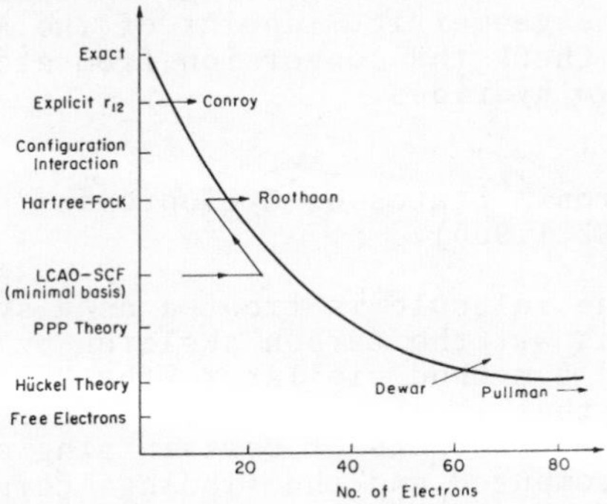

FIG. 1. A chart of quantum chemistry.

While in this issue, take a look at the paper on p. S244 and see if you like the quote the author uses to introduce the subject.

31. "Electric Dipole Moments for First- and Second-Row Diatomic Hydride Molecules, AH" P.E. Cade and W.M. Huo 45, 1063 (1966).

This exercises gives an example of another property other than the energy that can be usefully approximated by ab initio calculations. In addition to the really very good accuracy they have in Table I you can see that this type of calculation allows one to follow trends in physical properties where the experimental data are incomplete and may not even reveal the existence of the trend. Such a trend, which the authors discuss briefly, is the regular change in magnitude, including polarity, of the electric dipole moment of the hydrides. What is the trend in the experimental data for the second-row diatomic hydrides?

TABLE I. Electric dipole moment of first- and second-row hydrides, AH.[a]

AH	Configuration[b]	Term	R_e(Bohr)[c]	$-\langle\Psi \mid \sum_k z_k \mid \Psi\rangle$	$\mu(ea_0)$	μ(D)	μ(exptl)
LiH	$1\sigma^2\,2\sigma^2$	$X\,^1\Sigma^+$	3.015	0.6536	−2.3614	−6.002	−5.882[d]
BeH	$1\sigma^2\,2\sigma^2 3\sigma$	$X\,^2\Sigma^+$	2.538	3.6959	−0.1111	−0.282	
BH	$1\sigma^2\,2\sigma^2 3\sigma^2$	$X\,^1\Sigma^+$	2.336	5.3539	0.6819	1.733	
CH	$1\sigma^2\,2\sigma^2 3\sigma^2 1\pi$	$X\,^2\Pi_r$	2.124	5.9279	0.6179	1.570	1.46[e]
NH	$1\sigma^2\,2\sigma^2 3\sigma^2 1\pi^2$	$X\,^3\Sigma^-$	1.9614	6.5243	0.6401	1.627	
OH	$1\sigma^2\,2\sigma^2 3\sigma^2 1\pi^3$	$X\,^2\Pi_i$	1.8342	7.1202	0.7005	1.780	1.660[f]
HF	$1\sigma^2\,2\sigma^2 3\sigma^2 1\pi^4$	$X\,^1\Sigma^+$	1.7328	7.6951	0.7639	1.942	1.8195[g]
NaH	$KL\,4\sigma^2$	$X\,^1\Sigma^+$	3.566	15.0908	−2.7329	−6.962	
MgH	$KL\,4\sigma^2 5\sigma$	$X\,^2\Sigma^+$	3.271	17.3940	−0.5965	−1.516	
AlH	$KL\,4\sigma^2 5\sigma^2$	$X\,^1\Sigma^+$	3.114	18.7509	0.0669	0.170	
SiH	$KL\,4\sigma^2 5\sigma^2 2\pi$	$X\,^2\Pi_r$	2.874	18.7999	0.1189	0.302	
PH	$KL\,4\sigma^2 5\sigma^2 2\pi^2$	$X\,^3\Sigma^-$	2.708	19.1677	0.2117	0.538	
SH	$KL\,4\sigma^2 5\sigma^2 2\pi^3$	$X\,^2\Pi_i$	2.551	19.4714	0.3389	0.861	
HCl	$KL\,4\sigma^2 5\sigma^2 2\pi^4$	$X\,^1\Sigma^+$	2.4087	19.7404	0.4608	1.197	1.12[h]

[a] The sense is defined such that a positive dipole moment implies A^-H^+ and a negative dipole moment implies A^+H^-.
[b] $K \equiv 1\sigma^2$ and $L \equiv 2\sigma^2 3\sigma^2 1\pi^4$.
[c] R_e(exptl) values are from G. Herzberg except for LiH, BH, and HCl for which more recent values were used.
[d] L. Wharton, L. P. Gold, and W. Klemperer, J. Chem. Phys. **33,** 1255 (1960).
[e] Reference 2.
[f] F. X. Powell and D. R. Lide, Jr., J. Chem. Phys. **42,** 4201 (1965).
[g] R. Weiss, Phys. Rev. **131,** 659 (1963).
[h] C. A. Burrus, J. Chem. Phys. **31,** 1270 (1959).

Compute all of the $\mu(ea_0)$ entries given that the z axes are measured from the geometric midpoint of the A-H line pointing toward atom H. Check the conversion from atomic units to Debyes for one or two hydrides.

32. "Excited Electronic States of Cyclopropane" R.D. Brown and V.G. Krishna <u>45</u>, 1482 (1966).

The cyclopropane molecule is treated as a system of delocalized σ electrons wrt the carbon skeleton by both the simple Hückel MO method and a method similar to the PPP treatment for π electronic states that is shown as an ordinate of Fig. 1 of exercise no. 30. The positions of certain singlet and triplet excited states are computed and the findings compared with experiment. We will follow the treatment only through the Hückel stage including finding the wavefunctions of correct symmetry under the point group(?) from the first group of excited configurations. Fig. 1 shows the atomic orbital basis used.

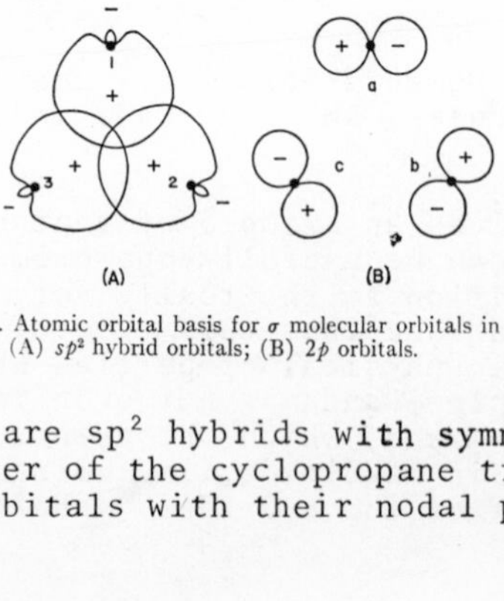

FIG. 1. Atomic orbital basis for σ molecular orbitals in cyclopropane. (A) sp^2 hybrid orbitals; (B) $2p$ orbitals.

χ_1, χ_2 and χ_3 are sp^2 hybrids with symmetry axes intersecting at the center of the cyclopropane triangle; χ_a, χ_b and χ_c are 2p atomic orbitals with their nodal planes intersecting along what symmetry element of C_3H_6? The six basis functions form the following symmetry orbitals, as you should check, belonging to three of the irreducible representations

$$\phi_1 = (1/3^{\frac{1}{2}})(\chi_1+\chi_2+\chi_3) \qquad a_1'$$

$$\phi_2 = (1/6^{\frac{1}{2}})(2\chi_a-\chi_b-\chi_c) \qquad e'$$

$$\phi_3 = (1/2^{\frac{1}{2}})(\chi_b-\chi_c) \qquad e'$$

$$\phi_4 = (1/2^{\frac{1}{2}})(\chi_2-\chi_3) \qquad e'$$

$$\phi_5 = (1/6^{\frac{1}{2}})(2\chi_1-\chi_2-\chi_3) \qquad e'$$

$$\phi_6 = (1/3^{\frac{1}{2}})(\chi_a+\chi_b+\chi_c) \qquad a_2'.$$

Here ϕ_2 and ϕ_3 mix together under the group operations as do ϕ_4 and ϕ_5. The application of the Hückel method gives the molecular orbitals as just these symmetry orbitals, and the energy level scheme of Fig. 2.

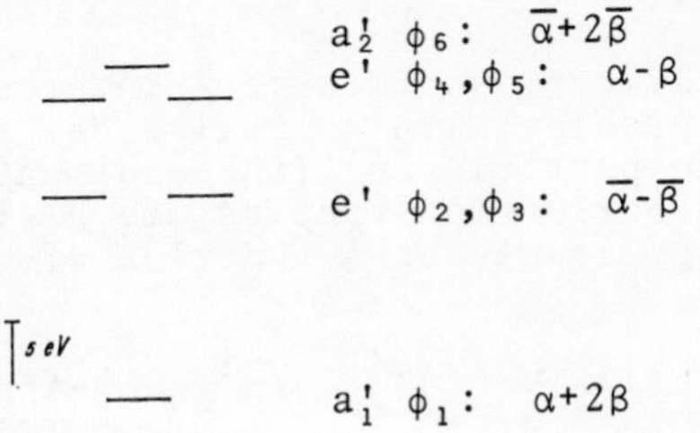

FIG. 2. Hückel molecular orbitals and their relative energies.

The coulomb integrals are α and $\bar{\alpha}$ for χ_1 and χ_a respectively; β and $-\bar{\beta}$ are the resonance integrals between χ_1 and χ_2 and χ_a and χ_b respectively. The negative sign is used with $\bar{\beta}$ to make it a negative quantity; is this correct? Look at Fig. 1 and the sign of the 2p lobes on adjacent carbons. From the energy scale in Fig. 2 determine values for β, $\bar{\beta}$, and $\alpha-\bar{\alpha}$; are all three quantities negative? Can we determine the values of the coulomb integral in Hückel schemes?

What similarities and what differences exist when you compare Fig. 2 and the corresponding diagram for benzene?

By considering the electronic promotions $\phi_2,\phi_3 \to \phi_4,\phi_5$ show that we obtain the four excited configurations

$$I = (\phi_1)^2\phi_2(\phi_3)^2\phi_4$$

$$II = (\phi_1)^2\phi_2(\phi_3)^3\phi_5$$

$$III = (\phi_1)^2(\phi_2)^2\phi_3\phi_4$$

$$IV = (\phi_1)^2(\phi_2)^2\phi_3\phi_5$$

of equal energy. From these construct the combinations belonging to the indicated symmetry species

$$\Theta_1[A_1'] = (1/\sqrt{2})(I-IV)$$

$$\Theta_2[A_2'] = (1/\sqrt{2})(II+III)$$

$\Theta_3[E'] = (1/\sqrt{2})(I+IV)$

$\Theta_4[E'] = (1/\sqrt{2})(II-III)$.

Show that the excitations $\phi_2,\phi_3 \to \phi_6$ lead to two degenerate configurations which are already symmetry species belonging to the E' representation:

$\Theta_5[E'] = (\phi_1)^2\phi_2(\phi_3)^2\phi_6$

$\Theta_6[E'] = (\phi_1)^2(\phi_2)^2\phi_3\phi_6$.

The authors then proceed to compute the energies of the singlet and triplet excited states by means of the PPP method and go on to include configuration interaction, but this is as far as we will go.

33. "Electron in Box Theory for Metal Atom Clusters" W.F. Libby 46, 399 (1967).

This article presents a very interesting application of the simplest problem with an analytical solution in quantum mechanics, the particle in a(three-dimensional) box. For an electron in a box of constant volume V with square cross section show the energy levels as a function of the quantum numbers n, o, and p are

$$E_{nop} = [h^2/(8mV^{2/3}\alpha^{4/3})](n^2+\alpha^2o^2+\alpha^2p^2)$$

with α the shape parameter giving the ratio of length of one edge to that of the other two which are taken to be equal to each other. Thus α=1 is a cube, α=2 is a rectangular parallelopiped with height twice the base-edge length, and α=½ is one with a height of half the base-edge length. The levels for these three α values are given in Table I.

TABLE I. Energy levels for electron in boxes of various shapes.[a]

Quantum numbers	Flat ($\alpha=\frac{1}{2}$)	Cube ($\alpha=1$)	Square column ($\alpha=2$)
111	3.8	3	3.6
211	11.3	6	4.7
121, 112	5.6	6	8.3
122	7.5	9	13.2
212, 221	13.1	9	9.6
222	15.0	12	14.4
123, 132	10.6	14	21.3
213, 321	16.3	14	17.7
312, 321	25.5	14	11.0
223, 232	18.2	17	22.4
322	27.5	17	16.4
114, 141	13.1	18	27.6
411	41.2	18	9.6
331, 313	28.9	19	19.6
133	13.8	19	29.2
142, 124	15.0	21	32.5
214, 141	20.6	21	18.8
421, 412	43.0	21	14.4

[a] $(h^2/8mV^{2/3})$ unit.

Plotting the energy levels against α, one can readily compute for any value of α the number of electrons required to fill each of the various levels and thus the total number of electrons for the various stable structures. These numbers are given in Table II.

TABLE II. Numbers of electrons to fill to successive energy levels.

Flat, two dimensional ($\alpha=0$)	Flat ($\alpha=0.5$)	Cube ($\alpha=1.0$)	Square column ($\alpha=2.0$)
2	2	2	2
6	6		4
8	8	8	8
12[a]	**12**		14
16	14	14	**18**
22	22	**16**	20
26	**24**		**26**
30	30	28	28
34	34	34	**32**
	38		
36	42	40	36
40	46	46	40
44			44
48			48

[a] The boldface numbers occur in the metal atom clusters and all known clusters are so included.

Read the rest of the article(quite short) and see if you agree with the arguments presented for the metal atom clusters.

Are all of the levels possible considered in Table I? Not in the sense of all possible integer combinations(∞) but as the energy increases beyond that of the (111) state are there any gaps in the states that should be considered that were not?

34. "Comment on 'Electron in Box Theory for Metal-Atom Clusters'" H. Müller <u>49</u>, 475 (1968).

This Comment answers the last question of the previous exercise in the affirmative. Table I shows the corrected numbers of electrons to fill to successive energy levels.

TABLE I. Numbers of electrons to fill to successive energy levels.

Flat ($\alpha=1/2$)		Cube ($\alpha=1$)		Square column ($\alpha=2$)	
This paper	Ref. 1	This paper	Ref. 1	This paper	Ref. 1
2	2	2	2	2	2
6	6	8	8	4	4
8	8	14	14	6	8
12	*12*	20	*16*	10	14
16	14	22	28	*16*	*18*
18	22	34	34	20	20
26	*24*	40	40	*24*	*26*
28	30	46	46	30	28
34	34			36	*32*
38	38			42	36
46	42			46	40
	46				44
					48

Can you verify Müller's results? Do you see which quantum number combinations(states) were omitted earlier? Are the arguments about cluster stabilities changed now? In what way?

The definition of a Comment is given e.g. in <u>46</u>, 413 (1967) and you are asked why Libby's rebuttal is not printed immediately following this Comment? Do you feel the matter is settled to the satisfaction of both men?

Special thanks are due to Professor Libby for allowing us to reproduce his Tables as part of the previous exercise.

35. "Separation of Rotational Coordinates from the N-Electron Diatomic Schrödinger Equation" R.T Pack and J.O. Hirschfelder 49, 4009 (1968).

"A formalism is presented for the exact separation of the center-of-mass and rotational coordinates from the Schrödinger equation for an arbitrary diatomic system. The use of operator arguments instead of the explicit representations of the rotation group simplifies the procedure. The coupled equations governing the internal motion of the system are obtained and expressed in a new and particularly simple form using angular-momentum raising and lowering operators. The formalism is carried out for three types of electronic coordinates (separated atom, center of mass of nuclei, and geometric center of nuclei) which differ as to the origin of the electronic coordinates."

Fig. 1 is the flow chart for the coordinate transformations and Fig. 2 shows the relative coordinates.

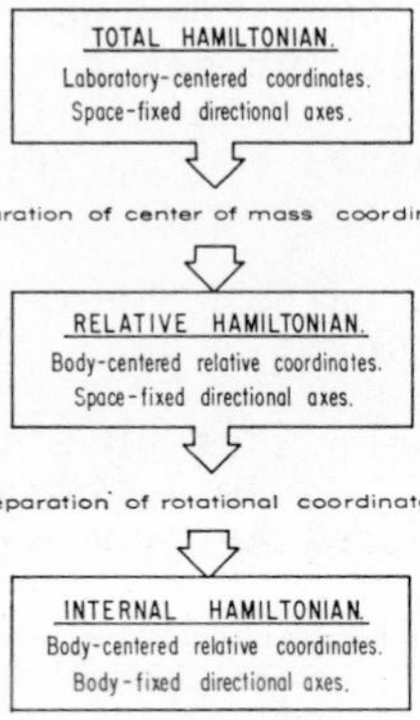

FIG. 1. Flow chart for the separation of center-of-mass and rotational coordinates to obtain the Hamiltonian governing the internal motion of the system.

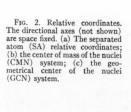

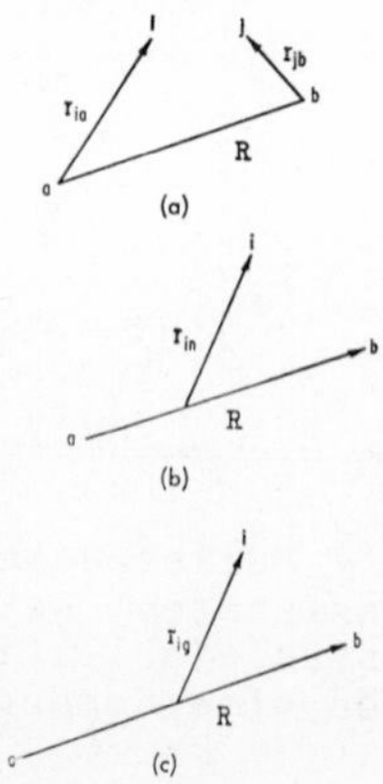

FIG. 2. Relative coordinates. The directional axes (not shown) are space fixed. (a) The separated atom (SA) relative coordinates; (b) the center of mass of the nuclei (CMN) system; (c) the geometrical center of the nuclei (GCN) system.

In this exercise we shall follow through a few of the steps in the transformation of the Hamiltonian in the coordinate systems and hopefully sharpen some of our mathematical skills in the process. We will allow the quoted abstract to summarize the work and not attempt to paraphrase the results as we proceed but merely try and pass from one equation to another for awhile. If we persist, however, we may increase our understanding of operators.

The Hamiltonian used throughout this paper is the Breit-Pauli approximation, which includes the Schrödinger Hamiltonian and also the relativistic and magnetic corrections through the order of $\alpha^2[\equiv O(\alpha^2)]$, where the fine structure constant $\alpha=2\pi e^2/hc \simeq 1/137$, and $\alpha^2 \simeq 5\times10^{-5}$. But because the Breit-Pauli Hamiltonian omits higher-order relativistic corrections such as the Lamb shift of $O(2\times10^{-6})$, one is only justified in calculating the relativistic corrections through $O(\alpha^2)$. Hence the authors consistently neglect all terms of higher order, as for example the terms containing nuclear spin, which are of $O(\alpha^2/\mu)$, where μ is the reduced mass of the nuclei in Hartree atomic units(mass of the electron taken as unity, charge on the proton and the unit of angular momentum $h/2\pi$ both also taken as unity, length in Bohr radii a_o, and energy in units of e^2/a_o). Since $10^3 \leq \mu \leq 10^5$ a.u. for all possible diatomic species, $O(\alpha^2/\mu) \lesssim 5\times10^{-8}$a.u.

Check the numerical values in the last paragraph.

To obtain the Hamiltonian in laboratory-fixed coordinates consider a system of N electrons and two nuclei(a and b) in the absence of any external field. Let $\hat{r}_q$, where q=a,b,1,...,N and ^ signifies a vector quantity, be the laboratory-fixed coordinate vectors of the particles. In these coordinates the Breit-Pauli Hamiltonian is given by

$$H_\tau = T_\tau + V + \alpha^2 H_\alpha \tag{2.1}$$

where τ stands for "total", and the total kinetic energy is

$$T_\tau = (2m_a)^{-1}\hat{p}_a{}^2 + (2m_b)^{-1}\hat{p}_b^2 + \tfrac{1}{2}\sum_{i=1}^{N} \hat{p}_i{}^2 \tag{2.2}$$

where $\hat{p}_q=-i\hat{\nabla}_q$, and m_a and m_b are the masses of the nuclei. Can a vector squared represent anything other than the dot product of it with itself? V is just the usual electrostatic potential energy

$$V = \frac{Z_aZ_b}{r_{ab}} - \sum_{i=1}^{N}\frac{Z_a}{r_{ia}} - \sum_{j=1}^{N}\frac{Z_b}{r_{jb}} + \sum_{j>i}^{N}\sum r_{ij}{}^{-1}. \tag{2.3}$$

H_α contains the relativistic and magnetic corrections and is given explicitly in their Appendix C.

We now proceed to express the Hamiltonian in terms of "separated-atom"(SA) coordinates(Fig. 2a) in which N_a electrons are arbitrarily centered on nucles a and $N-N_a$ are centered on nucleus b. $\hat{C}$ is the vector of the center of mass of the system and $\hat{R}$ is the internuclear vector. In the laboratory-fixed coordinate system we had N+2 vectors to locate N+2 particles; we will of course need N+2 vectors in the SA coordinate system. These will be defined below in equations (2.4) in terms of the N+2 laboratory-fixed set. Does this mean that we may think of the SA vectors as "ultimately" defined in the laboratory-fixed frame? When we separate the kinetic energy of the center of mass from the rest of the Hamiltonian we will see that the question is trivial as the remaining vectors all have their origins at one or the other of the nuclei. In atomic units $M=m_a+m_b+N$ is the total mass of the system. The new set of N+2 vectors is

$$\hat{C} = M^{-1}(m_a\hat{r}_a + m_b\hat{r}_b + \sum_{i=1}^{N} \hat{r}_i)$$

$$\hat{R} = \hat{r}_a - \hat{r}_b$$

$$\hat{r}_{ia} = \hat{r}_i - \hat{r}_a, \quad 1 \leq i \leq N_a$$

$$\hat{r}_{jb} = \hat{r}_j - \hat{r}_b, \quad N_a+1 \leq j \leq N. \tag{2.4}$$

We can obtain new momenta(operators) $\hat{p}_q = -i\hat{\nabla}_q$, where now q is any member of the set (2.4), by using the chain rule for derivatives. We find that

$$\hat{p}_a = (m_a/M)\hat{p}_C - \hat{p}_R - \sum_{i=1}^{N_a} \hat{p}_{ia}$$

$$\hat{p}_b = (m_b/M)\hat{p}_C + \hat{p}_R - \sum_{j=N_a+1}^{N} \hat{p}_{jb}.$$

$$\hat{p}_i = M^{-1}\hat{p}_C + \hat{p}_{ia}, \quad 1 \leq i \leq N_a$$

$$\hat{p}_j = M^{-1}\hat{p}_C + \hat{p}_{jb}, \quad N_a+1 \leq j < N. \tag{2.5}$$

Why are these equations written expressing the old momenta in terms of the new rather than vice versa? Suppressing the -i factors let us work with the x component of the first equation of the set (2.5). Applying the chain rule we find

$$\partial/\partial x_a = (\partial C_x/\partial x_a)\partial/\partial C_x + (\partial R_x/\partial x_a)\partial/\partial R_x + \sum_{i=1}^{N_a} (\partial x_{ia}/\partial x_a)\partial/\partial x_{ia}$$

where the partial derivatives may be evaluated from equations (2.4). Having verified that we do have the x component after all would you say that the first equation of (2.5) is 33% or 99+% confirmed? Is it necessary to follow the same route for the y and z components or may one merely introduce new subscripts? The rest of set (2.5) can be done in the same manner. Do it.

Is V easily written in SA coordinates? Does it depend on $\hat{C}$? Two of the required four relations among the vectors are supplied below; can you furnish the other two?

$$\hat{r}_i - \hat{r}_j = \hat{r}_{ia} - \hat{r}_{jb} - \hat{R}$$

$$\hat{r}_{ib} = \hat{r}_{ia} - \hat{R}$$

It turns out that H_α depends only on relative vectors(which are independent of $\hat{C}$) and on the electronic momenta. But in equations (2.5)

$$\hat{p}_i = \hat{p}_{id} + O(1/M), \quad 1 \leq i \leq N \text{ and } d = a \text{ or } b \tag{2.6}$$

so that the change in H_α is of $O(\alpha^2/M)$ and therefore negligible. Hence in H_α we just let $\hat{p}_i \to \hat{p}_{id}$ to express it in this relative coordinate system.

The translational kinetic energy of the center of mass can

be separated now from the rest of H_τ by substituting (2.5) into (2.1). The total kinetic energy term T_τ is the one that requires the real attention. One sees from (2.2) that we require the dot products of three vector operators with themselves, viz $\hat{p}_a$, $\hat{p}_b$, and $\hat{p}_i$. The first of these terms in T_τ is

$$[1/(2m_a)][(m_a/M)\hat{p}_C - \hat{p}_R - \sum_{i=1}^{N_a}\hat{p}_{ia}] \cdot [(m_a/M)\hat{p}_C - \hat{p}_R - \sum_{i=1}^{N_a}\hat{p}_{ia}]$$

or

$$m_a/(2M^2)\hat{p}_C^{\,2} + 1/(2m_a)\hat{p}_R^{\,2} + 1/(2m_a)(\sum_{i=1}^{N_a}\hat{p}_{ia}) \cdot (\sum_{i=1}^{N_a}\hat{p}_{ia})$$

$$-(1/M)\hat{p}_C\cdot\hat{p}_R - (1/M)\hat{p}_C \cdot \sum_{i=1}^{N_a}\hat{p}_{ia} + (1/m_a)\hat{p}_R \cdot \sum_{i=1}^{N_a}\hat{p}_{ia}$$

as you may show. Expand the second term, $(2m_b)^{-1}\hat{p}_b^{\,2}$, in the same way. The remainder of the total kinetic energy in laboratory-fixed coordinates is the sum of the individual electron momentum operators squared; this is not the same as the square of the sum of the operators, or is it?

We break the sum of the squares up into two parts - the first over electrons 1 through N_a and the second over electrons N_a+1 through N. Substituting from (2.5) we then have

$$\tfrac{1}{2}\sum_{i=1}^{N_a}[\{(1/M)\hat{p}_C + \hat{p}_{ia}\} \cdot \{(1/M)\hat{p}_C + \hat{p}_{ia}\}]$$

$$+ \tfrac{1}{2}\sum_{j=N_a+1}^{N}[\{(1/M)\hat{p}_C + \hat{p}_{jb}\} \cdot \{(1/M)\hat{p}_C + \hat{p}_{jb}\}].$$

Verify this step. Show that upon adding the kinetic energy terms and cancelling where appropriate one obtains for the total Hamiltonian H_τ the result

$$H_\tau = (1/2M)\hat{p}_C^{\,2} + (1/2\mu)\hat{p}_R^{\,2} + (1/2m_a)\hat{P}_a^{\,2} + (1/2m_b)\hat{P}_b^{\,2} + \alpha^2 H_\alpha$$

$$+ V + \hat{p}_R \cdot [(1/m_a)\hat{P}_a - (1/m_b)\hat{P}_b] + \tfrac{1}{2}\sum_{i=1}^{N_a}\hat{p}_{ia}^{\,2} + \tfrac{1}{2}\sum_{j=N_a+1}^{N}\hat{p}_{jb}^{\,2}.$$

Here $\mu=m_a m_b/(m_a+m_b)$ is the reduced mass of the nuclei. $\hat{P}_a$ and $\hat{P}_b$ are defined by

$$\hat{P}_a \equiv \sum_{i=1}^{N_a}\hat{p}_{ia}$$

$$\hat{P}_b \equiv \sum_{j=N_a+1}^{N}\hat{p}_{jb}. \tag{2.11}$$

This total Hamiltonian in the SA coordinate system, still relative to space-fixed directional axes, can now be written as

$$H_\tau = (1/2M)\hat{p}_C^{\,2} + H \tag{2.7}$$

where H is independent of $\hat{C}$. When we have a Hamiltonian given as a sum of noninteracting terms may we exactly write the total wavefunction as a product of functions of the noninteracting variables? Setting

$$\Psi_\tau = \Psi\psi(\hat{C})$$

the center of mass problem is separated off from the rest of the diatomic terms. In passing, to what standard problem in quantum mechanics is the former completely equivalent?

We are left with

$$H\Psi = E\Psi. \tag{2.8}$$

Demonstrate that H can be written in their form

$$H = (1/2\mu)\hat{p}_R{}^2 + (1/\mu)\hat{P}\cdot\hat{p}_R + H_e + (1/\mu)H_\mu(m) + \alpha^2 H_\alpha \tag{2.9}$$

where

$$\hat{P} = (\mu/m_a)\hat{P}_a - (\mu/m_b)\hat{P}_b \tag{2.10}$$

$$H_e = T_e + V \tag{2.12}$$

$$T_e = \sum_{i=1}^{N_a} \hat{p}_{ia}{}^2 + \sum_{j=N_a+1}^{N} \hat{p}_{jb}{}^2 \tag{2.13}$$

and

$$(1/\mu)H_\mu(m) = (1/2m_a)\hat{P}_a{}^2 + (1/2m_b)\hat{P}_b{}^2. \tag{2.14}$$

What physical interpretation can you attach to $\hat{P}_a$ and $\hat{P}_b$ defined in (2.11)?

The first term in (2.9) is the operator for the kinetic energy of two chargeless particles with interparticle distance variable both in length and orientation. The second couples the momenta of the electrons to the relative momentum of the nuclei. The third is the ordinary electronic Hamiltonian. The fourth is discussed in their ref. 18 and is essentially $(1/\mu)$ times the sum of the mass polarization operators or isotope corrections of the separated atoms. The last term has already been discussed.

Our exercise has now more or less covered the mathematics through section II C. There is much more but we are going to stop.

36. "Electronic Configuration of the O^{2-} Ion" D. Adler 52, 4908 (1970).

The author points out that the first assumption made upon analyzing the composition of any oxide is that all oxygen atoms are doubly ionized. This hypothesis is based upon the stability of the closed-shell, neon-type electronic configuration $2p^6$. Quoting his references 1 and 2, he states that the free O^{2-} ion is unstable by 9 eV. Restate this fact in terms of an exothermic reaction with O^- as one of the products. This instability of the free ion is usually not a problem in a crystalline solid, in which the oxide ions are stabilized by the Madelung potential of the order of 20 eV for a totally divalent material.

He demonstrates that there is a degree of s character to the oxygen 2p band in a Madelung-stabilized oxide, even when completely ionic. We will follow his reasoning. Table I shows

TABLE I. $2p \rightarrow 3s$ excitation energies for ions with the Ne electronic configuration.

Ion	Excitation energy (10^3 cm^{-1})	First differences (10^3 cm^{-1})	Second differences (10^3 cm^{-1})
Co^{17+}	6478	613	29
Fe^{16+}	5865	584	30
Mn^{15+}	5281	554	30
Cr^{14+}	4727	524	30
V^{13+}	4203	494	30
Ti^{12+}	3709	464	30
Sc^{11+}	3245	434	30
Ca^{10+}	2811	404	30
K^{9+}	2407	374	30
A^{8+}	2033	344	31
Cl^{7+}	1689	313	30
S^{6+}	1376	283	30
P^{5+}	1093	253	30
Si^{4+}	840	223	32
Al^{3+}	617	191	30
Mg^{2+}	426	161	30
Na^{+}	265	131	(30)
Ne	134	(101)	(30)
F^{-}	(33)	(71)	
O^{2-}	(−38)		

us the $2p \rightarrow 3s$ excitation energies, taken from the volumes by Moore referenced in the next exercise. The third and fourth columns would be dropped half of a line and a full line respectively if it were not such a nuisance for the typesetter. Check two or three of the excitation energies from Moore's volumes. Write(exothermic?) reactions, with ΔE in eV, for both the excitation shown in the Table and for the process $3s \rightarrow \infty$(ionization?). What is ΔE for the latter?

In the presence of the Madelung potential of the order of 20 eV, the O^{2-} is stabilized by about how much? From the above results the author might expect a $2p \rightarrow 3s$ excitation energy in the stabilized ion of the order of only 5 eV endothermic. This figure is sufficiently low that we can conclude that there is significant s character to the 2p band of the oxide ions, without the necessity of invoking covalency effects with the cations.

Make an energy level diagram showing all of the oxygen species mentioned, both free and stabilized in the lattice.

From Table I, what is the dependence of the $2p \rightarrow 3s$ excitation energy upon nuclear charge?

CHAPTER III

STATISTICAL MECHANICS

37. "Free Energy Functions for Gaseous Atoms from Hydrogen(Z=1) to Niobium(Z=41)" T.J. Katz and J.L. Margrave 23, 983 (1955).

Table I presents the free energy functions, $(F^\circ_T - H^\circ_0)/T$, for the atoms at four temperatures. These functions are seen to vary slowly enough so that reasonably accurate interpolation is possible on the temperatures.

TABLE I. Free energy functions for gaseous atoms.

Element	$-(F^0 - H_0^0/T)$ in cal deg^{-1} mole^{-1} 298°K	500°K	1000°K	2000°K
H	22.422	24.993	28.437	31.880
He	25.155	27.726	31.170	34.613
Li	28.173	30.744	34.188	37.631
Be	27.577	30.148	33.592	37.035
B	31.579	34.190	37.664	41.123
C	32.519	35.200	38.727	42.213
N	31.644	34.215	37.658	41.102
O	33.074	35.840	39.459	43.002
F	32.689	35.413	39.057	42.642
Ne	29.978	32.549	35.992	39.436
Na	31.744	34.315	37.759	41.202
Mg	30.534	33.104	36.548	39.992
Al	33.754	36.570	40.209	43.755
Si	34.064	37.102	40.928	44.587
P	34.010	36.581	40.024	43.469
S	34.746	37.538	41.276	44.928
Cl	34.426	37.060	40.688	44.338
A	32.012	34.583	38.027	41.471
K	33.326	35.897	39.340	42.785
Ca	32.023	34.594	38.037	41.481
Sc	35.120	37.691	41.134	46.260
Ti	37.021	40.103	44.033	47.793
V	37.205	40.448	44.663	48.840
Cr	36.666	39.237	42.681	46.151
Mn	36.523	39.094	42.538	45.981
Fe	37.618	40.532	44.504	48.368
Co	37.780	40.489	44.381	48.513
Ni	38.045	40.897	44.822	48.823
Cu	34.775	37.346	40.790	44.235
Zn	33.481	36.052	39.496	42.939
Ga	35.122	37.960	42.011	45.986
Ge	34.171	37.406	41.994	46.390
As	36.642	39.213	42.656	46.102
Se	37.242	39.817	43.334	47.024
Br	36.834	39.406	42.854	46.362
Kr	34.218	36.789	40.232	43.676
Rb	35.658	38.229	41.672	45.116
Sr	34.354	36.925	40.369	43.813
Y	37.371	40.284	44.221	48.015
Zr	37.850	40.798	44.989	49.264
Nb	37.792	41.308	46.015	50.512

For noninteracting atoms the only contributions to the thermodynamic functions are translational and electronic. The above Table was compiled making use of their ref. 1; C.E. Moore, "Atomic Energy Levels"(National Bureau of Standards, Washington D.C.), Circular 467, Vol. I (1949) and Vol. II, (1952). These volumes contain observed energy levels for atoms and ions with the corresponding degeneracies so that the electronic partition function can be evaluated as a numerical function of temperature.

The translational contribution can be evaluated in a straightforward fashion. Compute the free energy function for helium at the highest temperature ignoring the electronic contribution. Is the agreement good? Why is this?

Show that the contribution of the electronic excitations to the free energy function is -R ln Q_e where the electronic partition function

$$Q_e = \sum_i g_i \exp(-\varepsilon_i/kT).$$

Listed are the first few electronic levels ε_i and the corresponding degeneracies g_i for helium and carbon, taken from ref. 1.

He	term	i	$\varepsilon_i(cm^{-1})$	g_i
	$1s^2\ ^1S$	0	0	1
	$1s2s\ ^3S$	1	159850	3
	$1s2s\ ^1S$	2	166272	1
C	$2s^22p^2\ ^3P$	0*	0*	9*
	$2s^22p^2\ ^1D$	1	10194	5
	$2s^22p^2\ ^1S$	2	21648	1
	$2s2p^3\ ^5S$	3	33735	5
	$2s^22p3s\ ^3P$	4	60373	9

*ignoring spin-orbit splitting.
When we take this into account the first part of the carbon table becomes

$^3P_0$	0'	0	1
$^3P_1$	0"	16.4	3
$^3P_2$	0'''	43.5	5.

Demonstrate that the contribution of the excited He levels is negligible even at 2000°K.

Calculate the free energy functions for C at all temperatures using i) the partition function ignoring spin-orbit splitting and ii) including it. Would you expect the difference between calculations i) and ii) to be greatest at 298 or 2000°K? Was your expectation realized? Why is the electronic contribution appreciable for carbon at room temperature?

We will examine this area again in exercise no. 40.

38. "Monte Carlo Study of Coiling-Type Molecules. II. Statistical Thermodynamics" F.T. Wall, S. Windwer, and P.J. Gans <u>38</u>, 2228 (1963).

A non self-intersection random walk N steps long on a tetrahedral(diamond) lattice corresponds to an N-step polymer. After each polymer was generated it was scanned for nearest-neighbor pairs. The potential energy was infinite for double occupancy of a lattice site, equal to ε for nearest-neighbor pairs, and zero otherwise. Further, each polymer was assigned a weighting factor ω_i as configurations of a particular interest were sometimes generated. Finally many different polymers were examined for a given N; in this study N was a maximum of 120.

For fixed N and T the partition function becomes

$$Q = \sum_i \omega_i \exp(-\nu_i \xi)$$

with ν_i the number of nearest-neighbor pairs in the i'th polymer and $\xi=\varepsilon/kT$. The sum is carried out over all of the polymers generated for that value of N. This is then an "experimental" method of determining the energy level pattern with the associated degeneracies; $\nu_i\xi$ is the energy of polymer i and ω_i is its degeneracy.

How would you expect to gain a sharper picture of this pattern for a fixed N? Could it be done?

$\langle\nu\rangle$ and $\langle\nu^2\rangle$ are obtained by averaging ν_i and ν_i^2 through the partition function. If ΔE is the change in energy per polymer upon changing ξ from 0 to ξ, show that

$$\Delta E = kT^2[\partial \ln (Q/Q_\circ)/\partial T] = kT\langle\nu\rangle\xi$$

with $Q_\circ$ the value of Q for $\xi=0$. What variable(s) are constant in the partial derivative?

Also show from the last form of this equation that ΔC_v, defined as $\partial\Delta E/\partial T$, can be expressed as $k\sigma_\nu^2\xi^2$, with $\sigma_\nu^2 = \langle\nu^2\rangle-\langle\nu\rangle^2$.

A result of this study was that there is no basis for concluding that the thermodynamic behavior on the tetrahedral lattice differs in any essential detail from that of other two- and three-dimensional lattices; the various extensive thermodynamic functions are linear in N as expected.

39. "Equation of State for a Two-Dimensional Electrolyte" A.M. Salzberg and S. Prager <u>38</u>, 2587 (1963).

For a two-dimensional system of point charges confined to an area A these authors derive the partition function

$$Q(A,T,N^{(\ell)}) = A^{[N+\frac{1}{2}\sum_{i<j} q_i q_j/kT]}\, Q^*(T,N^{(\ell)}),$$

where $N^{(\ell)}$ is the number of charges of species ℓ, $N=\Sigma N^{(\ell)}$, and Q^* is a reduced partition function independent of A.

Show that the two-dimensional pressure P given by

$$P = kT(\partial \ln Q/\partial A)_{T,N^{(\ell)}}$$

takes the final form

$$P = [NkT/A]\,[1 - (4kT)^{-1} \sum_\ell x^{(\ell)} q^{(\ell)2}]$$

assuming overall electrical neutrality. Here $x^{(\ell)}$ and $q^{(\ell)}$ are the mole fraction and charge of species ℓ.

The sum term in this equation is twice the ionic strength. It is this term which governs departure from ideality.

What is the interesting suggestion that the authors advance about what might be happening to the ions when the term in square brackets begins to go negative at low temperatures?

40. "Thermodynamic Properties of a Monatomic Hydrogen Gas at High Temperatures" I. Oppenheim and D.R. Hafemann 39, 101 (1963).

These authors have made a study of atomic hydrogen at temperatures high enough so that the internal and intermolecular degrees of freedom are not separable. They derive expressions for the partition function and thermodynamic properties. We shall examine only their Introduction.

Upon being confronted for the first time with the statement that electronic partition functions must be evaluated by summing the observed terms, for example as in our exercise no. 37, one may perhaps wonder if approximations exist for the electronic energies in closed form corresponding to the rigid rotor and harmonic oscillator approximations for the rotational and vibrational energies. In particular what about the simplest case of atomic hydrogen for which the electronic levels vary as $-1/n^2$? We shall see below that the partition function for bound states of the hydrogen atom formally diverges for all temperature. The temperature range for which the divergence is of physical importance is examined and the high temperature region is discussed in the balance of the paper.

First some familiar equations are presented. For an ideal gas(defined as composed of point masses of noninteracting particles) of N atoms the canonical ensemble partition function is

$$Q_N = (Q_1)^N/N! \tag{1.1}$$

with Q_1 the single atom partition function

$$Q_1 = Q_t Q_e Q_n . \tag{1.2}$$

The translational partition function Q_t is

$$Q_t = V(2\pi mkT/h^2)^{3/2} \tag{1.3}$$

and the nuclear spin partition function Q_n is

$$Q_n = 2I + 1 \tag{1.4}$$

where I is the spin of the nucleus. By writing Q_n in this form are we restricting all of the atoms to be in their nuclear ground states? Is this rigorous? Is it realistic at all temperatures at which we conceive electrons to be still interacting with nuclei? The atoms begin to interact with one another, thus making the assumption of separability in equation (1.1) invalid, at temperatures far below those necessary to consider excited nuclear states.

The electronic partition function Q_e can be written

$$Q_e = \sum_n \omega_n \exp(-\beta\varepsilon_n) \tag{1.5}$$

where $\beta \equiv 1/kT$. The summation sign implies a sum over discrete states and integration over the continuum, ω_n is the degeneracy of the nth level, and ε_n is the energy of the nth level.

In nonrelativistic quantum mechanics, the bound-state energy levels for the hydrogen atom in an infinite volume are given by the expression

$$\varepsilon_n = Ry(1-1/n^2), \qquad n=1,2,3,\ldots \tag{1.6}$$

where n is the principal quantum number and Ry is the Rydberg energy 109,678 cm^{-1} for finite proton mass. Notice that by writing the energy levels ε_n in this form we are referring the energies to the ground electronic state as zero, the usual convention in statistical mechanics. The degeneracy of each of these levels, as every freshman chemistry student learns(is exposed to?), is

$$\omega_n = 2n^2. \tag{1.7}$$

Combining equations (1.5)-(1.7) we write

$$Q_e = \sum_{n=1}^{\infty} 2n^2 \exp\left[-\beta Ry(1-1/n^2)\right] + \int \text{continuum}. \tag{1.8}$$

The sum over discrete states must diverge for all finite values of $\beta(T>0)$. Can you prove this from simple considerations? Is there a simpler related series, obviously divergent, whose individual terms are smaller than ours on a matching basis?

In order to determine the temperature range in which this divergence is of physical importance we determine the behavior of the terms in (1.8) as a function of temperature. As do the authors we will suppress the factor 2 in the discussion.

The first term is then equal to 1 at all temperatures. Show that at 300°K the second term is O[exp (-400)], and the terms do not become appreciable(of order 10^{-3}) until $n=O(10^{113})$. We may, therefore, cut off the series after the first term and ignore the divergence entirely. At 1000°K evaluate both the second term and the value of n for which the nth term becomes as large as 0.001. Compare with the authors' values of O[exp-(100)] and $O(10^{33})$.

The situation changes at higher temperatures. Fig. 1 shows

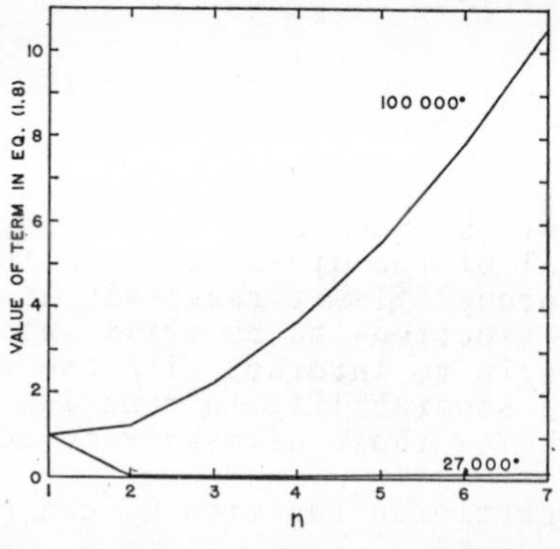

FIG. 1. Values of terms in Eq. (1.8) vs n at 27 000° and 100 000°.

that at 27 000° there is no clearcut cutoff, and at 100 000° there is no longer any indication of a cutoff.

Thus it is apparent that at high temperatures excited H atoms cannot be treated as isolated systems. Physically this is because the excited atoms have a less compact electron distribution(they are "larger") and therefore interact even at low densities. Furthermore, the continuum states correspond not to an atom at all, but to a pair of charged particles, an electron and a proton. The authors proceed to consider the mixture of charged particles and atoms and their interactions with expressions for the partition function and thermodynamic properties following.

Is the factor 2 suppressed in Fig. 1?

41. "Molecular Energy States and the Thermodynamic Properties of Solid Isotopic Methanes" J.H. Colwell, E.K. Gill, and J.A. Morrison 42, 3144 (1965).

This is one of a series of papers treating the low temperature thermodynamic properties of the methanes from both experimental and theoretical viewpoints.

Table IV contains calorimetric entropies and Table V comparisons between calorimetric and practical entropies for each

TABLE IV. Calorimetric entropies of the deuterated methanes (calories per mole·degree).

ΔS	CH_3D	CH_2D_2	CHD_3	CD_4
Solid ($0° \rightarrow T_{TP}$)[a]	14.17	14.89	15.39	16.02
Fusion (T_{TP})	2.49	2.44	2.43	2.42
Liquid ($T_{TP} \rightarrow T'$)[b]	···	···	···	1.11
Vaporization (T')	23.28	23.46	23.60	21.20
Correction to ideal gas	0.05	0.05	0.05	0.13
S_{cal}:	39.99±0.10	40.84±0.10	41.47±0.10	40.88±0.10

[a] Does not include the low-temperature anomalies of the partially deuterated compounds; for these, C_P was extrapolated smoothly to zero from 8°K.

[b] For CD_4, S_{cal} is calculated at T'=97.82°K and P=208.8 mm; for the other methanes, S_{cal} is calculated at the triple-point temperature T_{TP} and pressure as given in Table V of Ref. 4.

TABLE V. Comparison of calorimetric and practical entropies of the deuterated methanes (calories per mole·degree).

CH_3D (90.41°K, 84.50 mm)	CH_2D_2 (90.17°K, 82.00 mm)	CHD_3 (89.96°K, 80.20 mm)	CD_4 (97.82°K, 208.8 mm)
S_{spec}=42.82[a]	44.37	44.22[a]	40.84
S_{cal}=39.99±0.10	40.84±0.10	41.47±0.10	40.88±0.10
Δ= 2.83±0.10	3.53±0.10	2.75±0.10	−0.04±0.10

R ln4=2.76
R ln6=3.56

[a] Moments of inertia for CH_3D and CHD_3 were obtained from B. P. Stoicheff (private communication).

of the four deuterated methanes. In Table IV the correction to the ideal gas state, at the appropriate temperatures and pressures as shown in Table V, is positive in every instance. What is the most general explanation that you can give of this? Why is the correction for CD_4 over twice as large as the others?

The practical entropy is identical with the spectroscopic entropy which neglects the effects of nuclear spin. The value of the practical or spectroscopic entropy minus the calorimetric entropy, Δ in Table V, is called the residual entropy and should be a nonnegative quantity. Is this condition met within the experimental errors? The residual entropies are conventionally explained("exceptions to the third law") as being due to the freezing-in of nonequilibrium molecular configurations at low temperatures. Adopting this view for the moment what should the residual entropies of each of the four species be? In Table V, left to right, R ln 4, R ln 6, R ln 4, and R ln 1? Is the agreement within experimental error in each case?

The authors offer a different explanation for both the origin and the magnitudes of the residual entropies. Quoting from their Summary "We have shown that the anomalies(in heat capacities for T<8°K for the partially deuterated species) and residual entropies can be related and that they are not caused by the freezing-in of nonequilibrium molecular configurations at low temperatures. An interpretation of the results is advanced which ascribes the observed effects to the perturbation (by the lattice) of the molecular energy states which correspond to rotation in the free molecules."

The statistical entropy includes the effects of nuclear spin. The apparent zero-point entropy S_o is $S_{stat}-S_{cal}$ and thus includes both nuclear spin and residual effects; it is given by the authors' equation

$$S_o(\text{calc.}) = R\sum_i \chi_i \ln g_o(i) - R\sum_i \chi_i \ln \chi_i \qquad (8)$$

where i denotes the nuclear spin species, χ_i the mole fraction of species i in the normal(high-temperature) composition and $g_o(i)$ the effective degeneracy of the ground state available to species i. What assumption is made about conversion among different spin species by the use of the high temperature composition? The last term in (8) is the entropy of mixing.

Check the value of S_o for CD_4 of 8.69±0.10 e.u. with the degeneracies $g_o(A)=15$, $g_o(E)=12$, and $g_o(F)=54$ for the A, E, and F spin species. Why would the apparent zero-point entropy carry an experimental error?

42. "Low-Temperature Specific Heat of Gold, Silver, and Copper" L.L. Isaacs 43, 307 (1965).

Fig. 1 presents the specific heats C as functions of temperature in the form C/T vs T^2. What is the experimental temperature range? Empirical equations fit to the data are given alongside the solid curves. One see that a T^5 term(for C) was necessary to give good fits for the Au and Ag data while for Cu a T^3 term sufficed. Check a value for C/T for each metal in the middle of the temperature range with the appropriate equation. These are total specific heats, electronic plus lattice.

The Debye approximation for the lattice frequency distribution leads to the asymptotic expression for C_V in the limit as T→0

$$C_V = (12\ Nk\pi^4/5)(T/\theta_D)^3.$$

θ_D is the Debye temperature. See for example T.L. Hill "An Introduction to Statistical Thermodynamics" (Addison-Wesley Publishing Co., Inc. Reading, Massachusetts 1960), Chapt. 5.

From the coefficient of the cubic term in the equation

for C calculate the Debye temperatures for the metals. The author call them the zero-degree limiting Debye temperatures $\theta_C(0)$. Compare your values, within Isaacs' experimental errors, with Debye temperatures calculated from elastic constants $\theta_E(0)$.

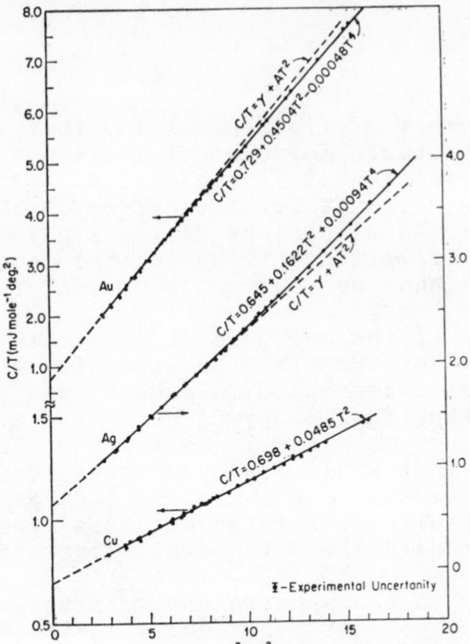

FIG. 1. The plots of C/T versus T^2 for gold, silver, and copper.

	$\theta_C(0)$		$\theta_E(0)$
Gold	?	±1.5°K	161.6°K
Silver	?	±2.0°K	227.3°K
Copper	?	±3.0°K	345.3°K

43. "Heat Capacity and Other Thermodynamic Properties of MoF_6 between 4° and 350°K" D.W. Osborne, F. Schreiner, J.G. Malm, H. Selig, and L. Rochester 44, 2802 (1966).

This exercise is taken from the same article as was no. 14. There some of the classical thermodynamic aspects were emphasized. Here we will compute the spectroscopic entropy from molecular data and see if the third law is valid for the crystal.

From electron-diffraction measurements the Mo-F distance is known to be 1.84±0.02 Å. The MoF_6 was assumed to have the configuration of a regular octahedron(point group?) with symmetry number σ=24. This is the case for all of the known hexafluorides with the possible exception of XeF_6 (ca late 1965); do you know of any definite determinations for xenon hexafluoride since the time of this article? The case of the MoF_6 molecule being octahedral or not can be settled quite neatly from the spectroscopic entropy as we shall see below.

Their vibrational frequencies are taken from the assignment

[redacted] [redacted]

[redacted]

of ref. 21 and are, with degeneracies in parentheses,

$\nu_1(1) = 741\ cm^{-1}$
$\nu_2(2) = 643$
$\nu_3(3) = 741$
$\nu_4(3) = 262$
$\nu_5(3) = 312$
$\nu_6(3) = 122$

Is this the right number of frequencies for this molecule? Application of the standard formulas, for example from Chapt. 9 of the book by Hill referenced in exercise no. 42, should enable you to duplicate their result for the entropy of MoF_6 at 298.15 °K and one atmosphere(ideal gas) of 83.765 e.u.. How does this compare with the calorimetric entropy for the same state given in exercise no. 14? What do you conclude about the third-law behavior of MoF_6?

If the symmetry of the molecule is less than octahedral how would the (decreased)symmetry number σ affect the entropy? Does it turn out that a decreased molecular symmetry and a finite residual entropy($S_{spec}-S_{cal} \geq 0$) both affect the spectroscopic entropy in the same manner? Is there any aspect of the molecular structure that would lead you to expect a residual entropy such as is present in the classical cases such as H_2O, CO, NO, and the partially deuterated methanes of exercise no. 41? If not may we regard the octahedral structure of MoF_6 as established?

One other point on the spectroscopic entropy. The triply degenerate ν_6 is an important contributor to the entropy of the gas; why?(two reasons). Other assignments have been proposed for this frequency ranging from 190-234 cm^{-1}. The 190 cm^{-1} value leads to $S^\circ_{298.15}(g)=81.303$ e.u. with the others leading to lower(or higher?) entropies yet. Still, could we possibly have the situation where one of the higher values for ν_6 could combine with a symmetry lower than octahedral to still yield an entropy close to the calorimetric value?

44. "Vibration Spectrum and Specific Heat of Lithium" R.P. Gupta 45, 4019 (1966).

The author has calculated the vibrational frequency spectrum and heat capacity, taking into account the effect of conduction electrons on the lattice vibrations, for crystalline lithium. The elastic-force model of earlier workers was employed.

For a crystal of N atoms or ions there are $3N-6 \simeq 3N$ vibrational frequencies. For a mole this becomes $\simeq 2\times10^{24}$ frequencies and so in practice one speaks of a distribution of frequencies $G(\nu)d\nu$, the number of frequencies in differential width $d\nu$ about ν. Also we must have

$$\int_0^\infty G(\nu)d\nu = 3N$$

the normalization condition. Fig. 1 presents the lattice vibrational spectrum of lithium. Notice it is given in terms of the angular frequency $\omega=2\pi\nu$. Will this make any difference in the ordinate $G(\omega)$? Notice that the factor 2π is in the abscissa ω.

[redacted]

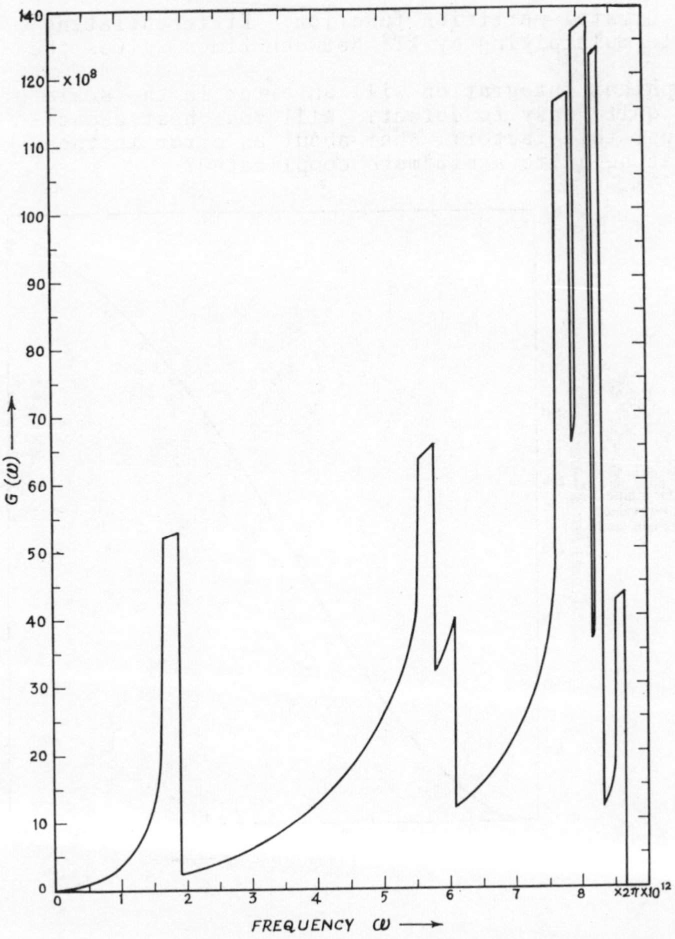

FIG. 1. The lattice vibrational spectrum of lithium.

Check, by a rough graphical integration, whether or not the distribution is normalized. If it seems to be off by a factor of ten would it most likely be in ω or G(ω)? Does 180, 1800, or 18 000 cm^{-1} seem the most likely upper limit for a lattice vibration? Recent Raman studies of graphite show the symmetric ring stretching mode at about 1500 cm^{-1}, though covalent anisotropic is by no means ionic isotropic. The question of normalization can be definitely settled when you calculate heat capacities at one or two temperatures to compare with those given in Fig. 2.

To do this you will need the relation between G(ν) and C_V

$$C_V = k\int_0^\infty [(h\nu/kT)^2 \exp (h\nu/kT)\, G(\nu)/\{\exp (h\nu/kT) - 1\}^2]d\nu.$$

This result can be obtained in about three steps by writing the crystal partition function as a product of 3N harmonic oscillator partition functions(in general all with different frequencies) then by replacing the sum of terms by an integral in the logarithm of the crystal partition function. Differentiating twice wrt T, while multiplying by kT^2 between times, gives the C_V equation.

In your graphical integration will an error in the scale of the $G(\omega)$ axis be quite easy to detect? Will your heat capacities be off by just this factor? What about an error in the ω scale? Would that be quite a bit more complicated?

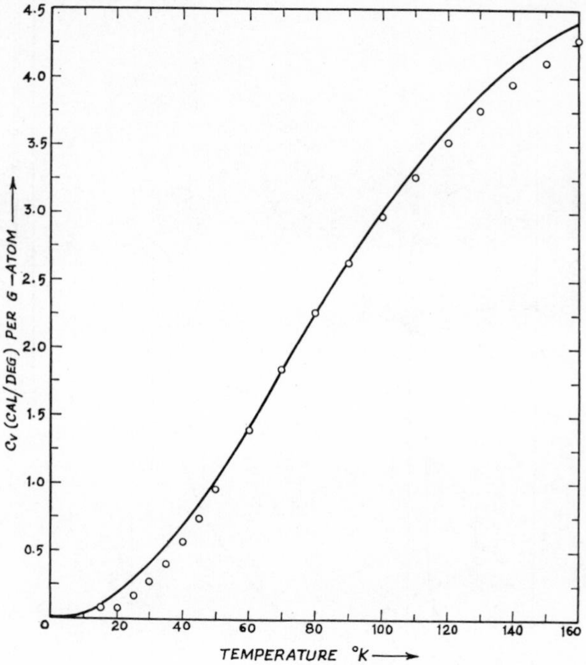

FIG. 2. The specific-heat curve for lithium. The circles represent experimental values due to the measurements of Simon and Swain.

45. "Significant-Structure-Theory Compressibility Coefficients, Thermal-Expansion Coefficients, and Heat Capacities of Alkali Metals in the Liquid Phase" R. Vîlcu and C. Misdolea 49, 3179 (1968).

Perhaps this exercise could just as well have appeared in Chapter I or Chapter VI. It was placed in the context of statistical mechanics because three of the quantities we will be working with, heat capacity at constant volume C_V, thermal expansion coefficient α, and isothermal compressibility coefficient β, are evaluated from the molar partition function Z which is written in the significant-structure form for liquid metals. We will follow this in outline below. You will then be asked to establish a thermodynamic identity for C_P-C_V and then check their values of C_P.

The partition function itself is given in the article. It is written for a canonical ensemble as we may determine by inspection; Z=Z(N,V,T). C_V is obtained from the energy E which itself is related to Z through

$$E = kT^2[(\partial \ln Z)/\partial T]_{N,V}.$$

From this express C_V in terms of T and Z. The basic result for the Helmholtz free energy or work function

$$A \equiv E - TS = -kT \ln Z$$

will be used to determine α and β. Show that the pressure P is given by

$$P = kT[(\partial \ln Z)/\partial V]_{N,T}.$$

Taking derivatives with respect to T and V and then making use of the basic result

$$(\partial V/\partial T)_P(\partial T/\partial P)_V(\partial P/\partial V)_T = -1$$

allows us to express

$$\alpha = \frac{(\partial V/\partial T)_P}{V}$$

and

$$\beta = -\frac{(\partial V/\partial P)_T}{V}$$

in terms of ln Z and its derivatives. Do this just as you did for C_V. We will leave these equations at this point knowing that given Z(N,V,T) we could evaluate C_V, α and β.

Next you are asked to derive the thermodynamic identity ("well-known equation")

$$C_P - C_V = \alpha^2 VT/\beta.$$

When you have done this check and see if you obtain $C_P - C_V = R$ for an ideal gas.

Table II gives us molar volumes at different temperatures for the five alkali metals. Are they for the liquid or the solid?

TABLE II. The molar volumes at different temperatures.

Li		Na		K		Rb		Cs	
T (°K)	V [a] (cc)	T (°K)	V [a] (cc)	T (°K)	V [a] (cc)	T (°K)	V [b] (cc)	T (°K)	V [c] (cc)
452	13.42	371	24.79	337	47.24	312	58.05	301	71.74
500	13.55	400	24.98	500	49.57	373	59.27	500	76.05
600	13.82	473	25.46	600	51.11	414	60.07	600	78.54
700	14.10	573	26.15	700	52.75	453	60.83	700	81.29
800	14.40	773	27.65	800	54.50	493	61.63	800	84.62

[a] E. Goltsova, Teplofiz. Vysokokh Temperatur, Akad. Nauk SSSR Inst. Poluprov. Sb. Tr. 1-90 2-90 Soveshch. po Termoelekt. **4**, 3 (1966).
[b] *Reactor Handbook Engineering* (U. S. Atomic Energy Comm. New York, 1955), p. 208.
[c] J. P. Stone, C. T. Ewing, J. R. Spann, E. W. Steinkuller, D. Williams, and R. Miller, J. Chem. Eng. Data **11**, 32 (1966).

Tables III, V, and VII present us with calculated and experimental results for molten lithium, kalium, and cesium respectively. Even if you don't know a little German could you determine the identity of the substance referred to in Table V from a knowledge of the periodic table and the assumption that the authors have presented their data in a logical fashion?

The units of β are $cm^2/dyne$; the dimension must be reciprocal pressure - right?

TABLE III. Heat capacities, thermal-expansion coefficients, and compressibilities for molten lithium.

Temperature °K	C_v (cal/g-atom·°K)		$\alpha \times 10^4$ (°K^{-1})		$\beta_T \times 10^{12}$ (cm^2·dyn Å^{-1})		C_p (cal/g-atom·°C)	
	Calc	Exptl [a]	Calc	Exptl [b]	Calc	Exptl [a]	Calc	Exptl [c]
452	6.01	5.35 [d]	6.68	1.80	3.76	8.8 [d]	23.2	7.26
500	5.96	···	3.62	···	2.32	···	15.1	7.20
600	5.93	···	3.44	···	2.64	···	14.8	7.06
700	5.92	···	3.96	···	3.48	···	16.5	6.93
800	5.94	···	4.57	···	4.49	···	18.7	6.92

[a] J. Lumsden *Thermodynamics of Alloys* (London, 1952), p. 79.
[b] C. Smithells, *Metals Reference Book* (New York, 1955), p. 636.
[c] R. Hultgren and R. Orr, *Selected Values of Thermodynamic Properties of Metals and Alloys* (New York, 1963).
[d] Orientative value for 298°K.

TABLE V. Heat capacities, thermal-expansion coefficients, and compressibilities for molten kalium.

Temp. (°K)	C_v (cal/g-atom·°K)		$\alpha \times 10^4$ (°K^{-1})		$\beta_T \times 10^{12}$ (cm^2 dyn Å^{-1})		C_p (cal/g-atom·°C)	
	Calc	Exptl [a]	Calc	Exptl [b]	Calc	Exptl [a]	Calc	Exptl
337	5.94	6.9	3.68	2.91	6.47	40	13.8	7.68 [d]
500	5.94	···	5.41	2.98 [c]	13.31	···	18.9	7.34 [e]
600	5.95	···	6.15	···	17.40	···	21.9	7.20 [e]
700	5.95	···	6.55	···	20.88	···	24.1	7.13 [e]
800	5.95	···	6.68	···	23.69	···	25.6	7.11 [e]

[a] See Table IV, Footnote a.
[b] J. Jarzynski and T. Litovitz, J. Chem. Phys. **41**, 1290 (1964).
[c] Value at 423°K.
[d] J. Wilson, Metallges. Rev. **10**, 539 (1965).
[e] D. Stull and G. Sinke, Advan. Chem. Ser. **18**, 1956.

TABLE VII. Heat capacities, thermal-expansion coefficients, and compressibilities for molten cesium.

Temp. (°K)	C_v (cal/g-atom·°K)		$\alpha \times 10^4$ (°K^{-1})		$\beta_T \times 10^{12}$ (cm^2·dyn Å^{-1})		C_p (cal/g-atom·°K)	
	Calc	Exptl [a]	Calc	Exptl [a]	Calc	Exptl [a]	Calc	Exptl [b]
301	5.97	6.7	3.55	3.7	8.49	67	13.62	7.6
500	5.95	···	5.51	···	19.84	···	19.85	7.6
600	5.95	···	6.13	···	25.40	···	22.66	7.6
700	5.96	···	6.45	···	30.21	···	24.68	7.6
800	5.96	···	6.64	···	35.17	···	26.21	7.6

[a] See Table IV, Footnote a. [b] See Table V, Footnote e.

With the calculated values for C_V, α and β compute C_P for each substance at the lowest temperature listed. Do you match the Table entries?

C_V and α, calculated values, are in reasonable agreement with experiment in every case shown but one. β and C_P on the other hand show rather large discrepancies. Are these discrepancies independent? Think about our thermodynamic identity.

Do you agree with the authors' conclusions "...the significant-structure theory gives good results as long as the quantities are expressed in regard to the variation of the partition function with the temperature. The quantities implying in their calculation the variation of the partition function with the volume(chiefly β and C_P) do not give a good enough agreement with the experimental data."?

They point out that α calculated from Z agrees much better with experiment for liquid alkali metals than for molten salts.

46. "Molecular Dynamics Studies of the Microscopic Properties of Dense Fluids" P.L. Fehder 50, 2617 (1969).

This exercise will serve to introduce us to a relatively new(ca 1959) line of attack for theories of dense fluids, that of computer calculations on various models.

The work reported in the paper is for dynamic calculations on a two-dimensional model system of 364 disk particles interacting through a Lennard-Jones potential. All dynamical variables entering into the calculations are reduced in units involving m, the particle mass, and σ and ε, the distance and energy parameters in the Lennard-Jones pair potential. Distances are expressed in units of σ, velocities in units of $(\varepsilon/m)^{\frac{1}{2}}$, and the reduced time increment $\Delta\tau$ corresponding to the time increment Δt is given by

$$\Delta\tau = (\Delta t)(\varepsilon/m)^{\frac{1}{2}}(\sigma^{-1}).$$

The pair potential is truncated at a "cutoff" radius $r_c = 2.50\sigma$, beyond which it is set equal to zero. A time increment equivalent to 10^{-14} seconds($\Delta\tau$ or Δt?) was used in the dynamics integration. If γ_i is the velocity of the i'th particle in units of $(\varepsilon/m)^{\frac{1}{2}}$, the two-dimensional temperature of the system is given by

$$T^* = (2N)^{-1} \sum_{i=1}^{N} \gamma_i{}^2.$$

Fig. 1 gives plots of velocity distribution at three different times after time zero, at which time the initial distribution was obtained by randomly assigning, positive or negative, a value of just over $1.0(\varepsilon/m)^{\frac{1}{2}}$ to the x and y coordinates of each particle. The density $\rho^* = N\sigma^2/A$, with A the true area, was 0.63 and T^* was 1.05.

What would the velocity distribution plot look like at zero time? Two delta functions? How many computational steps (time increments) have elapsed when "snapshot" (A) was taken? (B) and (C)? With this information make measurement of peak heights as accurately as possible and see if relaxation to equilibrium is a first-order process. Do the same for the growth of the center peak with velocity equal to zero. Same rate law? Can you account for the structure on the outside of the left-hand peak in (B)? Is (C) approximately gaussian? Is this what you would expect for a one-dimensional velocity distribution? Is a one-dimensional velocity distribution meaningful in this two-dimensional problem?

By fitting an approximate gaussian to (C) evaluate analytically the mean squared velocity $\langle\gamma^2\rangle$ and then use this with T* to see if the distribution in (C) is normalized.

The author comments in connection with Fig. 1 that reequilibration after an "instantaneous" temperature alteration is generally quite rapid. This figure is a very remarkable one and may well serve to help us have confidence in the three-dimensional Maxwellian distribution for velocities that we have studied. One final question: what will be the evolution of the distribution function in Fig. 1 at times later than (C)?

The relative density ρ_R is the ratio of the actual system density to that of closest packing and is given by

$$\rho_R = (N/A)(\sqrt{3}/2)\delta_{min}{}^2,$$

where δ_{min}=1.1132σ is the lattice parameter found computationally to give the lowest energy for a two-dimensional hexagonal closest-packed array of disks interacting with the truncated Lennard-Jones potential. What is the value of δ for the two-dimensional hcp array itself? The answer can be determined from books of course but try and determine it from a sketch of coins or even better use a coin to draw hcp circles.

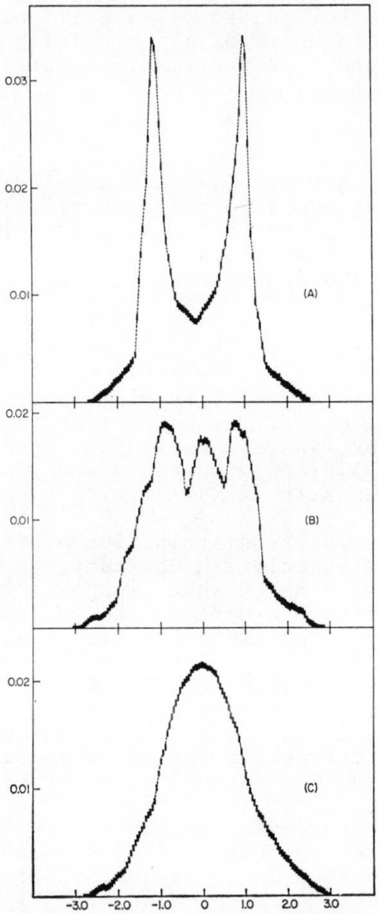

FIG. 1. Velocity distributions showing rapid equilibration of kinetic energy in the system at density $\rho^*=0.63$, nominal average temperature $T^*=1.05$. Initial distribution obtained by randomly assigning, positive or negative, a value of 1.0263 to the x and y velocity components of each particle. Plots show distribution after (A) 1.5×10^{-13}, (B) 3.0×10^{-13}, and (C) 2.0×10^{-12} sec.

Fig. 3 is a snapshot of the instantaneous configuration of the system at the temperature and density indicated. The most striking feature is the presence of "holes" or "vacancies"

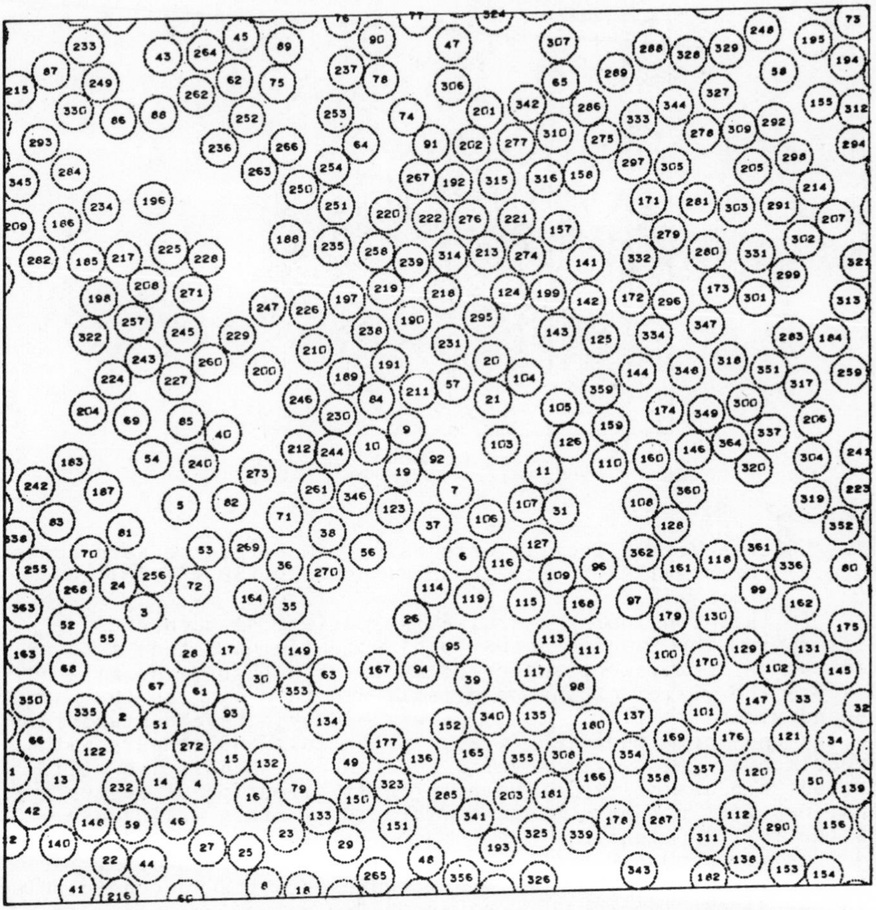

FIG. 3. Snapshot of instantaneous system configuration in state $T^*=0.927$, $\rho_R=0.6781$.

in the particle distributions. The vacancies are quite irregular and many encompass areas of several σ^2. How does this behavior differ from the experimental result interpretation quoted for liquid argon? Does his analogy of microscopic surface tension for explaining the relatively long life(? seconds) of the vacancies seem convincing?

Fig. 6 shows a plot of particle trajectories for periods of 2×10^{-12} seconds. The small circles mark the initial positions of the particle centers, and the irregular lines extending from the circles represent the paths of the centers during the remainder of the 2×10^{-12}-sec interval. The dashed lines enclose regions of chain and cluster diffusion into vacancies. Can you see which is which? Are there any particles which appear to be collision-free during this time interval? Single-particle uncorrelated motion was observed to be common only in the highest temperature, lowest density states examined.

There are more results and a good deal more discussion than that outlined here; do you agree that it is an intriguing area of research?

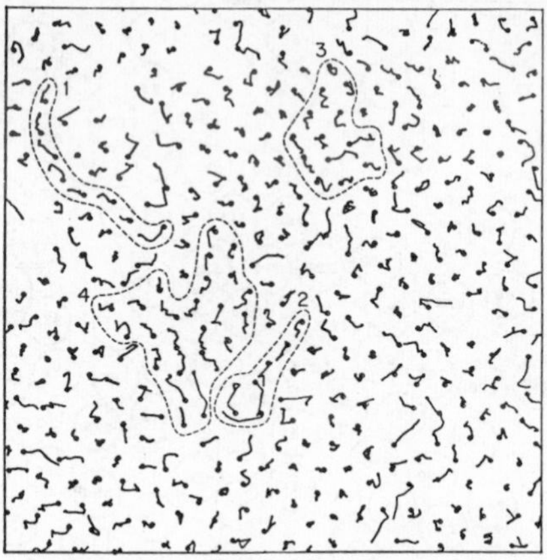

FIG. 6. Particle trajectories over a 2×10^{-12}-sec interval in the system in state $T^*=0.927$, $\rho_R=0.6781$.

47. "Second Virial Coefficients Using a Modified Lennard-Jones (6,n,γ) Potential" R.Y. Koo and H.W. Hsu 52, 2392 (1970).

In this concluding exercise in statistical mechanics we will review the Lennard-Jones (6,n) potential and then take a look at an improvement suggested by these authors. We will discuss the virial expansion briefly and then examine the second virial coefficients as determined by both the standard (6,12) and their modified (6,n,γ) potentials and compare with experimental values for several substances. Finally we will present a plausibility argument or nonrigorous proof suggesting that He_2 does not have a bound vibrational state making use of the (6,12) potential.

As you probably are aware the inverse sixth-power attractive term in the (6,n) accurately represents the induced-dipole-induced-dipole interactions of non-polar molecules; accurate in the sense of being a result derived from quantum mechanics. An interesting historical exercise might be to check the literature of the era and see if the attractive index -6 was first suggested on empirical grounds, the old quantum theory, or whatever. A place to start the literature search might be the article by J.E. Lennard-Jones, Proc. Roy. Soc.(London), A106, 463 (1924). What about the repulsive index -12? Is it rigorous also? Is it often chosen partially on the grounds of mathematical convenience since 2x6=12? Is twelve the only exponent used? Find one or two articles in the literature to support your claim. The (6,12) potential is given in the nomenclature of the present authors as

$$E(r) = 4\varepsilon[(\sigma/r)^{12} - (\sigma/r)^{6}]$$

where $\varepsilon>0$ is the well depth and σ is the collision diameter, that value of r for which E(r) is zero(other than infinity?). This curve is shown in Fig. 1 in reduced form. Here $r^*\equiv r/\sigma$. Can you determine by inspection how E* is defined? Show that $r_{min}=2^{1/6}\sigma$ for the (6,12) potential.

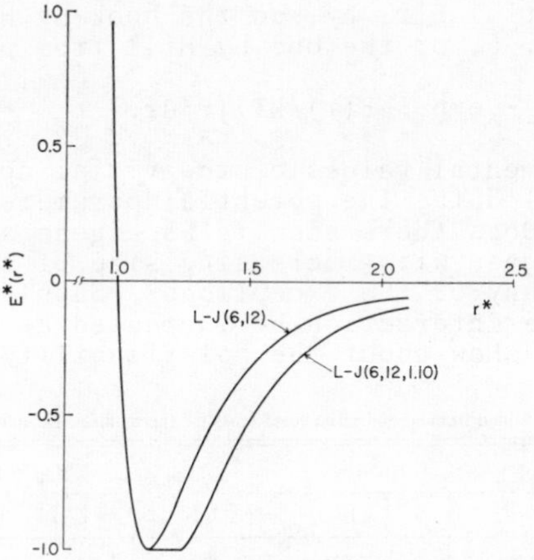

FIG. 1. Reduced Lennard-Jones (6,12, $\gamma = 1.10$) potential.

One sees immediately that the (6,12) potential is a two-parameter one, ε and σ. From a study of virial coefficients obtained from model potentials the authors' ref. 1 concluded that the bowl of the (6,12) potential is too narrow for the rare gases. If the inverse sixth power attraction is maintained and the bowl width is to increase then the repulsive exponent must be made less than 12, which in turn makes the potential too "soft". To overcome these deficiencies(really just one) the authors propose the three-parameter function shown in Fig. 1. If the repulsive exponent is allowed to vary the potential function becomes one of four parameters. Its analytic form is

$$E(r) = [n\varepsilon/(n-6)](n/6)^{6/(n-6)}[(\sigma/r)^n - (\sigma/r)^6] \quad \text{for } 0<r<r_{min}$$

$$= -\varepsilon \quad \text{for } r_{min}<r<\gamma r_{min}$$

$$= [n\varepsilon/(n-6)](n/6)^{6/(n-6)}[(\gamma\sigma/r)^n - (\gamma\sigma/r)^6] \quad \text{for } \gamma r_{min}<r<\infty,$$

where $r_{min}=(n/6)^{1/(n-6)}\sigma$. Demonstrate that the function is continuous at r_{min} and γr_{min}. What about the first derivatives at these two points; are they continous? Are your analytical results borne out by inspection of Fig. 1? This is an example of zeroth-order contact, but not first-order contact, in the sense of footnote 28 of the paper by D.D. Konowalow, J. Chem. Phys. 50, 12 (1969). For n=12 and γ=1 can you show that the above potential reduces to the (6,12)? By inspection you can see that l'Hospital's rule will be required.

For a one-component gas the equation of state in the form of the virial expansion is

$$(P/kT) = \rho + B(T)\rho^2 + C(T)\rho^3 + D(T)\rho^4 + \ldots,$$

with B(T), C(T), D(T),... called the second, third, fourth,... virial coefficients, functions of temperature only. A fast trick question might be what is the value of the first virial coefficient? The second virial coefficient is related to the intermolecular potential E(r) by(see the book by Hirschfelder et al mentioned on p. 6, or the one by Hill from p. 76)

$$B(T) = 2\pi N \int_0^\infty [1 - \exp\{-E(r)/kT\}] r^2 dr.$$

From the experimental values of the virial coefficients, determined from P-V-T data, the potential parameters of Table II were optimized. Does there seem to be a general trend for σ and ε to both increase with increasing size of the molecule? Can you rationalize any of the exceptions? Should the induced-dipole-induced-dipole interaction be increased as the number of electrons increases? How about the polarizability?

TABLE II. Potential parameters determined from second virial coefficients for the modified Lennard-Jones $(6, n, \gamma)$ potential.

	L-J$(6, n, \gamma)$				L-J(6,12)	
Gas	n	γ	σ [Å]	ϵ/K[°K]	σ [Å]	ϵ/K[°K]
N_2	12	1.01	3.635	97.98	3.694	96.26
CO_2	12	1.04	4.664	178.00	4.416	192.25
C_2H_6	15	1.04	4.615	258.96	5.220	194.14
n-C_4H_{10}	15	1.05	6.695	279.09	7.152	223.74
C_2H_4	15	1.11	4.256	239.65	4.433	202.52
C_6H_6	13	1.05	8.379	251.17	8.443	247.50

B(T) for both potentials are shown for six substances in Fig. 2. For which substance does the $(6,n,\gamma)$ potential make the biggest improvement in fitting B(T)? For which the least? At a given temperature which substance shows the greatest departure from the ideal gas law? Which the least? Notice that at lower temperatures B(T) is negative for all of the compounds. Does the attractive or repulsive portion of the potential predominate? Is the density less than or greater than that of the ideal gas?

At higher temperatures $B(T_B)=0$ where T_B is called the Boyle temperature. Why? At higher temperatures B(T) increases then passes through a maximum which is not shown in Fig. 2. Rationalize this maximum.

As a final activity consider the question of one or more bound vibrational states in the diatomic He_2. The simplest MO theory gives the configuration $1\sigma_g^2 1\sigma_u^2$ with two electrons in each of the bonding and antibonding orbitals and hence a slight net repulsion. Our effort here should reinforce this view from somewhat different considerations.

The Lennard-Jones (6,12) parameters for He are(Hill, op cit p. 484) $\varepsilon/k=10.22$°K and $\sigma=2.556$Å. Consider the (6,12) to be expanded in a Taylor series about $r=r_{min}$. Identify the terms in the expansion. Treating only the squared term, equivalent to assuming a harmonic potential and therfore a questionable procedure apart from any possible pedagogic merit, determine the force constant $(d^2E/dr^2)_{r_{min}}$ and hence the zero-point energy $\frac{1}{2}h\nu$. Is this the energy, above the bottom the well, of the first bound state in the harmonic oscillator approximation? It should be a multiple or submultiple of the well depth ε. Convincing?

In summary the original (6,12) potential can be improved upon somewhat, although it still maintains a very high ratio of usefulness to simplicity.

Notice that the authors have a Nomenclature section. This

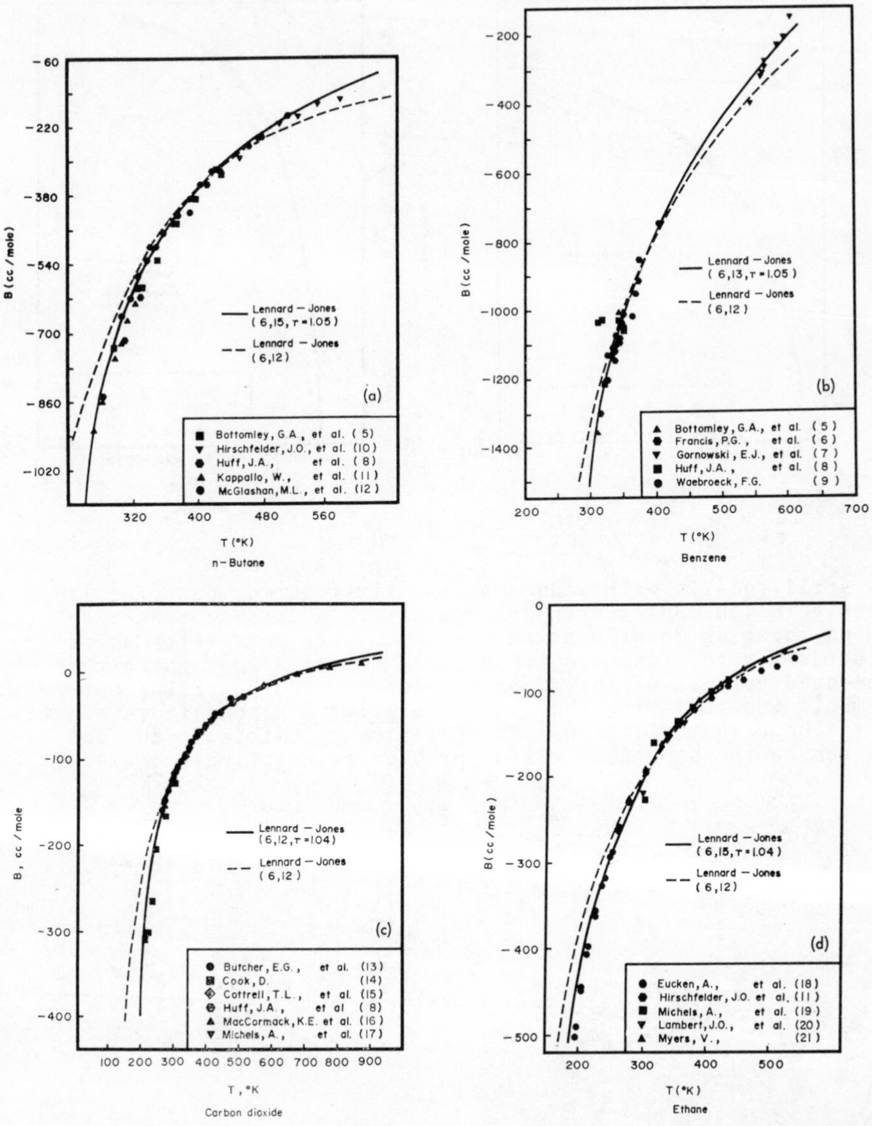

FIG. 2. Comparison of experimental and calculated values of second virial coefficients of Lennard-Jones (6, 12) and Lennard-Jones $(6, n, \gamma)$ gases a function of temperature. (a) *n*-Butane; (b) benzene; (c) carbon dioxide; (d) ethane; (e) ethylene; (f) nitrogen.

is not often found in the literature and can be extremely useful, especially to the new worker in an area. The book by Hirschfelder et al has a section on notation. The author of this exercise book, who happens to have a degree in petroleum refining engineering, might be willing to wager a small sum that the majority of articles carrying formal sections on nomenclature do originate in engineering departments or schools.

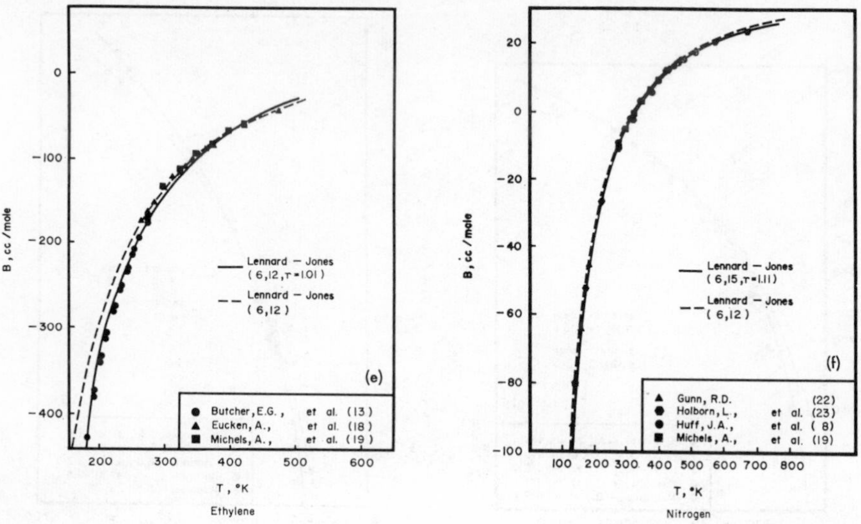

FIG. 2 (continued)

He still recalls with amusement the first summer out of college when a fellow employee in a large aircraft company, chairman of an engineering department in a rather large university and a visitor for the summer, grimly spent day after day going through selected volumes of the company's library red-penciling out symbols and replacing them with his private nomenclature scheme. Let's hope that all of us are a bit more flexible in our approach to the sometimes vexing problem of notation!!

CHAPTER IV

RATE PROCESSES

48. "The Rate Constant of the Reaction between Hydrogen Peroxide and Ferrous Ions" T. Rigg, W. Taylor, J. Weiss 22, 575 (1954).

It was shown earlier that reaction between hydrogen peroxide and ferrous ions proceeds in two consecutive steps

$$Fe^{2+} + H_2O_2 \rightarrow Fe^{3+} + OH^- + OH \qquad (k_1) \qquad (1)$$

$$Fe^{2+} + OH \rightarrow Fe^{3+} + OH^- \qquad (k_2) \qquad (2)$$

under certain conditions, which then correspond to the simple stoichiometry

$$2Fe^{2+} + H_2O_2 \rightarrow 2Fe^{3+} + 2OH^-. \qquad (3)$$

Apply the steady-state approximation(which species in aqueous solution?) to derive the rate law

$$-d[Fe^{2+}]/dt = d[Fe^{3+}]/dt = 2k_1[Fe^{2+}][H_2O_2]. \qquad (4)$$

It is thus possible to measure the rate constant k_1 directly; the method chosen was spectrophotometric measurement of the absorption at λ=304mμ due to the ferric ion formed. What can we determine about the rate constant k_2? Notice that it is not even identified in the article; do you see why?

Why was it necessary to have the reactants acidified?

The molar extinction coefficient ε is a mild function of temperature as shown in Fig. 1. This is in contrast to the assumption made in exercise no. 16, p. 28. From Fig. 3 show that the optical density when the reaction is complete, ε(Fig.2) and the concentrations shown(obvious misprint for $[Fe^{2+}]_o$) are all compatible with the 2-cm path length at the temperature indicated.

Table II and Fig. 2 present the results for k_1 as a function of temperature in different forms. From Fig. 2 determine the activation energy and pre-exponential factor in the Arrhenius equation to compare with their values

$$k_1 = 1.05 \times 10^8 \exp(-8460/RT) \text{ liter}\cdot\text{mole}^{-1}\cdot\text{second}^{-1}. \qquad (7)$$

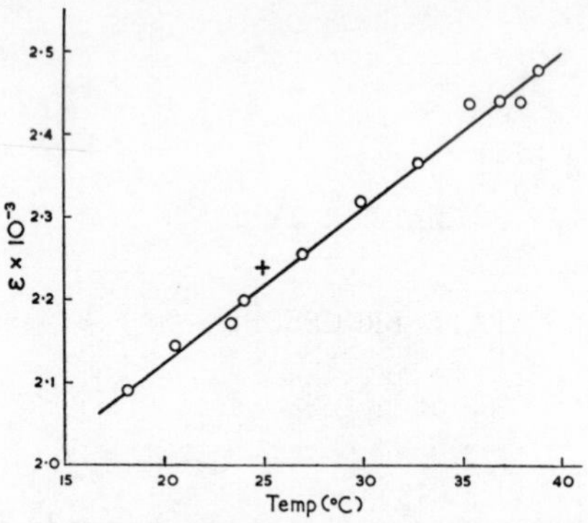

FIG. 1. Temperature dependence of the molar extinction coefficient at $\lambda = 304$ mμ. Ferric sulfate in 0.80 N sulfuric acid. (+) corresponds to the value obtained by Hochanadel and Ghormley (see reference 7).

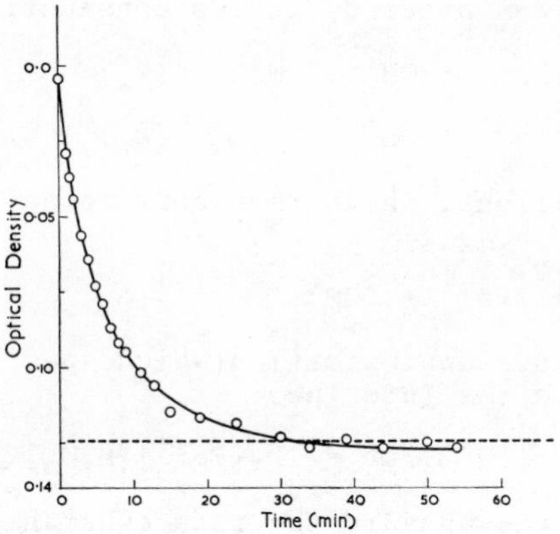

FIG. 3. Increase of the optical density at $\lambda = 304$ mμ. (2-cm silica cell) for initial concentrations. $[H_2O_2]_0 = 1.325 \times 10^{-5}$ moles/1 and $[Fe^{2+}]_0 = 5.15 \times 10^5$ moles/1 in 0.8 N sulfuric acid solutions at 23.8°C. The straight line corresponds to the theoretical amount of ferric salt produced according to the simple stoichiometry [Eq. (3)].

TABLE II.

Conc. of sulfuric acid (normality)	Temp. °C	k_1 (mean) (1 moles^{-1} sec^{-1})
0.5	15.1	40.5±2.0
0.8	18.6	48.9±2.5
0.8	19.75	52.8±2.5
0.5	20.0	51.0±2.5
0.05	20.6	58.8±3.0
0.8	23.8	64.4±3.0
0.5	25.1	63.4±3.0
0.8	29.1	81.2±4.0
0.8	33.7	102.8±5.0
0.5	35.5	102.5±5.0
0.5	40.0	125.2±6.0

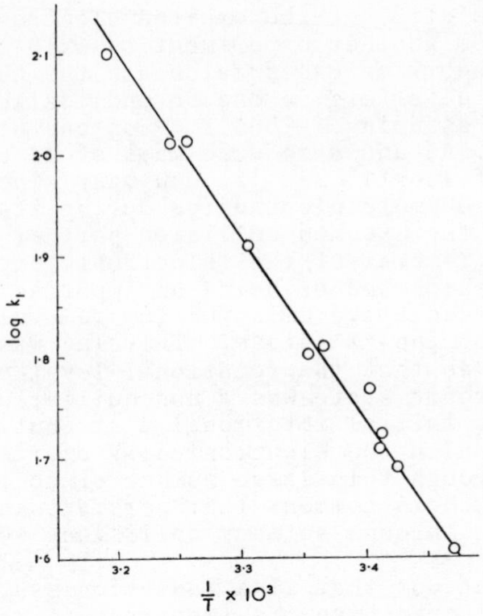

FIG. 2. Plot of $\log k_1$ *vs* reciprocal of absolute temperature.

49. "Fluorescence and Average Lifetime of Excited $OH(^2\Sigma^+)$ in Flames" H.P. Broida and T. Carrington 23, 2202 (1955).

Through chance we have juxtaposed two exercises involving the hydroxyl radical. What is the ground state of the gaseous species? What is the wavelength of the 0,0 band of the emission from the $^2\Sigma^+$ state to the ground state? As you are probably aware the volume "Spectra of Diatomic Molecules" by G. Herzberg (D. Van Nostrand Co. Inc. Princeton, New Jersey 1950) has in its Appendix a rich fund of information on diatomics and their states.

In this paper the authors show that the average lifetime of electronically excited OH in an atmospheric-pressure flame is of the order of 6×10^{-10} seconds, down by a factor of 1/1000 from previous estimates.

The apparatus used is shown schematically in Fig. 1. The

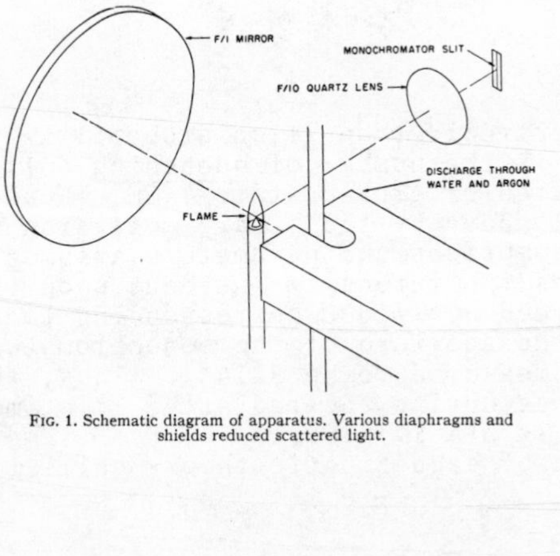

FIG. 1. Schematic diagram of apparatus. Various diaphragms and shields reduced scattered light.

normal radiative lifetime of the excited electronic state had been determined from another experiment as 6×10^{-7} seconds. From the kinetic theory of gases calculate the number of collisions an OH molecule suffers in one second, called the binary collision number, assuming T=2000°K, P of one atmosphere, a mean diameter of 2.7Å, and a reduced mass of 12 awu. Compare with their value of 1.6×10^{9} sec^{-1}. How many kinetic-theory collisions does an OH molecule undergo during its radiative lifetime? What is the assumed collision partner when μ=12 awu?

Quenching, or removal of the (electronic)excitation energy by collision, was suspected of being an important mechanism for competing with radiative emission for removal of the excitation energy from the $^2\Sigma^+$ state. The clue was that spectroscopic studies showed that the rotational level population of this excited electronic state was a nonequilibrium one. The assumption was that initial rotational distribution produced in the process in which the electronically excited OH molecules persisted, even through this large number of collisions at one atmosphere. The authors comment that persistence of the rotational distribution through so many collisions would be surprising.

They then point out that if a radiationless transition, such as collisional quenching, is important in removing excited OH molecules, the average lifetime of the state is decreased, and hence the molecules whose emission is observed will have undergone fewer collisions before radiating. Would this then make the observed nonequilibrium rotational state distributions more or less significant in implying long relaxation(in the sense of "relaxing" to equilibrium) times? In imparting information about the initial rotational distributions?

The measured fluorescence was about one part in 10^7 of the light from the discharge that was incident upon the flame. This intensity incident upon the flame is taken to be the intensity observed in the monochromator when the discharge is in the flame position, times a factor of 100. This factor is the simple quotient of the apertures of the mirror and the lens; what is the equation?

In this oxy-acetylene flame, just above the inner cone, the absorption of the two particular rotational lines of the (0,0) band had an absorption of about 25%. To calculate the importance of quenching the ratio of the observed fluorescence, I, to that in the absence of quenching, I_0, is needed and is given by

$$I/I_0 = A/(A+Z)$$

where A is the electronic transition probability for the $^2\Sigma^+$ state of OH and Z is the number of <u>quenching</u> collisions suffered per second by excited(or ground state?) OH. Would you expect Z to be bounded from above by the binary collision number if the kinetic theory assumptions and parameters assumed are valid? I_0 was estimated as the intensity incident upon the flame from the discharge reduced by 1/1600 corresponding to the absorption in the flame and the aperture of the monochromator lens. In this way I/I_0 was measured to be 8×10^{-4}. Thus, the number of quenching collisions during the radiative lifetime of the excited OH molecule is $Z/A=1.3\times10^3$.

What fraction of each kinetic theory collision leads to

quenching? Is this result reasonable? What is the average lifetime of electronically excited OH in a one-atmosphere oxyacetylene flame?

50. "On the Isomerization of Isobutyl Radicals" J.R. McNesby and W.M. Jackson 38, 692 (1963).

The authors study the competition between two reactions of deuterium-labeled isobutyl radicals, isomerization and straight decomposition. We will follow their arguments leading to the conclusion that the latter reaction is predominant at temperatures up to the highest obtained experimentally, 584°C.

The butyl radicals are formed in chain-propagating processes

$$R + (CH_3)_3CD \rightarrow RD + (CH_3)_3C \qquad (1)$$

$$R + (CH_3)_3CD \rightarrow RH + CH_2CD(CH_3)_2. \qquad (2)$$

The unimolecular reactions of $(CH_3)_3C$ include the known reaction

$$(CH_3)_3C \rightarrow H + i\text{-}C_4H_8 \qquad (3)$$

and the isomerization reaction in question

$$(CH_3)_3C \rightarrow CH_2CH(CH_3)_2. \qquad (4)$$

At the temperatures of the experiments, nearly all of the isobutyl radicals formed in (4) decompose by reaction (5)

$$CH_2CH(CH_3)_2 \rightarrow CH_2{=}CHCH_3 + CH_3. \qquad (5)$$

The unimolecular reactions of $CH_2CD(CH_3)_2$ include straight decomposition and isomerization

$$(CH_3)_2CDCH_2 \rightarrow CH_3 + CH_3CD{=}CH_2 \qquad (6)$$

$$(CH_3)_2CDCH_2 \rightarrow (CH_3)_2CCH_2D. \qquad (7)$$

The authors' objective is to assess the importance of reaction (7) relative to reaction (6). They tell us that reaction (4) is endothermic and that it has been shown not to contribute very much to the production of CH_3 radicals. Which C-H bond in reactant or product in (4) is weakened wrt the others if the reaction is endothermic?

Substantially all CH_3 radicals arise from reaction (6), and in such a chain reaction they may be regarded as always abstracting H or D from $(CH_3)_3CD$ to form CH_4 and CH_3D [by means of (1) and (2)?]. The occurrence of reactions (4) and (5) may cause CH_4+CH_3D to be too high by about 10%, but this effect is ignored. The H atom formed in reaction (3) nearly always forms H_2 or HD by abstracting H or D from $(CH_3)_3CD$. Why may the total concentration of $t\text{-}C_4H_9$ radicals produced be set equal to [HD]+ $[CH_3D]$?

To compare (6) and (7) the following procedure was employed. Let X be the fraction of $t\text{-}C_4H_9$ that decomposes in (3). The number of $t\text{-}C_4H_9$ radicals that decompose is equal to the number of $i\text{-}C_4H_8$ molecules formed; is the total amount of $t\text{-}C_4H_9$ formed equal to the number of D atoms abstracted from isobutane-2-d? Is this question identical with the last one? Same answer? Show that X is then given by

$$X = \frac{[i\text{-}C_4H_8]}{[HD]+[CH_3D]}. \tag{I}$$

It is now assumed that a fraction, 0.9X, of $t\text{-}C_4H_8D$ decomposes in (8)

$$(CH_3)_2CCH_2D \rightarrow H + i\text{-}C_4H_7D. \tag{8}$$

Show that

$$0.9X = \frac{[i\text{-}C_4H_8D]}{\text{Total } t\text{-}C_4H_8D \text{ radicals formed}} = \frac{0.9[i\text{-}C_4H_8]}{[HD]+[CH_3D]} \tag{II}$$

and that

$$\int_0^t k_7[(CH_3)_2CDCH_2]dt = \text{Total } t\text{-}C_4H_8D$$

$$\text{radicals formed} = \frac{\{[HD]+[CH_3D]\}[i\text{-}C_4H_7D]}{0.9[i\text{-}C_4H_8]}. \tag{III}$$

Thus the total number of CH_3 radicals and therefore the total number of $i\text{-}C_4H_8D$ radicals decomposing according to (6) is equal to(really proportional to?) $[CH_4]+[CH_3D]$.
Now demonstrate that

$$\int_0^t k_6[(CH_3)_2CDCH_2]dt = [CH_4]+[CH_3D] \tag{IV}$$

From (III) and (IV) show that the ratio sought is

$$\frac{k_7}{k_6} = \frac{[HD]+[CH_3D]}{[CH_4]+[CH_3D]} \times \frac{[i\text{-}C_4H_7D]}{0.9[i\text{-}C_4H_8]}.$$

The experimental results are collected in Table I.

TABLE I. Pyrolysis of isobutane-2-*d*.

t_{sec}	T°C	P mm	Percent conversion	Relative numbers of molecules H_2	HD	D_2	CH_4	CH_3D	i-butene-d_1 / i-butene	k_7/k_6
480	509	56.0	1.8	241	309	1.0	396	288	0.00	0.00
127	514	55.5	0.5	124	161	0.0	212	138	0.00	0.00
270	540	48.2	2.4	264	327	2.2	571	340	0.011	0.009
120	584	45.5	6.4	837	804	10.1	2295	1114	0.038	0.024

Verify the last column from the final equation.

Why would metal valves rather than stopcocks be used throughout the system? How were the analytical techniques of mass spectrometry and gas chromatography used? What experiment was used to establish that the deuterium atoms in isobutene do not exchange with H atoms in chromatographic column? Was the experiment a stringent test for exchange?

51. "Kinetics of Desorption. III. Rb^+, K^+, and Na^+ on Rhenium" M.D. Scheer and J. Fine 39, 1752 (1963).

When an alkali atom beam impinges on a metal surface whose work function is sufficiently large, positive ions are formed with high efficiency. If the surface coverage θ, the ratio of occupied sites to available sites, is small the processes which occur are

$$A(g) \underset{-1}{\overset{1}{\rightleftarrows}} A(ads)$$

$$A(ads) \underset{-2}{\overset{2}{\rightleftarrows}} A(ads)^+ + e_M$$

$$A(ads)^+ \overset{3}{\rightarrow} A(g)^+.$$

As long as the electron-transfer processes 2 and -2 are sufficiently rapid to insure equilibration of A(ads) and $A(ads)^+$ with the electrons e_M in the Fermi sea of the metal, the rate of positive-ion desorption is given by

$$-dn_+/dt = n_+/\tau_3 = i_+$$

where n_+ is the surface concentration of alkali positive ions, τ_3 is the mean lifetime of these ions on the surface, and i_+ is the current density of alkali ions emitted by the metal. Is the above equation kinetically first order? What is the relation between the mean lifetime and rate constant? Is the mean lifetime for any process defined as the time for the system to have some property "relax" to 1/e of its initial value? In exercise no. 49 the mean radiative lifetime was used in just this sense. What was the relaxing property in that instance?

Experimentally the alkali atoms were generated in a stainless steel effusion cell at about 700°C. The beam was pulsed and the mean lifetime τ_3 was the time constant for the ion current decay(relaxation?) during the "off-time" of the beam pulsing cycle. Ion beams with intensities of about $10^{12}/cm^2 \cdot sec$ were generated at the surface. What precautions were taken to insure surface cleanliness? For experimental lifetimes between 10^{-2} and 10^{-4} seconds and about 10^{14} available sites per square centimeter, what is the range of θ? Does it meet the requirement for the applicability of the equations above?

Speaking of θ, one of surface chemistry's finest(worst?) puns was dropped in the Pacific Northwest in 1959(possible earlier ref. also) when the fraction of vacant sites, 1-θ, was called θ_{bare}. The author will be identified alphabetically in the index only.

Back to our present authors. Their earlier work on Cs^+ (ref. 4) on rhenium gave millisecond lifetimes in the neighborhood of 1000°K. From the Arrhenius dependence of τ_3 upon temperature, the heat of desorption ℓ_+ was found to be 2.01 eV. This value is in approximate agreement with the electrostatic image energy calculated for the removal of a Cs^+ ion of radius 1.7 Å from the surface of an electrical conductor. One of their purposes in this study was to determine whether this simple electrostatic calculation of the desorption energy is valid for the three remaining alkali ions which can be readily sur face-ionized on rhenium.

Fig. 1 gives the experimental results including those for

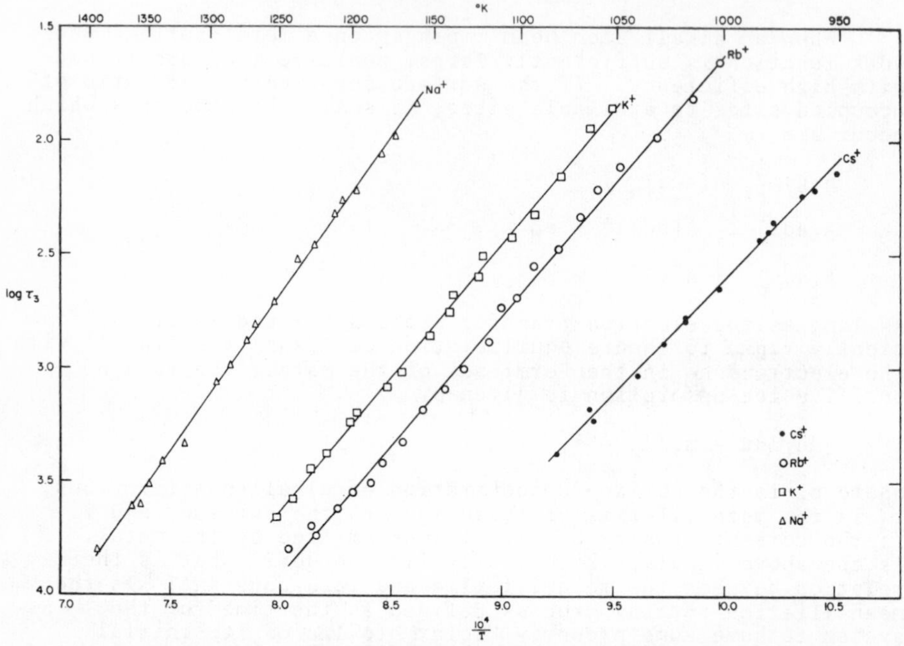

FIG. 1. Arrhenius plots of the mean adsorption lifetimes of the alkali ions on a clean polycrystalline rhenium surface.

Cs^+ for comparison. Table I collects the Arrhenius parameters ℓ_+ and τ_3^0. The latter is the inverse of the pre-exponential

TABLE I. Arrhenius desorption parameters for alkali ions on polycrystalline rhenium.

Alkali ion	r_i(Å)	$10^{13}\tau_3^0$(sec)	l_+(eV)	$e^2/4r_i$(eV)
Cs^+	1.69	1.9±0.9	2.01±0.04	2.13
Rb^+	1.48	0.8±0.3	2.28±0.03	2.43
K^+	1.33	1.0±0.3	2.33±0.03	2.71
Na^+	0.95	0.2±0.1	2.75±0.03	3.79

factor. Verify the Arrhenius parameters for one or two of the ions. Also shown are the ionic radii, taken from L. Pauling "The Nature of the Chemical Bond" (Cornell University Press, Ithaca, New York 1960). The last column in Table I will be discussed below.

It was assumed that the interaction of an isolated alkali ion with a metal surface can be approximated by the one-dimensional potential function shown schematically in Fig. 2. The x direction is normal to the surface and x=0 is defined by a point above the surface atom nuclei at which the electron density has decreased to a small fraction of that which exists in the interior of the metal. At large distances from the surface the ion is attracted to the surface by the electrostatic image force $e^2/4x^2$, while at small distances a strong repulsive force

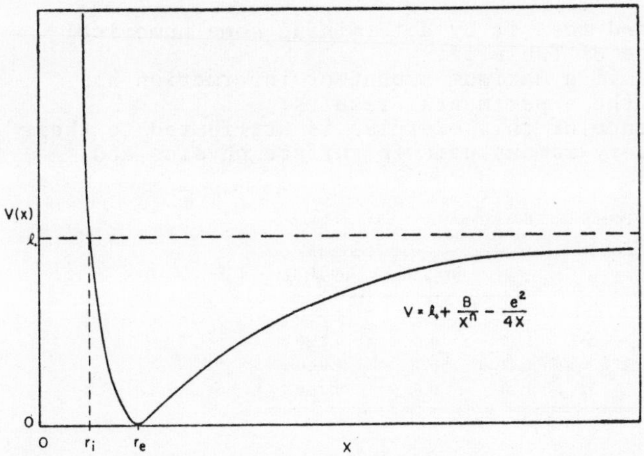

FIG. 2. Potential-energy function for alkali ions on a rhenium surface.

proportional to $x^{-(n+1)}$ predominates. What will be the dependence on n of the corresponding term in the potential energy function? The ionic radii are assumed to be the distance of closest approach to the surface (x=0) for an adsorbed ion whose energy is the heat of desorption ℓ_+. The potential function with these properties is written in Fig. 2.

At some equilibrium distance $r_e > r_i$, $dV/dx = V = 0$. Show that

$$B = e^2 r_e^n/(4r_e n)$$

and

$$\ell_+ = (n-1)e^2/(4nr_e).$$

V then become

$$V = \ell_+[1 + r_e^n/\{x^n(n-1)\} - nr_e/\{x(n-1)\}].$$

Now show that at $x=r_i$

$$r_i/r_e = (1/n)^{1/(n-1)}.$$

Prove that

$$\lim_{n\to\infty} (1/n)^{1/(n-1)} = 1$$

by taking a logarithm and applying l'Hospital's rule. Show that in the above limit $\ell_+ \to e^2/4r_i$. This limiting case is approximated by the ions of greater atomic number as Table I shows. Using the experimental values of ℓ_+ solve for n and r_e and compare your results with those shown in Table II. The next to last column gives $1/\nu_e$ as an order-of-magnitude estimate to the pre-exponential factor τ_3°. This $1/\nu_e$ is the time required for a single vibration in the region of the potential minimum. From the force constant and the ionic mass it is

$$1/\nu_e = [4\pi/e] \cdot [mr_e^3\{e^2/(4r_e\ell_+) - 1\}]^{\frac{1}{2}}.$$

Derive this result and test it by determining some numerical values to check those of Table II.

Does it seem as if a maximum amount of information has been extracted from the experimental results?

The first sentence of this exercise is attributed to their ref. 1(1933) and a very famous name in surface physics and chemistry.

TABLE II. Potential-function parameters for alkali ions adsorbed on rhenium.

Alkali ion	n	r_e(Å)	$10^{13}/\nu_e$(sec)	$10^{13}\tau_3^0$(sec)
Cs^+	100	1.77	0.9_2	1.9±0.9
Rb^+	85	1.56	0.6_7	0.8±0.3
K^+	30	1.49	0.7_1	1.0±0.3
Na^+	12	1.20	0.6_4	0.2±0.1

52. "Vibrational Relaxation of Carbon Monoxide by Ortho- and Parahydrogen" R.C. Millikan and L.A. Osburg 41, 2196 (1964).

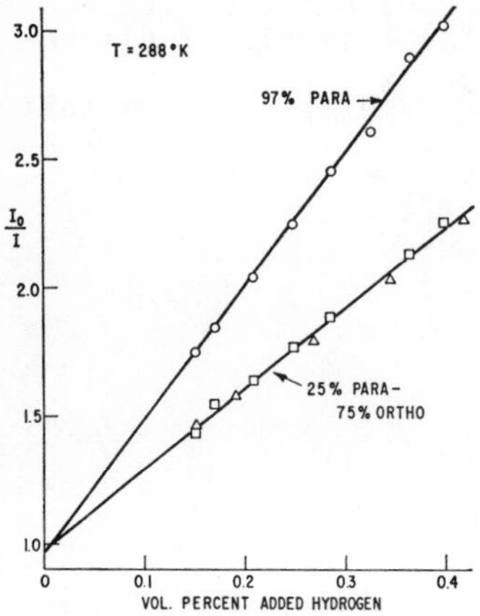

FIG. 1. Stern–Volmer quenching of vibrationally excited CO by mixtures of ortho- and parahydrogen. I_0 is the intensity of fluorescence without foreign gas; I is the intensity of fluorescence with a given amount of hydrogen added. ○, Mix 1 (97% para); □, Mix 1 after conversion to normal H_2 by Pt catalyst. △, Mix 2: normal H_2 component taken from cylinder.

Fig. 1 presents the fluorescence ratio I_o/I, in the absence and presence of a foreign quenching gas, as a function of the added gas which in this case is separately 97% parahydrogen and normal(25% para-) hydrogen. The significant and even remarkable thing is that ortho- and parahydrogen should vary so markedly in the quenching efficiency.

The two competing processes for the transfer of CO vibrational energy are(their ref. 2)

$$CO^* + M \rightarrow CO + M + K.E. \qquad k_1[M] = 1/\tau_1 \qquad (1)$$

$$CO^* \rightarrow CO + h\nu \qquad k_2 = 1/\tau_2 \qquad (2)$$

where the asterisk denotes a vibrationally excited molecule in its ground electronic state, and in our case M is either ortho- or parahydrogen. τ_2 was determined from absorption intensity studies and is 0.033±0.001 sec. We will still check some values given in the article under consideration, but will merely refer to their references rather than attempt complete derivations.

From ref. 2 the Stern-Volmer relation

$$I_o/I = 1 + (\tau_2/\tau_1)p \tag{6}$$

can be used to determine the quenching vibrational relaxational time τ_1 as a function of the fluorescence quotient and the partial pressure of quenching gas p expressed in atmospheres. What about the relative rates of processes (1) and (2) when the fluorescence quotient has the value 2? Show that in this instance for a total pressure of one atmosphere τ, the collisional relaxation time for infinitely dilute CO in the quencher M at one atmosphere pressure when $I_o/I=2$) is given by(ref. 2)

$$\tau(\text{atm}\cdot\text{sec}) = f\tau_2 \tag{7}$$

with f the fraction of foreign gas required for half-quenching.

From Fig. 1 determine the values of τ and compare with the authors' $6.6\text{x}10^{-5}$ atm·sec for 97% parahydrogen and $1.1\text{x}10^{-4}$ atm·sec for normal hydrogen. To obtain the values for pure para- and orthohydrogen of $6.5\text{x}10^{-5}$ and $1.43\text{x}10^{-4}$ atm·sec respectively, obviously some kind of extrapolation with mole fraction is called for. Determine whether it is the relaxation time or its reciprocal the reaction rate that is assumed to be linear in the mole fractions.

The value measured for τ_2 is independent of pressure and temperature. The presents experiments were carried out at a temperature of 288°K. What special precautions were necessary in the handling of the 97% parahydrogen? At temperatures above say 1000°K would you expect processes (1) or (2) to dominate? Assume fixed densities of both CO and M and think about the usual Arrhenius behavior of rate constants. Does this last consideration apply to k_1? to k_2?

Notice that τ for the two forms of hydrogen differs by a factor of greater than 2. To quote the authors "We hope these data on such a simple system will stimulate renewed theoretical attacks on this problem."

53. "Halogen-Atom Reactions. II. Luminescence from the Recombination of Chlorine Atoms" L.W. Bader and E.A. Ogryzlo 41, 2926 (1964).

Earlier work had suggested that studies of light emission accompanying recombination of halogen atoms would be difficult to carry out since the electrical discharge was not a useful source of halogen atoms because of a very rapid wall recombination. In Part I of this series, by the second author, it was shown that large concentrations of chlorine and bromine atoms could be formed with an electrodeless microwave discharge if the walls were coated with a suitable nonvolatile oxyacid.

What was the acid used in the present set of experiments? How did they determine that the impurities present in tank chlorine would not affect their results? How did they measure the absolute concentration of the atoms along the reaction tube?

Under typical operating conditions with about 30 micromoles/sec of chlorine flowing through the microwave discharge the gas stream coming from the discharge was seen to emit a weak orange-red glow. The glow could quite easily be made to extend a distance of 1 to 2 m along a 2-cm-i.d. reaction tube that was properly coated. Show that at the pressures mentioned in Fig. 2 and at room temperature the transit time through the reaction tube is the order of a second with typical flow rates.

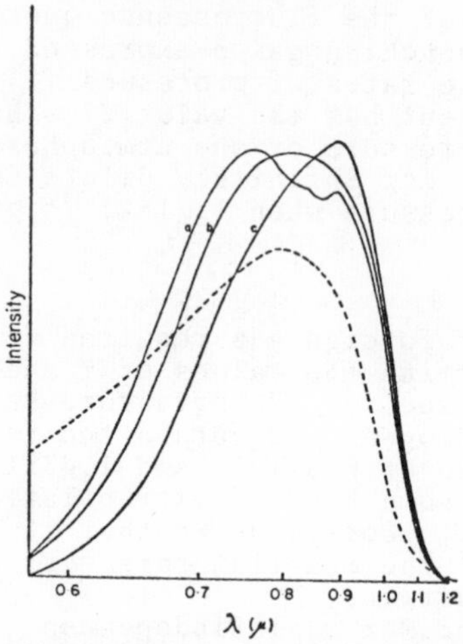

FIG. 2. Pressure dependence of chlorine emission obtained with a wide slit to remove vibrational structure. (a) 0.3 mm Hg, (b) 1.3 mm Hg, (c) 1.6 mm Hg. Spectral sensitivity of photomultiplier is given by dotted curve.

Fig. 2, obtained with a low resolution glass-prism spectrophotometer, shows the marked shift of the intensity maximum toward longer wavelength with increasing pressure. To positively identify the electronic transition involved the spectrum shown in Fig. 3 was obtained. As you probably know the conven-

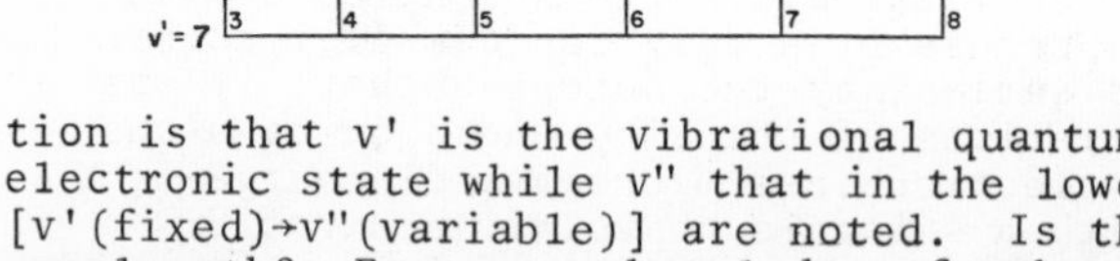

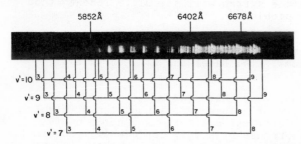

FIG. 3. Chlorine emission spectrum photographed on a Kodak 103 a F plate with a Jarell-Ash $f/6.3$ grating spectrograph. Slit=50 μ, exposure time=1 h. Neon calibration lines superimposed.

tion is that v' is the vibrational quantum number in the upper electronic state while v" that in the lower. Four progressions [v'(fixed)→v"(variable)] are noted. Is the scale linear in wavelength? From your knowledge of anharmonicity in vibrational levels does the scale appear to be linear in wavenumber? Bands originating between the 6th and 12th vibrational level in the upper electronic state were the most intense and no bands were observed for levels above v'=13. The electronic transition was identified as $^3\Pi_{o^+u} \rightarrow {}^1\Sigma_g^+$(ground) for Cl_2. Compute the energy of the v'=13 level above that of v"=0. From the volume by Herzberg referenced on p. 93 one finds that ν_{oo}, the wavenumber of the (0-0) band, is 18147 cm^{-1} and that ω_e and $\omega_e x_e$ for the upper electronic state are 239.4 and 5.42 cm^{-1}. From this value and the dissociation energy of the ground state, 20 000 cm^{-1}, we will follow the arguments presented in Mechanism 4.

This exercise should give us a workout in mechanism deduction. Before that we will examine their determination of that mechanism-independent quantity, the empirical rate equation.

To determine the order of the reaction wrt the atom concentration [Cl] and the molecule concentration [Cl_2] the rate expression was written in the form

$$I = d(h\nu)/dt = k[Cl]^n[Cl_2]^m$$

with I the emission intensity. At a fixed pressure([Cl_2] fixed)

$$\log I = n \log [Cl] + \text{constant}.$$

Hence a plot of log I vs log [Cl] should yield a straight line with slope n. Such a plot is shown in Fig. 4; the slope was determined to be 2.02. Verify this as closely as you can. The order of the reaction wrt Cl was thus taken as 2. The order of the reaction wrt Cl_2 can be obtained by rewriting the rate law in the form

$$\log (I/[Cl]^2) = m \log[Cl_2] + \text{constant}.$$

The appropriate plot is given in Fig. 5 for data at two different pressures. The departure of the slope, 0.9, from 1.0 was attributed to the redshift of the peak emission with increasing pressure. Does increasing pressure actually blueshift the peak? Is the slope affected in the way suggested? Within the accuracy of the results the rate law was therefore written

$$I = k[Cl]^2[Cl_2].$$

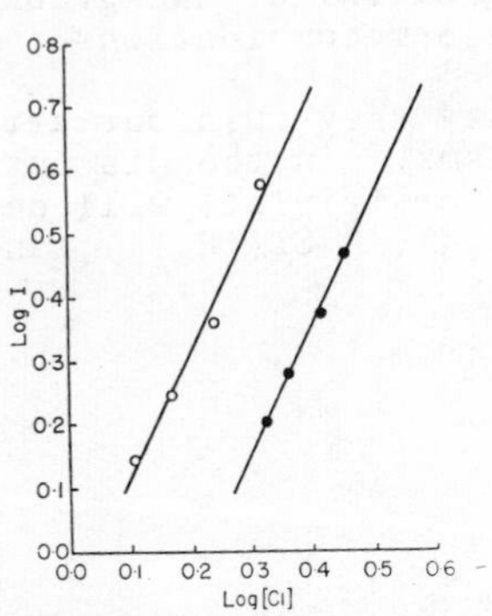

FIG. 4. A log-log plot of the emission intensity, measured with Kodak infrared film, against the chlorine-atom concentration at constant total pressure: open circles $p=1.61$ mm Hg; solid circles $p=0.65$ mm Hg.

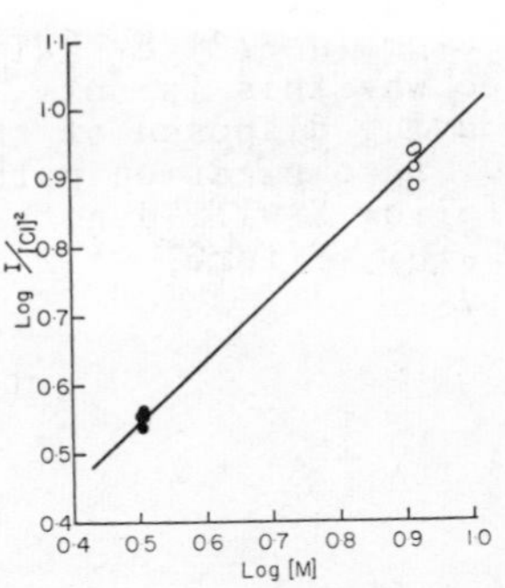

FIG. 5. The pressure dependence of the chlorine-emission intensity. The third-body concentration [M] is assumed equal to [Cl_2].

The balance of the discussion will be concerned with their four possible mechanisms. Some relevant potential-energy curves

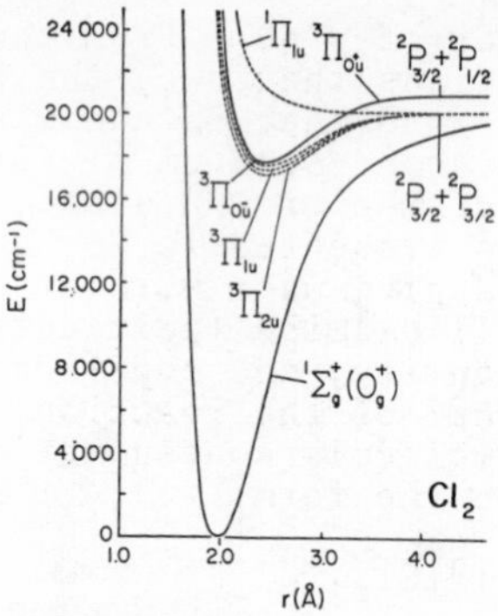

FIG. 6. Potential-energy diagram of Cl_2 showing the six lowest states. Solid curves—spectroscopically observed states; broken curves—extrapolated or estimated.

are presented in Fig. 6. Notice that the upper state responsible for the emission, $^3\Pi_{o^+u}$, does not correlate with the ground state of both atoms, $^2P_{3/2}+^2P_{3/2}$, but rather with $^2P_{3/2}+^2P_{1/2}$. The excitation energy of the latter state is 881 cm^{-1} as may be determined from Moore's Tables referenced on p. 70. Notice that the doublet is inverted; the state with the higher J value lies lowest. This is a general rule when the incomplete shell is over half filled. Notice also that the spin-orbit splitting is much greater than that for carbon(p. 71).

The lifetime of the upper electronic state for Cl_2 toward radiative transitions is less than 10^{-3} sec. Comparing this with the transit time estimated earlier can we discard the possibility that the excited molecules originate in the discharge? Why? The authors suggest that it is conceivable that large concentrations of $^2P_{1/2}$ atoms are produced in the discharge, yet would find it difficult to believe that the process remains second order in [Cl] in every experiment since to do so would require that the deactivation of this species(J=1/2) takes place at exactly the same rate as the removal of the ground state atoms(J=3/2) by wall and gas-phase recombinations. Do you see why this is so?

Having disposed of this possibility they turn our attention to four recombination paths or mechanisms. In the discussions that follow X will denote $Cl_2(^1\Sigma_g^+)$, T(for triplet) will denote $Cl_2(^3\Pi_{o^+u})$, S(for?) will denote $Cl_2(^1\Pi_1u)$, A $Cl(^2P_{3/2})$, and A* $Cl(^2P_{1/2})$.

Mechanism 1

$$A + X \underset{2}{\overset{1}{\rightleftarrows}} A^* + X$$

$$A + A^* + X \xrightarrow{3} T + X$$

$$T \xrightarrow{4} X + h\nu$$

This mechanism assumes that because of the small excitation energy of A* an equilibrium concentration A*/A≃1/140, at room temperature, of this species will be maintained in the atom stream, and then three-body recombination into T can occur directly. Compute this concentration quotient remembering the degeneracies of the two states involved. The experimental kinetic order

$$I = (k_1/k_2)k_3k_4(A)^2(X)$$

can be obtained by assuming reactions 1,2, and 4 are fast and that reaction 3 is rate controlling. We will sketch out the kinetic derivation for this mechanism though not for all of them; that will be an activity for you.

The rate constants for the elementary steps, k_i, are given a second subscript in the article to indicate to which mechanism they belong. We have suppressed them in our discussion - just remember that when we move on to a new mechanism it is also to a fresh set of k's.

The concentration of A* is constant in time with the assumption of thermal equilibrium so that

$$d(A^*)/dt = k_1(A)(X) - k_2(A^*)(X) - k_3(A)(A^*)(X) = 0.$$

Solving for (A*)

$$(A^*) = k_1(A)/[k_2+k_3(A)] \simeq k_1(A)/k_2$$

with the approximation a result of considering the relative rates. The rate of production of T, the rate-controlling step, is

$$d(T)/dt(\text{production}) = k_3(A)(A^*)(X) = (k_1/k_2)k_3(A)^2(X)$$

and the rate of destruction of T, the emission rate, is just this production rate times the rate constant for the emission process. Think about this last step; is it legitimate?

Since the mechanism postulates only a small equilibrium concentration of A*, the authors suggest that the recombination into the emitting state would probably be about two orders of magnitude slower than that into the ground state, and that a measurement of the total absolute emission intensity might provide evidence for or against this mechanism. However the mechanism cannot easily account for the experimental observation that emission from levels above v'=13 is absent. Furthermore, a similar emission is observed for bromine(ref. 6) where this mechanism is not applicable. Why isn't it so? What is the magnitude of the spin-orbit splitting in the ^2P lowest term and the resultant Boltzmann ratio for bromine?

Mechanism 2

When the rotational constant B' for T is plotted against v' a change of slope is observed(ref. 7) at about v'=13. Such perturbations are normally attributable to the crossing or near approach of another state. Referring to Fig. 6, the $0u^-$, 1u, and 2u components of the $^3\Pi_u$ state would be expected to lie below the $^3\Pi_{0^+u}$ or T state(ref. 9). The $^1\Pi_{1u}$ state, which should lie just above T, is held responsible for the continuous absorption centered at 30 000 cm^{-1}, and the authors suggest that a reasonable extrapolation of this curve could cross the T curve at the latter's 13th level as shown in Fig. 6. The curve for the S state could predissociate the T state above v'=13 either alone(less likely), with the inhomogeneous electric field of a colliding molecule(Mechanism 2) or of a perturbing species in a doublet state such as A(Mechanism 3). For Mechanism 2

$$A + A \underset{2}{\overset{1}{\rightleftarrows}} S$$

$$S + X \xrightarrow{3} T + X$$

$$T \xrightarrow{4} X + h\nu.$$

Follow through with all of the details as we did for Mechanism 1. Show that the same expression for I results.

Since T would be formed initially in the 13th level, this mechanism is consistent with the experimental observation that the emission begins at v'=13 and though it is very strong from levels just below v'=13, it is too weak to be detected from levels above v'=13.

Mechanism 3

Notice the differences between this mechanism and the last.

$$A + A \underset{2}{\overset{1}{\rightleftarrows}} S$$

$$S + A \underset{4}{\overset{3}{\rightleftarrows}} T^{\dagger} + A$$

$$T^{\dagger} + X \xrightarrow{5} T + X$$

$$T \xrightarrow{6} X + h\nu$$

$$I = (k_1/k_2)(k_3/k_4)k_5k_6(A)^2(X).$$

$T^{\dagger}$ indicates a level of T which can be predissociated by a chlorine atom, and the absence of $\dagger$ indicates that a collision has moved T to a level which is metastable only wrt radiation.

Derive the result for I. The correct kinetic order follows when the S concentration is chiefly determined by reactions 1 and 2, when the $T^{\dagger}$ concentration is chiefly determined by reactions 3 and 4, and when a steady-state concentration of T forms. Show that at very high pressures the reaction should become third order in A.

Mechanism 4

The authors quote increasing evidence(ref. 13,14) that recombination of halogen atoms into the ground state occurs chiefly via weakly bound Cl_3, Br_3 and I_3 complexes in the absence of added gases. This last mechanism shows how T could be formed by such a path:

$$A + X + X \underset{2}{\overset{1}{\rightleftarrows}} Cl_3 + X$$

$$Cl_3 + A \overset{3}{\rightarrow} T + X$$

$$T \overset{4}{\rightarrow} X + h\nu$$

$$I = (k_1/k_2)k_3k_4(A)^2(X).$$

Once again derive the result for I. The correct kinetic order can be obtained for the overall reaction by assuming an equilibrium concentration of Cl_3, governed by reactions 1 and 2, and a steady-state concentration for T. Mechanism 4, like Mechanisms 2 and 3, could result in emission only from the 13th and lower levels of T. What argument(s) could you give in defense of this statement? What about the energy of v'=13 relative to v"=0 versus the dissociation of the ground state of the diatomic?

Mechanisms similar to 2 had been used to explain chemiluminescence from the recombination of N+N, N+O, Br+Br, S+S, and S+O. In each case it was necessary to assume the presence of a spectroscopically unobserved predissociating state that crosses the emitting state at, or just below, the dissociation limit.

"It is beginning to seem like too much of a coincidence that there is always a convenient predissociating state just at the dissociation limit. It is for this reason that we favor Mechanism 4, which is generally applicable to any system".

Show that Mechanism 4 predicts a change in the kinetic order at low pressures to

$$I = k_1(A)(X)^2.$$

The authors propose a test of this mechanism when their experimental techniques are more refined. Specifically what form would this test take?

They note that no need was found to invoke collisional deactivation

$$T + X \rightarrow X + X.$$

If this reaction did remove T at a rate comparable with the radiative reaction, what would happen to the dependence of the glow on (X) at higher pressures? Why? The dependence of the emission intensity observed on the concentration of X remained first order at the higher pressures.

Is the shift in the emission peak to longer wavelength, observed as the pressure was increased, consistent with a more rapid vibrational relaxation of T? Why? Is the Franck-Condon principle the only effect in operation here? Look carefully at Fig. 2 again.

Do you see how the experimental results and the application of theories are woven together? What advances in the general understanding of gaseous recombination kinetics would you say might result from this paper? Do the authors proclaim them as such or rather do they tentatively extrapolate from their system to similar ones? Did you learn a bit about molecular states?

54. "Vacuum-Ultraviolet Photochemistry. III. Formation of Carbon Atoms in the Photolysis of Carbon Suboxide at 1470 Å" L.J. Stief and V.J. DeCarlo 43, 2552 (1965).

In this short exercise we will examine the authors' path of reasoning insofar as identification of reactants and products formed when C_3O_2 is photolyzed at 1470 Å. They conclude that carbon atoms are formed which then react with methane to form both ethylene and acetylene.

Earlier photolysis experiments carried out at wavelengths greater than 2200 Å indicated that the reactive intermediate might be the C_2O radical. With the heat of formation of the C_3O_2 molecule taken as -25 kcal/mole it is shown that at these wavelengths it is energetically impossible to produce free carbon atoms according to the reaction

$$C_3O_2(g) + h\nu \rightarrow C(g) + 2CO(g).$$

Table I summarizes the calculations. The three terms ob-

TABLE I. Wavelengths below which it is energetically possible to produce the indicated carbon atom by the reaction: $C_3O_2 \rightarrow C(g) + 2CO$.

State of carbon atom	Energy of state[a] (kcal/mole)	ΔH of reaction (kcal/mole)	Wavelength (Å)
$C(^3P)$	0	143	1998
$C(^1D)$	29	172	1661
$C(^1S)$	62	205	1394

[a] Energies of excited states of carbon are from Natl. Bur. Std. Circ. No. 467 (1949).

tained from the $1s^2 2s^2 2p^2$ configuration are shown. The energies in wavenumbers are given on p. 71. Do you recognize the reference of Table I's footnote? Calculate ΔH for the above reaction for each of the three states of carbon and compare with the third column. Do you see how the wavelength column is determined? Verify the numbers. Is ΔH endothermic?

The two xenon resonance lines at 1470 and 1295 Å were excited for use as a source; a sapphire window eliminates the latter line. Which states of carbon will be produced with the former line? Which with both? All of these experiments were done at 25°C and a total pressure of 90 mm Hg. Acetylene and ethylene, along with CO, were found to be major reaction products for mixtures of C_3O_2 with methane. What would be the obvious tool for analysis of the products? What steps would be necessary to insure that the reaction(s) were not taking place in the absence of photolysis? In agreement with earlier results photolysis at 2537 Å(or λ2537 as the astronomers would say) did not result in acetylene formation. Two these results are offered:

(I) C atoms are formed in the photolysis of C_3O_2. Insertion of a C atom into a C-H bond of methane results in a highly excited C_2H_4 molecule. The latter may be collisionally quenched or may decompose to acetylene. The first step is already given above. Write the products for the other three:

$$C + CH_4 \rightarrow \tag{1}$$

$$C_2H_4{}^* + M \rightarrow \tag{2}$$

$$C_2H_4{}^* \rightarrow \tag{3}$$

Compute ΔH or ΔE for (1) assuming that the product is in its ground state as is the C atom. We should find about 141 kcal/mole for ΔE. The authors say that the ethylene produced in (1) is excited at least to the extent of 142 kcal/mole. The numbers are close enough so that we won't quarrel, but do they refer to the same quantity? Explain. Why is the qualifying "at least..." used? How could there be greater excitation? Now that we have allowed that $141 \simeq 142$ how about the 143 from Table I? Is this the same quantity as the other two?

The asterisk is used to indicate excitation of internal degrees of freedom. Sometimes different symbols are used to differentiate between vibrational and electronic excitation. Can the distinction be made in the present case? Is it possible to have electronic excitation in ethylene with 142 kcal/mole? cf G. Herzberg "Electronic Spectra and Electronic Structure of Polyatomic Molecules"(D. Van Nostrand Co. Inc. Princeton, New Jersey 1966), p. 629.

(II) Photolysis at $\lambda 1470$ is the same as at $\lambda 2537$

$$C_3O_2 + h\nu \rightarrow C_2O + ? \tag{4}$$

$$C_2O + CH_4 \rightarrow C_2H_4 + ? \tag{5}$$

but secondary photolysis of C_2H_4 occurs resulting in the formation of acetylene

$$? + ? \rightarrow ? + ?. \tag{6}$$

If (II) occurs to the exclusion of (I), could this be determined? For conversions greater than 2% it was found that the rate of ethylene production decreased, the rate of acetylene production increased, but the sum of the two rates remained constant. What does this suggest about the relative importance of (I) and (II) at smaller and larger conversion fractions?

We see again how thermodynamic arguments underlie kinetic ones. Is the converse statement true?

55. "Sites for the Dehydrogenation of Formic Acid on Gold" M.A. Bhakta and H.A. Taylor <u>44</u>, 1264 (1966).

As well as determining rate constants, Arrhenius parameters and a probable mechanism, this study examines the one case of heterogeneous catalysis in depth. Yet the system chosen appears to be so clean-cut, in terms of absence of extraneous effects, that understanding of other cases of heterogeneous catalysis could be enhanced.

Gold, in the form of spongy polycrystalline cohesive pellets, was assayed at 99.99%. What was the ultimate source, other than Mother Nature?

Earlier work on the <u>decomposition</u> of formic acid, catalyzed by other metals, was not free of the possibility of oxidation of the metal. Is this objection met by the use of gold? The temperature range for the contact with formic acid is 130-200°C although higher temperatures were used in the study of enhancement of catalytic activity. Gold had been used before, of course, for catalytic studies. The authors' ref. 1 found the very interesting and pertinent results that

i) no hydrogen-deuterium exchange on gold at temperatures below 200°C,
ii) no exchange between hydrogen and deuterated formic acid under similar conditions, but
iii) full exchange between HCOOH and DCOOD, or in the decomposition of HCOOD.

Would this indicate the existence of adsorbed hydrogen atoms on the gold surface which can only arise from the formic acid?

The dehydrogenation(as a special case of decomposition) reaction

$$HCOOH \rightarrow H_2 + CO_2$$

was shown to be the only one occuring by mass-spectral analyses of the products. What would you expect to be the ratio of the two products? Traces of water, in the absence of any carbon monoxide, were regarded as an accidental impurity. The decomposition was brought about only by the catalyst. How would this be established?

You might make a simple sketch of the system from their description. The surface area of the fixed amount of gold was found to be 9.01×10^4 cm^2 by the B.E.T. method of nitrogen absorption. The purity of the formic acid was improved from 90.6%(reagent grade!) to >99% by drying followed by distillation. For a dehydrogenation run, the system was evacuated at 10^{-5} mm Hg for at least two hours. Formic acid vapor at the desired pressure was introduced into the reaction chamber, and pressure and pressure changes with time were read on an Octoil manometer.

Examination of the pressure-time plots suggests that the dehydrogenation proceeds by a fractional order which tends to zero as the pressure of the monomer vapor increases. The rate law may be written

$$r = -dP_M/dt = aP_M/(1 + bP_M)$$

where a and b are constants and P_M is the pressure of monomer vapor. Write an equation relating P_M at any instant to the total pressure P at the same instant and P_o, the initial pressure of monomer vapor. Is there any other method of determining P_M from pressure measurements? How about from spectroscopy? Infrared spectroscopy would seem to be a natural; the times involved are such that very effective monitoring of the HCOOH or CO_2 bands could take place. Some basic questions arise quickly. What about the stability of the salt windows and their sealing material at the temperatures involved?

The empirical constants a and b were obtained by least-squares analysis as described in the paper. They are given in Table II for the temperatures used. Table I presents the pressure measurements for two different runs(different values of P_o) at 160°C.

TABLE I. Dehydrogenation of formic acid on gold at 160°C, Series 1.[a]

Time (min)	P(mm Hg) Obs	P(mm Hg) Calc	P(mm Hg) Obs	P(mm Hg) Calc
0	5.99	5.99	15.61	15.61
1	6.53	6.34	16.93	16.24
2	6.72	6.68	17.06	16.86
3	7.04	7.00	17.50	17.46
4	7.39	7.31	18.09	18.04
5	7.71	7.60	18.71	18.62
6	7.97	7.88	19.25	19.19
7	8.23	8.14	19.91	19.75
8	8.47	8.39	20.34	20.29
9	8.68	8.63	20.77	20.81
10	8.91	8.85	21.24	21.31
15	10.01	9.79	23.21	23.63

[a] $a=0.080\ \text{min}^{-1}$; $b=0.063\ \text{mm}^{-1}$.

TABLE II. Values of a and b as functions of temperature.

Series I Temp (°C)	Series I a(min^{-1})	Series I b(mm^{-1})	Series II Temp °C	Series II a(min^{-1})	Series II b(mm^{-1})
145	0.045	0.063	130	0.089	0.064
160	0.080	0.063	145	0.298	0.122
176	0.305	0.175	160	0.700	0.171
191	0.690	0.240	170	0.890	0.163
201	0.858	0.165	180	1.520	0.199

For the temperature of Table I and the appropriate Series I values of a and b from Table II check the rate law at several of the total pressures listed, both for P_o of 5.99 and 15.61 mm Hg. You will need to obtain P_M from your equation using P and P_o. Approximate the slope at time t with $\Delta P/\Delta t$ where the increments could span a two-minute interval centered on t.

In Table II notice that the rates in Series II are appreciably higher than those of Series I at the same temperature. This was because of the activity enhancement of the catalyst by extended heating between the two series. The discovery of this effect can best be related in the authors' own words:

"The dehydrogenation, Series I runs were performed at 145°, 160°, 176°, 191° and 201°C in that order. For Series II, runs were made at 130°, 145°, 160°, 170° and 180°C in that order. Before Series I, the catalyst had never been heated to a temperature above 160°C and had been exposed only to helium and nitrogen during the B.E.T. measurements. However, between the two series several experiments with ethylene, alone or mixed with hydrogen or with hydrogen and formic acid vapor, had been performed. The data of these hydrogenation experiments are not included in the present discussions. However, it must be stated that periodically conducted formic acid dehydrogenation runs showed no evidence of the catalyst being poisoned by ethylene. In fact, it was only during such checking runs that the tendency of the gold to be activated by heating to high temperature was first sensed(our emphasis). Since there was no suspicion of such temperature dependence of activity during the Series I runs, the catalyst had been heated at each temperature for not less than 4 h before performing dehydrogenation runs. However, in Series II the heating at each temperature was for about 16 h." The original thrust of the research was actually in hydrogenations.

Before considering their annealing experiments and further discussions of catalyst activity, we will examine their mechanism.

From the rate equation it is seen that as $P \to \infty$, $r \to a/b = r^\circ$, the zero-order rate of reaction: as $P \to 0$, $r \to aP$, the first-order rate of reaction. The constant a is the specific first-order rate in $\min^{-1}$. The ratio a/b gives the zero-order rate r° in mm Hg/min. Verify these statements from the experimental rate law. Arrhenius plots of k°(zero-order rate in molecules per square centimeter per second) and a as functions of T gave for Series I

$$k^\circ = 10^{18.23} \exp(-12\ 700/RT)$$

$$a = 10^{21.94} \exp(-22\ 120/RT)\ \text{sec}^{-1},$$

and for Series II

$$k^\circ = 10^{18.13} \exp(-11\ 420/RT)$$

$$a = 10^{21.01} \exp(-18\ 780/RT)\ \text{sec}^{-1}.$$

Do we see the effect of the increased catalytic activity on the a values? Try these equations out with some values from Table II.

As the authors point out a mechanism should be able to account for (1) the fractional order observed, (2) the presence of adsorbed hydrogen atoms on the gold surface shown by ref. 1, and (3) the existence of formate ions or formyl radicals on metal catalysts shown by ref. 4.

The following mechanism was suggested:

$$2S + HCOOH \underset{k_{-0}}{\overset{k_0}{\rightleftarrows}} SH + SCOOH \qquad \text{(i)}$$

$$SCOOH \xrightarrow{k_1} SH + CO_2 \qquad \text{(ii)}$$

$$SH + HCOOH \xrightarrow{k_2} SCOOH + H \qquad \text{(iii)}$$

where S, SH, and SCOOH represent a surface site, an adsorbed hydrogen atom, and a formate ion or formyl radical that is adsorbed. Step (i) represents dissociative adsorption; (ii) represents decomposition of the relatively unstable gold formate and is assumed to be a fast process; (iii) represents an adsorbed H atom being struck by an acid molecule and is assumed to be slow and rate-controlling. Do you see the chain reaction at the surface? Explain. Show that the net overall reaction as given on p. 110 is exothermic by 7 kcal/mole.

Treating gold as a poor adsorbent, the sum of [SH] and [SCOOH] is small. If we assume that the fast reactions in (i) are in equilibrium does it the follow that [SH]+[SCOOH]=constant? Then the time derivative of the sum is zero. It is an additional reasonable assumption that the time derivative of each term is zero. To what standard kinetic approximation is this equivalent?

Taking $k_o/k_{-o}=K_S$, k k k K_S/k 1, and writing $[S]^\circ=[S]+[SH]+[SCOOH]$, where concentrations of total available sites and empty sites are denoted by $[S]^\circ$ and [S], show that after making the appropriate approximations

$$-d[HCOOH]/dt = k_2[S]^\circ[HCOOH]/(1+k_2[HCOOH]/k_1).$$

Identify the empirical a and b in terms of the specific rate constants and $[S]^\circ$. Can you show that step (iii) is pseudo first order? If the steady-state approximation tells us that [SH] quickly attains its steady-state value, should it remain independent of the formic acid pressure?

Fig. 1 shows the decrease of the initial rate r_o(measured at $P_o=P_M$ and all adjusted to P_o=15 mm Hg) of dehydrogenation at 130°C after heating at the temperatures indicated and then air-quenching to 130°C. The abscissa shows the time in hours that the catalyst is annealed at this temperature, under vacuum, befor the formic acid vapor is admitted and the run begins. Do you see that each point on each curve then represents a complete and separate experimental determination of pressure vs time? The gold was held for three of the four elevated temperatures at times varying from 3.7 to 4.6 h at those temperatures. For the highest enhancement temperature of 400°C the gold was kept at that temperature for 24 h. From Fig. 1 does the length of time the catalyst is held at the enhancement temperature, beyond 3-4 h, appear to be significant? These four curves show a parallel decay of activity with annealing time in spite of the differences in enhancement temperature. For the first 4 or 5 h the decay rate is faster but subsequently becomes approximately first order. The dashed curves represent a rough extrapolation of the first-order curves to zero time. From the proposed mechanism is the initial dehydrogenation rate a measure of [S]? How does [S] vary with the enhancement temperature? Regardless of the en-

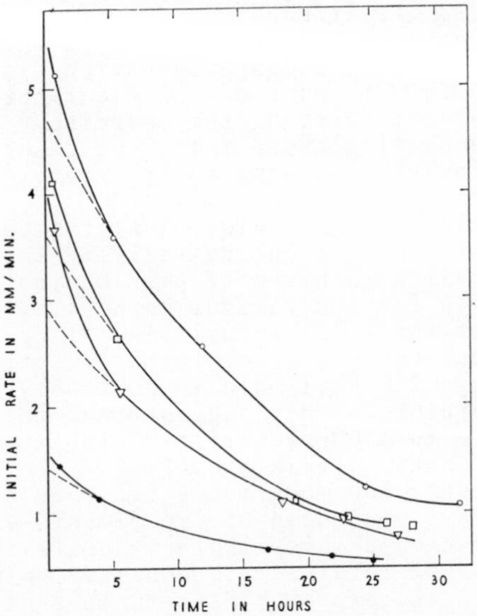

FIG. 1. Initial rate of dehydrogenation at 130°C vs annealing time after quenching from: ○, 400°; □, 365°; ▽, 305°; ●, 215°C.

hancement temperature, the site density [S] appears to decrease with time to a common value which approximates that for gold long annealed at 130°C. Fig. 2 gives us information that allows

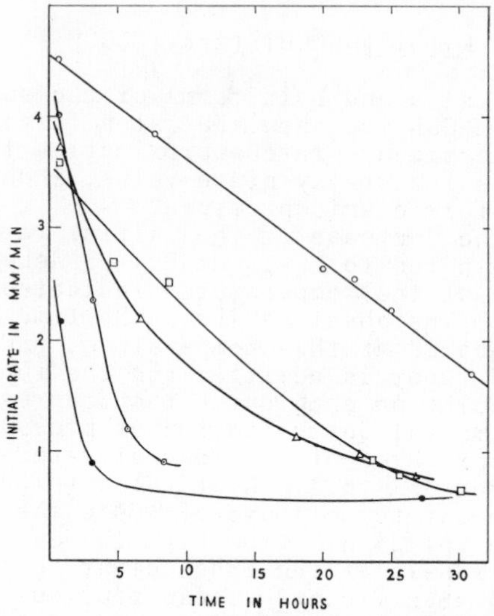

FIG. 2. Initial rate of dehydrogenation at 130° vs annealing time, after heating at: ○, 315° (27 h); □, 315° (2 h); △, 305° (4.5 h); ⊙, 315° (3 h); ●, 315°C (5 h).

us to answer the earlier question about the length of time that the catalyst is held at a given enhancement temperature being significant. Pick out zero-, first-, initially second- and subsequently first-, and even higher-order decay kinetics for the enhanced activities. Is it apparent from these two figures that the decay of enhanced activity is dependent on time alone and little, if any, upon the dehydrogenation processes catalyzed in between times. This is part of the clear-cut behavior of the system mentioned at the first of the exercise. Finally, after each heating and annealing experiment, the catalyst is left with slightly, but definitely and apparently irreversibly increased activity.

These decay kinetics bear close qualitative resemblance to the decay of extra electrical resistivity in quenched gold.

Further discussion is given of possible active sites, energies of activation for the various mechanistic steps, and possible effects of the 0.01% impurities. They feel that there is little doubt that the energy of formation of vacancies in gold is in the range 2-6 kcal/mole in contrast to the 1 eV/atom quoted in the literature. They suggest that the nearness of the 2-6 kcal figure to a characteristic property of gold may not be a coincidence. What is this property?

At $35/troy ounce how much money did they have tied up in their catalyst? Have you known of experiments in which the cost of the chemicals exceeded this figure by quite a bit? Would you say that this catalyst retained its monetary value after the experiments and could therefore be "spent" in a different sense than the usual exhausted catalyst?

56. "Mechanism of the 'Bimolecular' Hydrogen-Iodine Reaction" J.H. Sullivan 46, 73 (1967).

The mechanistic step

$$H_2 + I_2 \rightarrow 2HI$$

was thought to occur in the hydrogen-iodine reaction, and was one of the very few, if not the only, examples of a homogeneous gas-phase bimolecular reaction. The number of teachers of courses in physical chemistry who have had to rewrite this part of their notes on kinetics must have been added to daily at the time this article appeared! See also R.M. Noyes, 48, 323 (1968).

As the present paper indicates there are two concurrent mechanisms, (A) and (B), which are each both experimentally and theoretically consistent with the rates of reaction of hydrogen with iodine:

(A)

$$H_2 + I_2 \overset{k_1}{\rightarrow} 2HI \qquad (1)$$

$$H_2 + 2I \overset{k_2}{\rightarrow} 2HI \qquad (2)$$

(B)

$$I + H_2 \rightarrow HI + H$$

$$H + I_2 \rightarrow HI + I$$

$$H + HI \rightarrow H_2 + I$$

$$I + HI \rightarrow I_2 + H.$$

Although it was recognized in the earlier experimental work by the author that the reaction occurring in Mechanism (A) might be (2), the data were actually interpreted on the assumption that the reaction was (1). In thermal systems(iodine atoms in equilibrium with iodine molecules) show that

$$d[HI]/dt = k_2[I]^2[H_2] = k_1[I_2][H_2]. \qquad (3)$$

What is the relation between k_2 and k_1? How would one distinguish between (1) and (2) in thermal systems?

What is needed is a scheme for producing iodine atoms at a temperature low enough so that (1), if it occurs at all, is greatly suppressed because of the shortage of I atoms. By taking a range of (relatively low)temperatures one could determine Arrhenius parameters for k_2. If these parameters match those obtained from interpreting the higher-temperature thermal data in terms of (2), it would be be taken as evidence that the bimolecular reaction (1) does not occur. Is this reasoning sound?

The author suggests the scheme and produces iodine atoms photochemically at the lower temperatures. The photostationary iodine atom concentration is determined by

$$I_2 + h\nu \rightarrow 2I$$

$$2I + I_2 \overset{k_{r1}}{\rightarrow} 2I_2$$

$$2I + H \xrightarrow{k_{r2}} I_2 + H_2.$$

Could the r stand for recombination? If the hydrogen iodide is produced by (2) above, the local reaction rate is

$$d[HI]_\ell/dt = 2k_2[H_2][I]_\ell^2$$

where the subscript ℓ indicates a local concentration. The factor of 2 on the rhs of this equation, absent in (3), is due to the fact that the rate of iodine atom production is equal to twice the rate of absorption of light quanta. What does this statement say about fluorescence from the excited($B^3\Pi_{0^+u}$) iodine molecule? This excited molecule, which undergoes induced predissociation to ground state atoms, could just as well have been shown as an intermediate in the quantum absorption step above.

Since the rate of HI production is directly proportional to the first power of the (constant)rate of energy absorption, E_a, and the I_2 and H_2 concentrations are almost constant during a run, integration over the reaction space and the time interval of a run gives for the total amount of HI produced during a run

$$HI = 2k_2E_at[H_2]/(k_{r_1}[I_2] + k_{r_2}[H_2]), \tag{5}$$

where the concentrations are average(essentially initial) over the time t of the run. Can you derive this equation? It may be easier to work backwards. The recombination rate constants were calculated from the experimentally determined formulas of ref. 13 and 14 and are both given in liter^{2}mole$^{-2}\cdot$sec^{-1}:

$$k_{r_1} = 10^{4.65}T^{3/2} \exp (5320/RT)$$

$$k_{r_2} = 6.6\times10^8 \exp (1220/RT).$$

Rationalize the dependence on T, which is not the usual one for rate processes. Table I gives the results for k_2 in the photo-

TABLE I. Experimental data.

$10^5[I_2]$ (moles/liter)	$10^3[H_2]$ (moles/liter)	10^7 HI (moles)	10^7E_a (einsteins/sec)	10^3t (sec)	$10^{-5}k_2$ (liters^2 mole$^{-2}\cdot$sec^{-1})
			$T=417.9°K$		
6.42	10.96	8.35	4.71	33.6	1.113
6.80	10.96	9.96	4.99	41.46	1.04
8.96	4.66	12.73	5.18	80.88	1.11
14.67	4.66	12.12	6.46	79.20	1.20
15.75	2.585	7.63	6.78	85.5	1.11
22.00	4.50	6.08	4.13	93.6	1.11
17.53	5.715	7.30	3.96	79.2	1.16
			$T=480.7°$		
6.16	4.98	17.45	5.80	22.74	2.56
8.80	1.77	9.36	5.57	30.60	2.33
13.60	2.18	12.81	6.54	37.80	2.61
16.33	2.18	14.40	7.22	41.10	2.82
22.43	2.955	5.08	2.52	42.42	2.80
14.98	6.21	6.60	2.22	29.34	2.72
22.48	2.40	6.96	4.24	41.10	2.39
			$T=520.1°$		
7.73	9.34	49.4	3.08	57.42	4.06
5.71	9.34	19.51	2.67	24.48	3.90
13.96	2.495	6.47	6.63	9.36	3.79
23.07	2.335	7.39	2.42	40.50	4.20
5.67	11.90	12.32	2.66	15.60	3.85

chemical temperature regime. Using the previously developed equations compute a value for k_2 at each of the temperatures shown. Fig. 1 combines these values of k_2 with those previously (ref. 2, same author) determined in the thermal system temperature regime. Do the rate constants from the two regimes have the same Arrhenius parameters? What are they? Is the ordinate in Fig. 1 logarithmic? For the column headings in Table I and the ordinate label in Fig. 1, which of the two systems of labeling is used? viz i) heading, including power of 10, = entry, or ii) heading = entry times power of 10 from heading. There are almost always "obvious" checks you can make to determine which system is used. The einstein is surely a familiar unit.

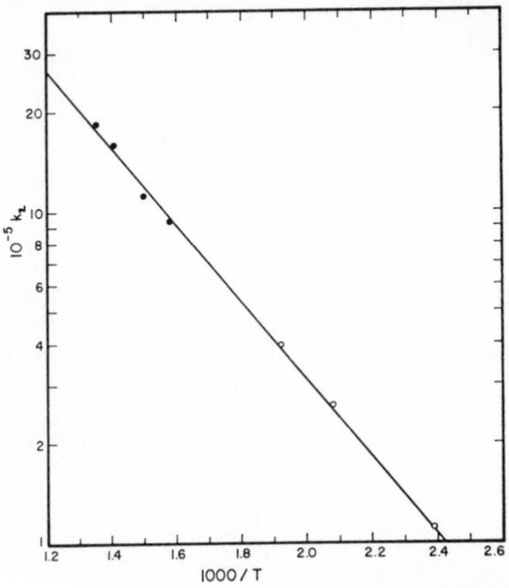

FIG. 1. Rate constants for reaction $H_2+2I\rightarrow 2HI$ from thermal and photochemical data: ●, thermal results; ○, photochemical results.

Compare your values of the Arrhenius parameters with his:

$$k_2 = (6.66\pm 0.5)\times 10^7 \exp(-5310\pm 85/RT)\ \text{liter}^2\text{mole}^{-2}\cdot\text{sec}^{-1}.$$

Write equation (5) in a form such that a plot of $2E_at/HI$ vs $[I_2]/[H_2]$ should be a straight line with slope k_{r_1}/k_2, intercept k_{r_2}/k_2, and ratio of slope to intercept, k_{r_1}/k_{r_2}. The data are plotted in Fig. 2. Demonstrate that the least-squares lines, with slope to intercept ratios of 79, 44 and 38, give quite good agreement with the values of k_{r_1}/k_{r_2} from the previous equations, at each of the three photochemical temperatures.

The reverse reaction is $2HI\rightarrow 2I+H_2$. This reaction has an activation energy of 43.7 kcal(ref. 2). By how much is it endothermic? Show that the H_2 produced is necessarily in the v"=0 level.

Note that his independently determined values of k_{r_1}/k_{r_2} agree with those from the flash-photolysis experiments of ref. 13 and 14. Why is this evidence that the rate-determining step in the photochemical formation of HI involves species which react in the rate-determining step of iodine-atom recombination?

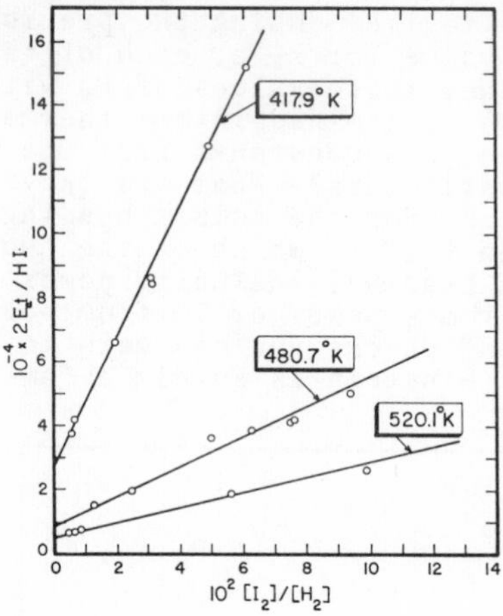

FIG. 2. Determination of ratio of recombination rate constants, k_{r1}/k_{r2}.

Is this significant? In what way? Since the rate-determining step in recombination is not vibrational relaxation, the rate-determining step in HI formation cannot be a reaction of H_2 with vibrationally excited I_2 in the latter's ground electronic state.

Table II shows a comparision of the experimental and two transition-state theoretical values of k_2. What is the difference between the two theoretical values in terms of assumptions?

TABLE II. Experimental and transition-state theory values of k_2.[a]

T (°K)	$10^{-5}k_2$ (Experimental)	$10^{-5}k_2$ (Theory I)	$10^{-5}k_2$ (Theory II)
417.9	1.12±0.02	0.87	0.64
480.7	2.60±0.17	1.86	1.45
520.1	3.96±0.15	2.79	2.18
633.2	9.38	6.90	5.48
666.8	11.5	8.65	6.9
710.3	16.1	11.2	8.9
737.9	18.54	13.3	10.4

[a] Units are square liters per square mole·second.

Finally, reconcile values of k_1 from your physical chemistry text(or possibly from footnote 2 of ref. 3) with the k_2 values reported here making use of the relation between k_1 and k_2 that you derived at the first of this exercise. His ref. 3 contains values of the equilibrium constant for the dissociation of I_2 or you should be able to easily approximate it from readily available data.

Those who subscribe to Chemical & Engineering News might recall that this research was written up there as well. That article should have caught the rest of the teachers of physical chemistry.

57. "Recoil Tritium Reactions: Studies on the System T + SiH_4" G. Ceteni, O. Gambino, M. Castiglioni, and P. Volpe 46, 89 (1967).

Write the equation for the production of "hot" T atoms by the neutron bombardment of ^3He. In these experiments ampules of quartz were filled with monosilane, ^3He, scavenger(I_2), and moderators in various proportions by means of a vacuum line and then irradiated. What similarity is thought to exist between the process of neutron thermalization and that whereby the hot T atoms are able to engage in chemistry?

The radio gas-chromatographic analyses of the products are shown in Fig. 1 and 2, where the former was done in the absence and the latter in the presence of the I_2 scavenger.

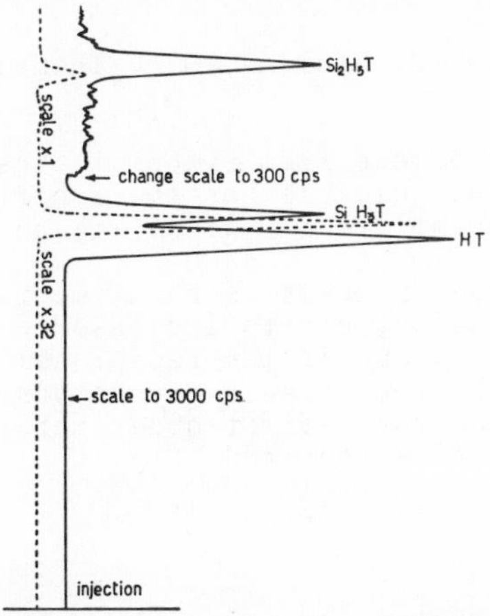

FIG. 1. Radio gas-chromatographic analysis of the irradiated mixture of SiH_4 and He. Column: 12 m long; partition liquid: silicon oil 702 20% on celite. Carrier gas: He at 50 cc/min. Mixture for proportional counter: He and CH_4 (1:1) 100 cc/min.

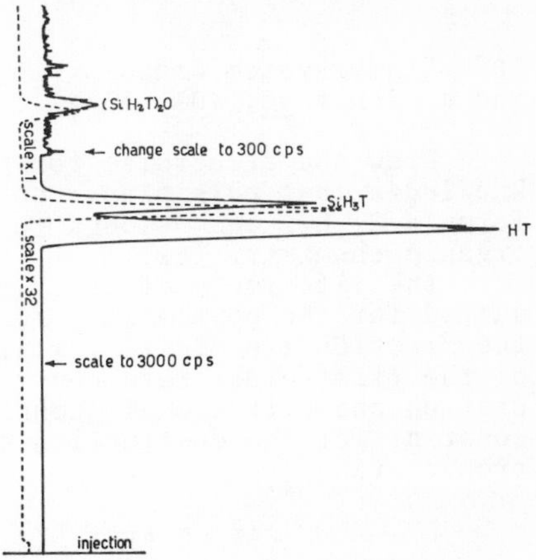

FIG. 2. Radio gas-chromatographic analysis of the irradiated mixture of SiH_4 and He (scavenger present). Column: 12 m long; partition liquid: silicon oil 702 20% on celite. Carrier gas: He at 50 cc/min. Mixture for proportional counter: He and CH_4 (1:1) 100 cc/min.

Which direction is that of increasing time? Does the order in which the products come off the column seem to be in accord with your empirical(what else?) knowledge of chromatography? Make measurements of areas under the curve as accurately as possible for each compound and compare with the relative abundances shown in Tables I and II. The siloxane probably occurs from the

TABLE I. Reactions of hot tritium with SiH_4 (I_2 absent, $T\sim40°C$).[a]

Composition		Yields in % of total activity observed			
SiH_4 press. (cm of Hg)	^3He press. (cm of Hg)	HT	SiH_4	Si_2H_6	Others
20	0.3	68	28	3.5	Trace

[a] Neutron flux $4.2\times10^{13}/cm^2$ sec—irradiation time 4 min.

TABLE II. Reactions of hot tritium with SiH_4 (scavenged with I_2, $T\sim40°C$).[a]

Composition		Yields in % of total activity observed			
SiH_4 press. (cm of Hg)	^3He press. (cm of Hg)	HT	SiH_4	$(SiH_3)_2O$	Si_2H_6
20	0.3	56	26	1	Trace

[a] Neutron flux $4.2\times10^{13}/cm^2$ sec—irradiation time 4 min.

silyl iodide and water adsorbed on the ampule walls. Write the reaction.

Notice that all of the species in Tables I and II have at least one T atom per molecule. Would those molecules with two or more T atoms per molecule be relatively abundant? Why or why not? From Fig. 1 and 2 does it seem as though their retention times could be resolved from the monotritiated species if indeed these di- and higher-tritiated species were present in measureable concentrations?

The authors continue with discussions of hot and thermal abstractions and substitution reactions, then compare results with methane reactions with T by means of theory developed for the latter case. Our interest was in the method of analysis only; let us ask, however, if the function of the scavenger is apparent from Tables I and II?

58. "Intersystem Crossing in Pyrene" B. Stevens, M.F. Thomaz, and J. Jones 46, 405 (1967).

Draw the structural formula of pyrene from either i) the knowledge that this planar conjugated molecule has the empirical formula $C_{16}H_{10}$ and belongs to the point group D_{2h}, or ii) an organic chemistry text.

The lifetime τ of any first-order process is the time required for the population, concentration, etc to decrease to the fraction 1/e of its initial value, and is the reciprocal of the first-order rate constant. In the absence of photoassociation and diffusional quenching the total first-order rate constant for the destruction of the first excited singlet electronic state is

$$1/\tau = k_F + k_{IC} + k_{IS}$$

where the subscripts stand for fluoresence emission(F), internal conversion(IC), and intersystem crossing(IC); you may need to refresh your memory on the meaning of the last two processes. Is it proper to just add rate constants up like this? Are the processes occuring in series or in parallel? Is each process a first-order one? At room temperature, where the quantum yield of molecular fluorescence $\gamma_F = k_F/(k_F + k_{IS}) = 0.65$(ref. 2), it turns out(ref. 3,4) that internal conversion is negligible, i.e.

$$k_{IC} << k_F + k_{IS},$$

although it may become significant at higher temperatures if it requires an activation energy. Now let us assume that k_{IS} goes to zero in the limit of lower temperatures. Practically this means at about 80°K and below. Since k_F, the reciprocal of the radiative lifetime, is temperature independent, the total first-order rate constant takes the form

$$1/\tau = k_F + k_{IS} = k_F + k_{IS}^{\circ} \exp(-E_{IS}/RT),$$

and the solid curve in Fig. 1 is drawn with

$$1/\tau_F(\text{sic}) \equiv 1/\tau = 1.33\text{x}10^6 + 3.3\text{x}10^7 \exp(-2270/1.987T) \text{ sec}^{-1}$$

where the first form of the equation recognizes that one determines an effective fluorescence lifetime in an experiment. Is it correct that this effective lifetime is the harmonic mean of the true fluorescence and intersystem crossing lifetimes? From the solid curve evaluate k_F as the intercept at 0°K. Try two or three other points on the curve to check the Arrhenius parameters. What is the best method to plot the data to check these values? Now check our assumption that k_{IS} vanishes at low temperatures by computing the fluorescence yield γ_F at room temperature. Was the assumption justified? Is IC eliminated as a significant competing process?

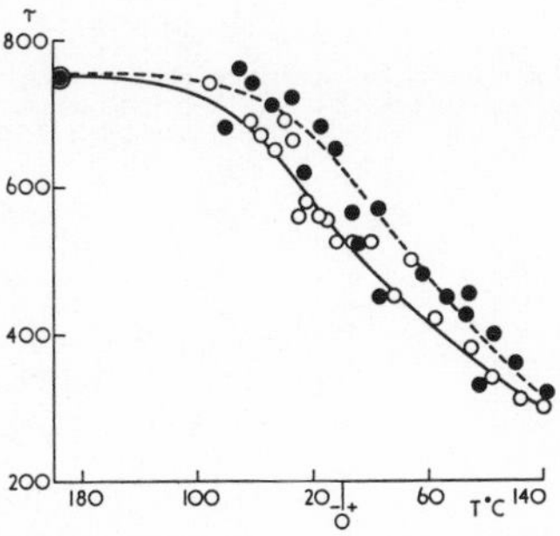

FIG. 1. Temperature dependence of fluorescence lifetime τ (nanosecond) of pyrene, ○, in ethanol ($2\times10^{-6}M$); ●, in liquid paraffin ($10^{-4}M$).

Why is the ethanol solution so dilute in pyrene? Since E_{IS} is close to the activation energy for viscous flow(determined how?) for this solvent of ∿3 kcal/mole, similar measurements of effective fluorescence lifetimes were made with pyrene in liquid paraffin, whose viscous-flow activation energy is ∿12 kcal/mole. Why was this precaution taken? What were the results?

These results are similar to those of ref. 5, on mono- and disubstituted anthracenes, whose authors conclude that "a significant temperature dependence of fluorescence quantum yield(or lifetime) is diagnostic of an intersystem crossing to a virtually degenerate triplet state...".

59. "Energy Transfer in Solution between UO_2^{2+} and Eu^{3+}"
J.L. Kropp 46, 843 (1967).

This article presents results of a study of intermolecular energy transfer in solution to an Eu^{3+} ion using a completely inorganic system. We will concentrate on just two aspects of the work; the experimental apparatus shown in Fig. 1 and the energy level diagram of Fig. 5.

Why is the monochromator M_2 set with its entrance at right angles to the beam coming from M_1? Of course it is so that M_2 might see something other than the primary beam. Do a little calculation with the Einstein A and B coefficients and the density of the radiation field. Take a 3000°K source and a frequency of 20 000 wavenumbers to show that the ratio of spontaneous emission to induced emission in this case is ∿ e^{10}. Does the fact that induced emission is in the direction of the

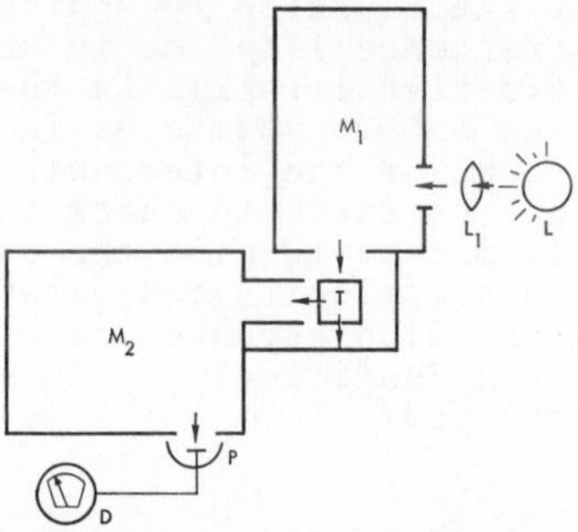

FIG. 1. Apparatus for measuring relative intensity of fluorescence as a function of wavelength. M_1, M_2, monochromators; L, light source; L_1, lens, T, sample; P, photomultiplier; D, electrometer.

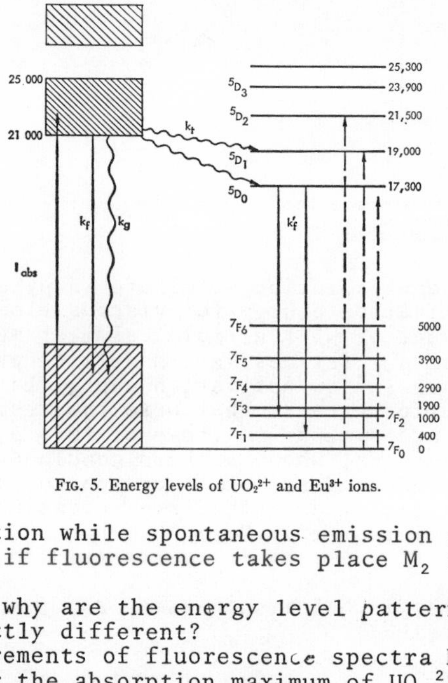

FIG. 5. Energy levels of UO_2^{2+} and Eu^{3+} ions.

inducing radiation while spontaneous emission is isotropic guarantee that if fluorescence takes place M_2 is in the right place?

In Fig. 5 why are the energy level patterns for the two ions so distinctly different?

For measurements of fluorescence spectra M_1 is set at a wavelength near the absorption maximum of UO_2^{2+} (should this be about 23 000 cm^{-1} from Fig. 5?) and M_2 is then dialed through the wavelength region of fluorescence; should the resulting fluorescence intensity show up as the electrometer current? For excitation spectra M_2 is set at the wavelength of the fluorescence maximum, 19 000 cm^{-1} for UO_2^{2+} (why not 23 000?), and at what two values for Eu^{3+}? The fluorescence intensity is then recorded as M_1 is varied. The fact that the excitation spectra for both ions were always identical implies what? Eu^{3+} absorbs only weakly, as the dashed transitions indicate. There are two separate errors connected with these absorptions which are completely negligible as far as the main arguments go, but which might be instructive to rectify. The arrow begins on the wrong level in one case and the J values are switched for another as described on p. 847 of the paper. Notice the magnitude of the spin-orbit splittings for the two terms. Can you confidently assign a label to the uppermost member of the ^5D term from your knowledge of L-S coupling? Is L-S coupling suspect on physical grounds for this heavy atom? Is it still legitimate as a classification scheme?

It turns out experimentally that adsorption by $UO_2{}^{2+}$ and fluorescence by Eu^{3+} are the dominant processes with the converse processes rather weak. From Fig. 5 do you see that there is a distinct cutoff in Eu^{3+} levels above which emission cannot originate? The process shown as k_g should be k_q for quenching. This is a bimolecular process returning the $UO_2{}^{2+}$ to its ground electronic state. Under what condition of excitation of the Eu^{3+} ion will k_q be identical with k_t(for transfer?) shown in Fig. 5? The latter does not seem to be identified in the article. Say something about the magnitudes of k_f and k_f'. What distinct color is the emission of the europium ion? "bright ?".

Does it seem as though we have a high ratio of uv spectroscopy to kinetics in this exercise? May we take refuge behind the statements in the Preface?

60. "Hydrogen-Chlorine Explosion Laser. II. DCl" P.H. Corneil and G.C. Pimentel 49, 1379 (1968).

Part I(ref. 1) concluded that the H_2-Cl_2 laser emission comprised 1→0 vibrational transitions. The work summarized in the present paper shows that the transitions are rather 2→1. Refer to Table I. "While the evidence permits some contribution from $P_1(10)$ and/or $R_3(0)$, there is no need to assume emission from other than 2→1 transitions." Is this an excellent general procedure?

TABLE I. Observed laser emission from the H_2–Cl_2 explosion, ν(cm^{-1}).

Spectrometer position	Calculated[a] 1→0	2→1	3→2
	$P_1(8)$ 2702.97		
2697.5	...	$P_2(4)$ 2697.55	$R_3(0)$ 2697.99
	$P_1(9)$ 2677.71		
2675.0	...	$P_2(5)$ 2674.97	...
			$P_3(1)$ 2659.13
2652	$P_1(10)$ 2651.96	$P_2(6)$ 2651.86	...
			$P_3(2)$ 2639.02
2628	...	$P_2(7)$ 2628.21	...
	$P_1(11)$ 2625.74		$P_3(3)$ 2618.20
2604	...	$P_2(8)$ 2604.05	...
	$P_1(12)$ 2599.06		
			$P_3(4)$ 2596.66

[a] See Ref. 9.

Notice the labeling on these vibration-rotation lines. The subscript is the upper vibrational level, while the J value of the lower vibrational level is in parentheses. Should we be tempted to interpret the J values otherwise, the presence of $R_3(0)$ should stop us. Write the full set v', v", J', and J" for $P_2(4)$.

Table II, not shown here, shows that the three lines of the DCl emission can likewise be assigned to the 2→1 scheme; the lines are $P_2(7)$, $P_2(8)$, and $P_2(9)$. How do they now interpret the evidence which earlier(ref. 1) had caused them to make the assignment 1→0?

The effects of varying partial pressures, ratios, total pressure, flash duration and total energy, and temperature were studied. Their laser output continues to increase as the temperature increases to 530°K, above which no laser emission is obtained at all. Why is this? Fig. 3 shows the evolution in time of the individual HCl laser lines at two temperatures. The

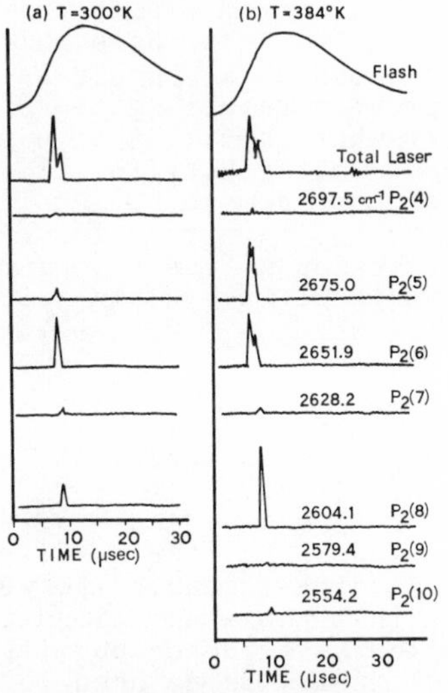

FIG. 3. Intensities of individual HCl laser lines. (a) $T=300°K$, 2000-J flash, 3 torr Cl_2, 6 torr H_2; slitwidth 1 cm^{-1}. (b) $T=384°K$, 1800-J flash, 4 torr Cl_2, 8 torr H_2; [except 12 torr H_2 for $P_2(9)$]; slitwidth, 2 cm^{-1}.

most striking feature is the greatly diminished intensities of the $P_2(5)$ and $P_2(7)$ lines at 300°K and of the $P_2(7)$ and $P_2(9)$ lines at 384°K. Lack of rotational equilibration must be involved. Show that at 10 torr H_2 and 2 torr Cl_2, an average HCl molecule makes about 190 H_2 collisions and 12 Cl_2 collisions/μsec. Since rotational exchange characteristically requires only a few collisions, should the 6 μsec prior to laser threshold suffice to establish rotational equilibration at laser onset? Maintaining equilibrium is, however, not to be expected over the short time interval between successive laser emissions. Hence, when $P_2(6)$ begins to emit, it depletes the v=2, J=5 level and overpopulates the v=1, J=6 level. During the microsecond of $P_2(6)$ emission, collisions will tend to repopulate the v=2, J=5 level preferentially from adjacent states, v=2, J=4 and 6. Furthermore, the v=1, J=6 level will transfer molecules preferentially to adjacent states v=1, J=5 and 7. Such repopulation detracts, then, from transitions immediately adjacent to the first line to emit, i.e. from $P_2(5)$ and $P_2(7)$. Apply the same argument, mutatis mutandis, to explain the "cannibalization" process weakening the $P_2(7)$ and $P_2(9)$ lines at 384°K. Should higher pressures facilitate rotational equilibration so that the alternation in intensity disappears? Check your expectation against Fig. 4. These are the only lines observed for DCl.

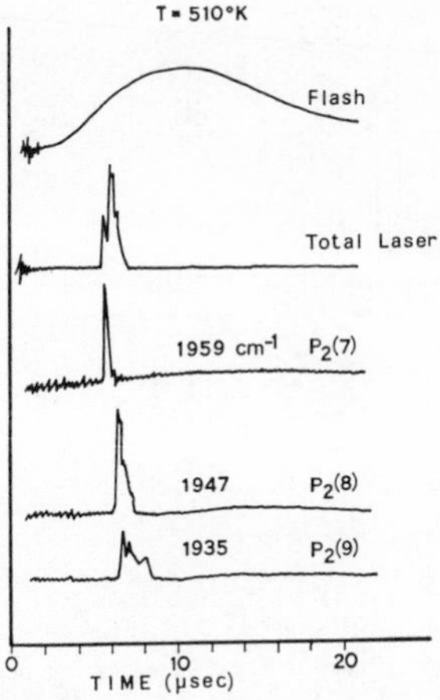

FIG. 4. Intensities of individual DCl laser lines. $T=510°K$, 2000-J flash, 18 torr Cl_2, 175 torr D_2, slitwidth 2.2 cm^{-1} except at 1959 cm^{-1}, 1.3 cm^{-1}.

The following reactions occur in this chemical laser:

$$Cl_2 + h\nu \rightarrow 2Cl$$

$$Cl + H_2 \xrightarrow{k_a} HCl + H \qquad \Delta H=+1 \text{ kcal} \qquad \Delta H^{\S}=4.4 \text{ kcal} \qquad (a)$$

$$H + Cl_2 \xrightarrow{k_b} HCl^{\dagger} + Cl \qquad \Delta H=-45 \text{ kcal} \qquad \Delta H^{\S}=2 \text{ kcal.} \qquad (b)$$

What must § and † denote? Why isn't HCl from (a) shown with vibrational excitation? Is the absolute value of the energy -45 kcal in (b) an upper or lower bound to the actual energy released? Would vibrational energy show up on the kcal scale? By how much for v'=2? Rapid production of a high concentration of chlorine atoms is desireable to initiate a high concentration of free-radical chains. Therefore raising Cl_2 pressure, flash energy, or flash power is beneficial. Why must a sufficient pressure of H_2 be retained? What is the effect of increasing the partial pressures, say in a constant ratio, indefinitely?

Fig. 6 shows concurrent HCl and DCl laser emissions of Cl_2 with H_2 and D_2 and with H_2 and HD. In the last line of the caption read HD for D_2 as per the second line and the label on (b). The two peaks shown in (a), $P_2(6)$ for HCl and $P_2(8)$ for DCl, have the same relative intensity, at least approximately, as they do in (b). It is necessary to consider the authors' statement concerning (b): "There is an initial HCl emission spike at $P_2(6)$ followed 1 μsec later by at least three DCl peaks some but not all of which emission is at $P_2(8)$."

For reaction (a) the ratio $k_a(HD)/k_a(H_2)$ is 0.49 at 510°K (ref. 13) where the species in parentheses are not concentrations but labels. The rate constant $k_a(HD)$ is the sum of those for the two competing reactions that cannot be distinguished through classical kinetic studies:

$$Cl + HD \rightarrow HCl + D, \quad k_a(H); \quad Cl + HD \rightarrow DCl + H, \quad k_a(D).$$

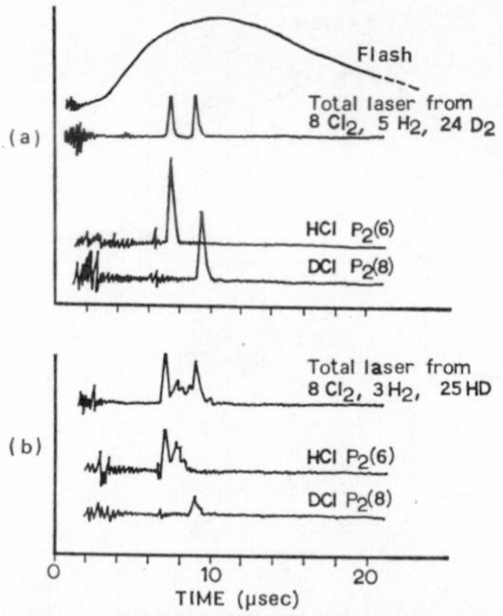

FIG. 6. Concurrent HCl and DCl laser emissions from mixtures of Cl_2 with H_2 and D_2 and with H_2 and HD; $T=510°K$, 2000-J flash. (a) 8 torr Cl_2, 5 torr H_2, 24 torr D_2; slitwidths 5 cm^{-1}. (b) 8 torr Cl_2, 3 torr H_2, 25 torr D_2; slitwidths 4 cm^{-1}.

The DCl laser action from HD can arise only when a(H) occurs, freeing a D atom for the pumping step b(D). Similarly, HCl laser action from HD requires the succession of reactions a(D) and b(H):

$$D + Cl_2 \rightarrow DCl^{\dagger} + Cl, \quad k_b(D); \quad H + Cl_2 \rightarrow HCl^{\dagger} + Cl, \quad k_b(H).$$

Do you see why these statements can be made? The authors then argue that the ratios of the HCl and DCl population inversions must have been the same in the two experiments portrayed in Fig. 6. In the reaction mixtures(could there have been H-D atom exchange, e.g. on the walls, and the present considerations still apply?) the HCl and DCl population inversions are determined by the rates of production of HCl and DCl in the appropriate reaction (b). Why are these rates limited by (a)? When only H_2 and D_2 are present, the relative population inversion is given by the ratio

$$\frac{k_a(H_2)\cdot[H_2]}{k_a(D_2)\cdot[D_2]} .$$

In a mixture of H_2 and HD show that $HCl^{\dagger}$is produced by two processes whose rates are $k_a(H_2)\cdot[H_2]$ and $k_a(D)\cdot[HD]$ for the rate-determining steps. On the other hand the $DCl^{\dagger}$ comes only from the process with rate fixed by $k_a(H)\cdot[HD]$. The relative population inversion here, part (b) of Fig. 6, is the ratio

$$\frac{k_a(H_2)\cdot[H_2] + k_a(D)\cdot[HD]}{k_a(H)\cdot[HD]} .$$

Equating these ratios(why?) and given that $k_a(H_2)/k_a(D_2)=4.35$ from ref. 13, show that the ratio $k_a(H)/k_a(D)=1.9$. "...the unique possibilities of chemical lasers are revealed by the ability to distinguish for the first time the relative reaction rates of chlorine atoms with the different ends of HD."

The authors make an interesting suggestion for this difference in terms of the classical rotation motion. Most appealing!

61. "Some Formal Results in a Theory of Molecular Rearrangements: Photoisomerism" W.M. Gelbart and S.A. Rice 50, 4775 (1969)

It is quite inaccurate to label this exercise with the title of the Journal article as we are only going to examine their rather complete discussion in Part II. Experimental Background, and ignore their theoretical approach completely.

Photoisomerism(cis-trans) in the stilbene molecule is considered as a prototype of the general rearrangement reactions induced by absorption of light. Of the two forms of 1,2-diphenylethylene which is the most stable? Irradiation with uv light converts the one form to the other. Earlier work had suggested that these photochemical rearrangements might be understood in terms of the excited electronic states of the molecule and their properties. Their ref. 2 seems to have been primarily concerned with determining whether or not molecular isomerization can take place in the lowest-lying singlet state reached directly by optical excitation. See Fig. 1. They observed that the trans-

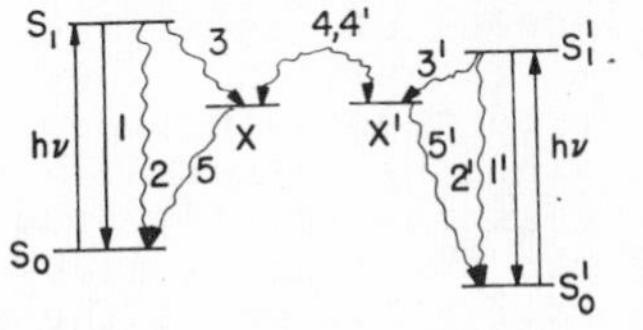

FIG. 1. A modified energy level scheme for stilbene (based on the SCF–CI results of Borrell and Greenwood).

stilbenes fluoresced strongly and that the cis isomers showed no $S_1 \rightarrow S_0$ emission whatever. These are the processes shown as 1' and 1 in Fig. 1. Why is this result a proof that conversion must proceed through lower-lying excited states? Their ref. 8 finds an activation energy of 2-3 kcal from temperature variation of the rate constant and argues that this energy cannot be attributed to a potential barrier in the S_1^1 state, again for the trans→cis reaction, for two reasons: i) an activation energy of this magnitude is improbable for a reaction which must occur in $<10^{-9}$ sec(how would one measure the lifetime of S_1^1?); and ii) the fact that the singlet absorption bands of the trans-stilbene lie to the red of the cis isomer bands implies something about the energy difference S_1-S_1^1 vs S_0-S_0^1; which of these will be larger? If the isomerization took place through a barrier in the excited singlet state(s) would an activation energy of >6 kcal exist? The ground state energy difference is 6 kcal/mole.

The nature of the one or more intermediate states involved directly in the photochemical reaction is now considered. Fig. 1 displays the simplest scheme compatible with the experimental results sketched above. Identify, by the numbers, the processes of $S_1 \rightarrow S_0$ internal conversion, $S_1 \rightarrow X$ intersystem crossing, "isomerization", and deactivation to the ground state. X and X' may represent a common state. Run down their ref. 20 and 21 and see if this the case for the twisted triplet state(lowest-lying?) in ethylene. Is the behavior of ethylene of some interest to us in connection with stilbene? The authors point out that it was just in 1962(ref. 7) that the 2-3 kcal activation energy discussed above could be attributed to a potential barrier in the $S_1^1 \to X^1$ intersystem crossing. The reasoning was that: i) T_1^1 (which was tentatively associated with X^1) had been determined to have an energy of ∿51 kcal/mole relative to S_0^1 from the oxygen perturbation studies on the 0-0 $S_0^1 \to T_1^1$ transition that was measured in absorption; ii) the activation energy of the thermal cis-trans isomerization in solution had been measured to be ∿37 kcal/mole - to be contrasted with the value quoted above of 2-3 kcal/mole for the photochemical isomerization; and iii) from this their ref. 7 concludes that the T_1^1 and S_0^1 potential surfaces intersect upon twisting which causes the excited-state electronic energy to be lowered. Let's try and separate cause and effect. Do the surfaces intersect because twisting lowers the T_1^1 energy, or is the energy of T_1^1 lowered upon twisting because of the intersection? What then would be the potential barrier of the twisted ground-state configuration? What was the reason for the oxygen used in the study establishing point i)? Why do these postulates imply a shortlived triplet state? Is this consistent with the fact that no phosphorescence had been observed in rigid(why?) media, at liquid-nitrogen temperatures (why?), for either cis- or trans-stilbene? Define the quantum yield $\phi(t \to c)$ as the molecules of cis-stilbene produced per photon absorbed in the $S_0^1 \to S_1^1$ transition. Thus it was argued that the experimentally observed temperature dependence of the quantum yield could be explained by a thermally activated radiationless crossing(which process in Fig. 1?) from the directly excited singlet state. How can the radiationless transition compete with a radiative one $S^1 \to X_1^1$? What about the omitted adjective "intersystem"?

Their ref. 10 found that $\phi(t \to c)$ as a function of temperature is characteristic of the compound in question, varying only slightly with the viscosity(diffusional quenching?) and chemical nature of the medium. Specifically it was found in all cases to increase from very low values(viz ∿0.002) at -180°C to higher values at higher temperatures, eventually approaching a limiting value(e.g. ∿0.50 at -30°C for stilbene, and ∿0.40 at -160°C for p-bromostilbene). What is the activation energy from these two temperature-dependent values for stilbene? Should it be compared with the photochemical or thermal values given earlier? Is the comparison encouraging?

In this high-temperature limit it is argued that equilibrium between the excited isomers should be established quickly, and one should observe $\phi(t \to c) + \phi(c \to t) \to 1$. Why should this be so? It was in fact the case for stilbene and p-chlorostilbene, but not for p-bromostilbine. What is the difference between bromine and chlorine atoms that they might give different limiting sums of quantum yields for the monosubstituted halostilbenes? Their ref. 24 explains these results by suggesting that there are at least two electronic states(called T_1^1 and I_2^1) below S_1^1 which participate in the isomerization scheme. Comment briefly on the reason(s) that might possibly have lead the present authors to label this state by I. Mnemonic for intermediate? Noncommittal about spin pairing? Misprinted? Ref. 24 further suggests that the directly reached S_1^1 state can pass into the manifold of intermediates in two ways - one predominating in the unsubstituted stilbene and the p-chloro derivative, and the other in the bromine compound. Only the $S_1^1 \to I_2^1$ transition is assumed to be thermally activated. In the case of the bromine compound the temperature-dependent $S_1^1 - I_2^1$ potential surface intersection and

and consequent radiationless decay to the T_1^1 triplet state are bypassed by the direct $S_1^1 \rightarrow T_1^1$ transition. What difference in properties, bromine vs chlorine, is invoked to explain the enhancement of the singlet-triplet mixing in p-bromostilbene?

Support for the hypothesis of the direct $S_1^1 \rightarrow T_1^1$ transition was found in the experimental determination of the fluorescence quantum yield Y_f(defined how?), as a function of temperature. It was found that $Y_f(T)$ for trans-stilbene increases dramatically upon cooling. How would this fact affirm the role of the thermally activated $S_1^1 \rightarrow I_2^1$ transition? Similarly, the observation that $Y_f(T)$ for trans-p-bromostilbene does not rise significantly at low temperatures is said to attest to the increased importance of the temperature-independent $S_1^1 \rightarrow T_1^1$ transition. Why is this so? Thus there seems to be good evidence for the thermally activated mixing between the singlet and triplet manifolds as postulated.

This argument for the bromine derivative is shown in Fig. 2 as the temperature-independent, spin-orbit-coupling-enhanced step 5' bypasses steps 3' and 4'. Notice that the excited electronic state I_2^1 invoked above is now labeled T_2^1. Notice also that X and X^1 are now shown as a common state as for the ethylene molecule. Thinking about the twisting angle as the single pertinent independent structural variable in the Born-Oppenheimer sense, do you see why the potential energy surfaces need be thought of only as curves?

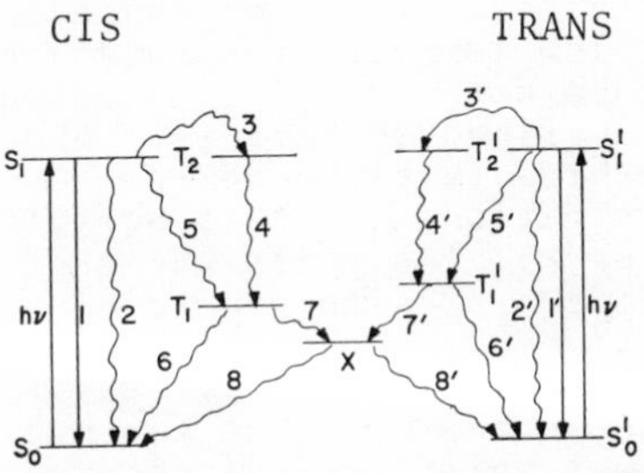

FIG. 2. A schematic plot of electronic energy vs angle of twist (about the double bond) in unsubstituted stilbene.

They feel their theoretical approach is naturally implied in the report of ref. 18 who suggests the fruitfulness of treating the isomerization process as simply another route for radiationless transitions in large molecules.

CHAPTER V

SOLIDS

62. "The Nature of the Thermal Color Change in Zinc Oxide" C.K. Coogan and A.L.G. Rees 20, 1650 (1952).

Pure zinc oxide was known to undergo a reversible color change from white to yellow upon heating. If the heating had not been prolonged at temperatures $>0.25\ T_m$(melting point in °absolute-°K or °R?) the color change was completely reversible upon cooling to room temperature. The absence of lag in the color changes at such low temperatures suggested that solid and surface diffusion do not take place and that the color change should not be ascribed to defects. Should their adjective "pure" really be "relatively pure" or "quite pure"?

Two possible explanations for this behavior existed at the time of the authors' work. The first was that the ZnO partially decomposes at high temperatures to give a zinc-rich nonstoichiometric oxide having a defect absorption band at 4000-4300 Å. Upon cooling the nonstoichiometric oxide was thought to recombine with oxygen. The second was that the near-uv absorption band undergoes a thermal shift towards the red. Can you rationalize the direction of this proposed shift? Think about groups of vibrational levels associated with both the upper and the lower electronic states, or groups of states, of the transition. Would heating populate the vibrational levels of the ground electronic states in such a fashion as to cause the red shift? Would the solid then experience a (relative)blue shift in the absorption band upon cooling? Earlier work for a single crystal of ZnO heated in air showed a red shift of 1.4 Å/°C in the range 26-77°C. A long extrapolation would place the absorption edge at about λ4700(from where at room temperature?) at 425°C($\leq 0.25\ T_m$?). Would this produce a yellow appearance of the solid?

ZnO was heated in vacuum to 400-500°C, a barium getter was fired to "get" any evolved oxygen, and the temperature was allowed to fall. The yellowed color changed to white immediately Which of the above mechanisms is proven? Only after prolonged heating under nonoxidizing conditions(reducing or inert atmospheres or vacuum?) did ZnO attain a permanent yellow color due to stoichiometric excess of zinc.

The authors suggest heating-induced color changes in other oxides, such as CeO_2 and In_2O_3, may be of similar origin.

63. "Optical Determination of the Compressibility of Solid Argon" B.L. Smith and C.J. Pings 38, 825 (1963).

The isothermal compressibility $k_T = -(1/V)(\partial V/\partial P)_T$ of solid argon was determined by measuring as a function of pressure the angle of minimum deviation D of light from a sodium lamp which was refracted from a prism-shaped specimen of solid argon. The prism angle A was 45°1.69'. A, D, and the refractive index n are related by(see their ref. 7)

$$n = \sin[\tfrac{1}{2}(A+D)]/\sin\tfrac{1}{2}A.$$

D is given as a function of pressure, to 75 atmospheres, in Fig. 1.

According to dielectric theory(e.g. ref. 4), the polarizability α of a nonpolar dielectric is related to the refractive index, and the density ρ, by the relation

$$\frac{(n^2-1)}{(n^2+2)\rho} = \frac{4\pi N_0\alpha}{3M} = L,$$

where N_0 is Avogadro's number, M the molecular or atomic weight, and L is called the Lorentz-Lorenz function; it is found to remain nearly constant over a wide range of temperature and pressure.

Show that(trivially) $k_T = (1/\rho)(\partial\rho/\partial P)_T$ and also that

$$k_T = \frac{6n}{(n^2-1)(n^2+2)}(\partial n/\partial P)_T$$

$$= \frac{3[1-\cos A]\sin(A+D)}{[\cos A - \cos(A+D)][3-\cos(A+D) - 2\cos A]}(\partial D/\partial P)_T.$$

From the slope in Fig. 1 and the value of D at 0-1 atmosphere see if you can match their value of k_T at 78°K, $(6.65\pm0.19)\times10^{-11}$ cm^2/dyn.

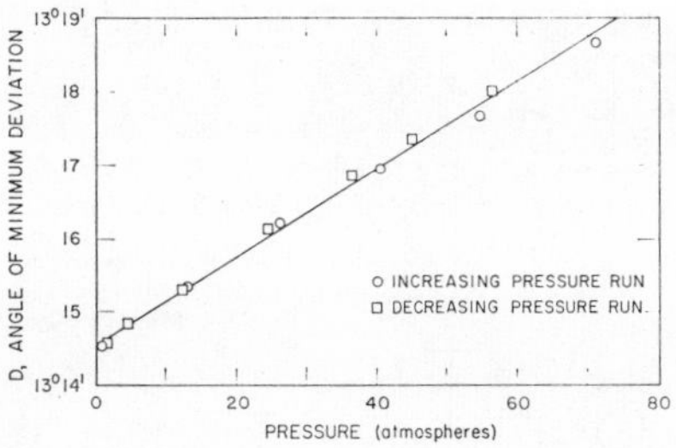

FIG. 1. Angle of minimum deviation vs pressure at 78.10°K.

64. "Melting and Polymorphism in $SbCl_3$ and $SbBr_3$" B.F. Bowles, G.J. Scott, and S.E. Babb, Jr. 39, 831 (1963).

The law of corresponding states, while forming a convenient and reasonably accurate means of predicting PVT properties of low-density fluids, was thought not to be generally useful for solids. Their ref. 2 showed that for similar molecules where the peripheral atoms are dominant in determining the intermolecular interactions, a law of corresponding states should be expected to hold even in the solid phase. In that study CCl_4 and CBr_4 were chosen because the tetrahedral molecular structure gave the halogens equivalent positions, and their large volume was thought to screen the effects of the carbon atom.

In the work under discussion here the molecules have a pyramidal structure, with the Sb-halogen distance somewhat greater than the C-halogen distance in CCl_4 or CBr_4. What effect should this have on the screening of the Sb vs the C atoms? Would a larger difference in the reduced phase diagram now be expected?

Experimentally the volume change when passing between the solid and liquid phases was quite large and thus the equilibrium pressure could be established within narrow limits as shown by the open circles in Fig. 1 and 2. With the solid-solid transi-

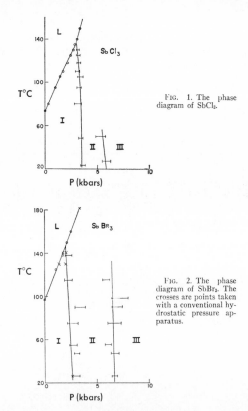

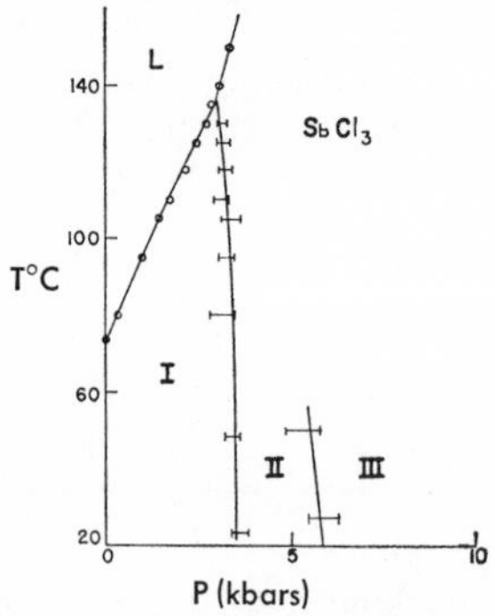

FIG. 1. The phase diagram of $SbCl_3$.

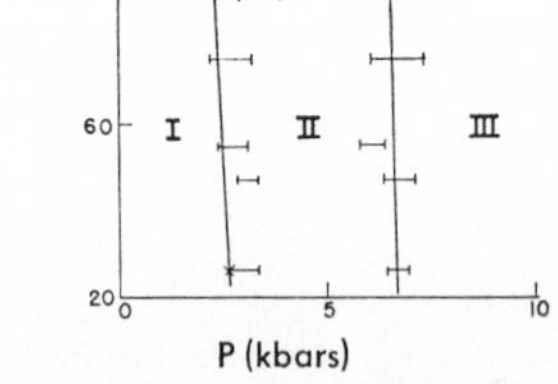

FIG. 2. The phase diagram of $SbBr_3$. The crosses are points taken with a conventional hydrostatic pressure apparatus.

tions this was not possible due to the small volume changes(show from the slopes and the Clapeyron equation that these are small in all four cases). These transitions were detected either by a break in the P-V curve itself or by a discontinuity in its slope. The transition pressures are given in Table I.

TABLE I. Transition data.

Temperature (°C)	L–I	Pressure[a] (bars) I–II	II–III
	Antimony trichloride		
25	...	3530±250	6000±500
50	...	3500	5800
75	70	3475	...
100	1210	3360	...
125	1910	2490	...
	L–II		
150	3310	...	...

L–I–II triple point: Approximately 136°C and 2900 bars.

Temperature (°C)	L–I	Pressure[a] (bars) I–II	II–III
	Antomony tribromide		
25	...	2700±350	6675±300
50	...	2570	6600
75	...	2415	6575±650
100	115	2300	6500
125	1160	2185	6475
	L–II		
150	2150	2010	...
175	3100	...	...

L–I–II triple point: approximately 146°C and 2000 bars.

[a] The uncertainties apply to all values below them in the table.

Where is the vapor region in Fig. 1 and 2? Are there possibilities for other triple points? Where and for what phases? Table II lists critical and reduced temperatures and pressures

TABLE II. Critical and reduced parameters.

Substance	T_c (°K)	P_c (bars)	T_m (°K)	T L–I–II °K	P L–I–II bars	T_m^*	T^* L–I–II	P^* L–I–II
$SbCl_3$	791	180	346.6	409	2900	0.438	0.518	16.1
$SbBr_3$	904	157	369.8	419	2000	0.409	0.466	13.1
$AsCl_3$	629	125	255.0			0.405		
$AsBr_3$	769	153	305.8			0.398		

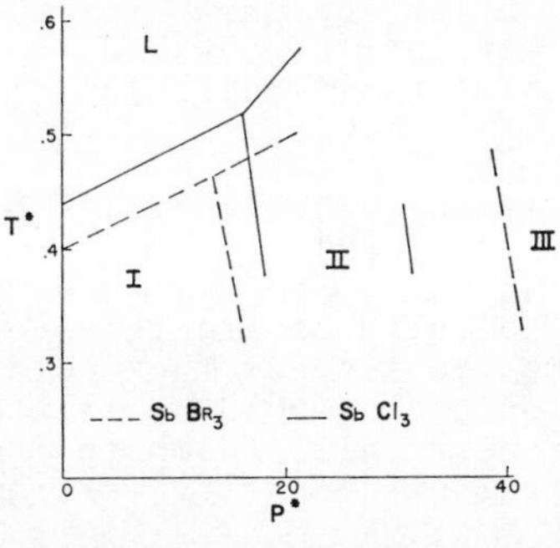

FIG. 3. The phase diagram of $SbCl_3$ and $SbBr_3$ in reduced units.

so that the corresponding state phase diagram of Fig. 3 could be constructed. You might want to check Fig. 3 against the one given by ref. 2. Are all of the phase boundaries placed correctly in Fig. 3? $AsBr_3$ has the same crystal structure as the antimony compounds(all phase I at atmospheric pressure?), though apparently that of $AsCl_3$ is not known. However the mere coincidence of T_m^*, coupled with chemical similarity, is not sufficient evidence to conclude that the phase behavior is similar. Follow their arguments for AsI_3 in this regard. What modification of the law of corresponding states do they suggest might lead to better agreement in reduced phase diagrams?(final paragraph).

65. "Single-Crystal X-Ray Diffraction Study of β-Fluorine" T.H. Jordan, W.E. Streib, and W.N. Lipscomb 41, 760 (1964).

A solid-solid phase transition exists for fluorine at 45.55°K. The β phase is stable from this temperature to the melting point of 53.54°K. Compare this last temperature to the more recently determined triple point(β-L-V), on p. 39 herein. From the fluorine vapor pressure equation given on that same page what would be the pressure inside sealed bulbs that were filled with F_2(probably at room temperature?) and then stored in liquid nitrogen? In addition to the yellow liquid observed at 77°K and the pale yellow crytals grown at a lower temperature (obtained how?) there appeared a small amount of white frost. To what was this probably due?

From the x-ray data various trial structures were assumed. In the third paragraph of p. 761 the process of "research" is characterized quite frankly. The sharp humor of the senior author was apparent on another occasion in 1962 at Washington State University while speaking on structures of boron hydrides, at a molecular structure conference. The preceding speaker had discussed NMR spectra of phosphorous compounds and, probably because of his affiliation at the time, had remarked two or three times that millions of dollars worth of these compounds were being produced.* Our author's opening remark, which can be fully appreciated with a knowledge of the government's truly massive program to take advantage of the higher heating value of the boron hydrides and thus incorporate them into jet fuels, was that "Recently millions of dollars worth of contracts for boron compounds have been canceled!".

The structure deemed most reasonable is shown in Fig. 1. The cubic unit cell has a lattice parameter a of 6.67±0.07Å. The molecules that are not part of the body-centered lattice have for the coordinates of their molecular centers, in units of a, as (x,y,z):

(1/4,1/2,0), (3/4,1/2,0), (0,1/4,1/2), (0,3/4,1/2), (1/2,0,1/4), and (1/2,0,3/4).

Check and see if these coordinates are those of the "rotating" diatomics in Fig. 1. Do you see how the coordinates of the body-centered molecular centers would be written? How many molecules are there per unit cell? Compute the crystal density to compare i) with their liquid density, extrapolated to the

* annually?

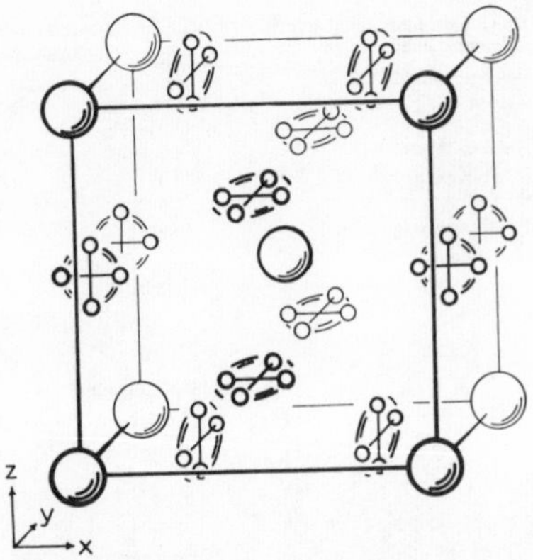

FIG. 1. Structure of β-F_2. The balls forming the body-centered part of the unit cell represent freely rotating or nearly isotropically disordered F_2 molecules. Each of these F_2's has 12 nearest neighbors at 3.7 Å. The disorder (24*k* positions[13]) of the other F_2's is mainly constricted to a plane as shown. Each of these F_2's has two neighbors at 3.3 Å, four at 3.7 Å, and eight at 4.1 Å. All distances are between molecular centers, and can be compared with van der Waals contacts of 2.7 Å if two F_2's have their molecular axes perpendicular to the line of contact or 4.1 Å if their axes are along the line of contact.

freezing point, of 1.67 g/cm^3, and ii) with the more precise density of liquid fluorine at or near this temperature, determined from exercise no. 20. Is the slight difference about what you would expect for a liquid→solid transition?

Verify the statements made in the caption of Fig. 1 about distances of neighbors, both the numbers and the distances between molecular centers. Sketch two fluorine molecules in the alignments mentioned showing the van der Waals contacts in each case. Use the internuclear distance of the free diatomic, 1.42 Å, to make your sketches to scale.

Why are the terms "nearly freely rotating" and "nearly isotropically disordered" used interchangeably?

66. "Magnetic Characteristics of Lanthanide-Silver Compounds Having the CsCl Structure" R.E. Walline and W.E. Wallace <u>41</u>, 3285 (1964).

The magnetic characteristics of 11 intermetallic binary compounds of the formula LnAg were studied, between temperatures of 2° and 300°K. Ln stands for those members of the lanthanide series listed in Table II, including yttrium. Stoichiometric proportions of the two components were fused together in a magnesia crucible by induction heating in an argon atmosphere. Powder x-ray digrams showed all compounds crystallized in the CsCl structure. The magnetic measurements consisted of measuring magnetization or susceptibility as a function of temperature at constant applied field.

In this exercise we will examine several of the columns of Table II, computing magnetic moments for the free tripositive ions, determining the effective magnetic moments for two or three compounds from plots of inverse susceptibility vs temperature, and check a few of the values obtained for the Weiss constants θ and the Néel temperatures.

TABLE II. Magnetic characteristics of the LnAg compounds.

Compound	Neel temp. (°K)	Θ[a] (°K)	Maximum temperature of Curie–Weiss behavior (°K)	Effective moment (μ_B/formula unit) μ_{eff}	Free Ln^{3+} ion
YAg	Pauli paramagnetism[b]			...	
CeAg	Ferro.		Temperature dependent		...
PrAg	Ferro.	2	165[c]	3.44	3.58
NdAg	22	−3	170[c]	3.64	3.62
SmAg	...		Not determined		...
GdAg	138	−84	291	8.57	7.94
TbAg	106	−11	297	9.40	9.72
DyAg	55	−23	297	10.58	10.60
HoAg	32	−4	281	9.93	10.60
ErAg	15		Temperature dependent		...
TmAg	10	−4	284	7.53	7.60

[a] Weiss constant.
[b] A slight temperature-dependent paramagnetism was observed for YAg. This paramagnetism yielded an effective moment of less than 10^{-3} μB/formula un It is thought that this small moment is due to a slight amount of impurity and that pure YAg is a Pauli paramagnet.
[c] Above this temperature, κ is too small for accurate measurement and the data are not used to obtain the effective moment. In the other cases, the temperatu listed is the maximum temperature at which data were taken.

The units of the magnetic moments in the last two columns are Bohr magnetons(hence the subscript B) equal to $eh/4\pi mc>0$, where e and m are the magnitude of the charge and the mass of an electron, h is Planck's constant, and c the velocity of light keeps the units in the gaussian system. This cgs system has electrostatic quantities in esu and magnetic quantities in emu, leading to the value of 0.927×10^{-20} erg/gauss for the Bohr magneton. The magnetic moment of the free ion will be the total angular momentum in units of $h/2\pi$, $[J(J+1)]^{\frac{1}{2}}$, times the Landé g factor

$$g = 1 + \frac{J(J+1) - L(L+1) + S(S+1)}{2J(J+1)}$$

This factor is present because the magnetic moment contribution due to electron spin is twice what one would expect from its spin angular momentum on classical grounds. See J.C. Slater, "Quantum Theory of Atomic Structure" Vol. I (McGraw-Hill Book Co., Inc. New York 1960) pp. 250-251. Derive the expression for the Bohr magneton, showing that it needs only to be multiplied by the <u>magnitude</u> of the angular momentum $g[J(J+1)]^{\frac{1}{2}}$ to obtain the magnetic moment, from the law of Biot and Savart that the magnetic moment is the product of the current in a ring and the ring area. Use an electron in a circular path for ease of derivation, though with Kepler's second law one could show that the path need not be circular. Show that for singlet atomic or ionic states, when the anomalous magnetic moment due to spin can be ignored, g takes the value 1 as expected.

In order to perform the calculation of the free ion moments you will need terms for the ground states of Ln^{3+} with the exception of yttrium which forms a Pauli paramagnet(paramagnetism due to ?). The ground state terms given below are taken from B.G. Wybourne, "Spectroscopic Properties of Rare Earths" (Interscience Pulishers, John Wiley & Sons, Inc. New York 1965) p. 3. Notice that this article precedes the book in time; the latter is a very convenient source. The terms are Ce^{3+} $4f(^2F_{5/2})$, Pr^{3+} $4f^2(^3H_4)$, Nd^{3+} $4f^3(^4I_{9/2})$, Pm^{3+} $4f^4(^5I_4)$, Sm^{3+} $4f^5(^6H_{5/2})$, Eu^{3+} $4f^6(^7F_0)$, Gd^{3+} $4f^7(^8S_{7/2})$, Tb^{3+} $4f^8(^7F_6)$, Dy^{3+} $4f^9(^6H_{15/2})$, Ho^{3+} $4f^{10}(^5I_8)$, Er^{3+} $4f^{11}(^4I_{15/2})$, Tm^{3+} $4f^{12}(^3H_6)$, Yb^{3+} $4f^{13}(^2F_{7/2})$, and Lu^{3+} $4f^{14}(^1S_0)$. How many of these terms could you have predicted from your knowledge of atomic structure? Are the multiplets regular or inverted when the f shell is less or more than half filled? What about when exactly half filled?

From the linear portions of the inverse susceptibility vs temperature plots in Fig. 5 and 6 determine the observed values of μ_{eff} for comparison with Table II. The Weiss constant θ is the value of the temperature for which the paramagnetic susceptibility χ goes to infinity. Determine θ values for the three compounds of Fig. 5 and 6. The Néel temperature, or antiferromagnetic transition temperature, is marked by a maximum in the

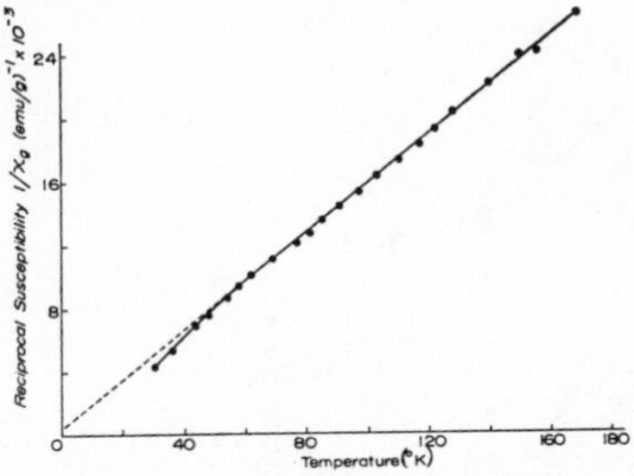

FIG. 5. Reciprocal-susceptibility–temperature curve for NdAg.

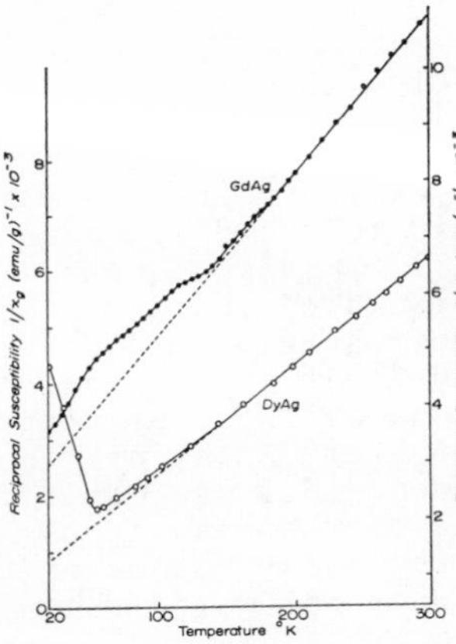

FIG. 6. Reciprocal-susceptibility–temperature curves for DyAg (left scale) and GdAg (right scale).

χ vs T plot. Show that the Néel temperatures for DyAg and GdAg from the χ-T plots of Fig. 3&4 are strikingly and barely apparent respectively.

From Fig. 3 the low-temperature upturn which occurs below 20°K for DyAg is also found for the Nd, Tb, Ho, and Er silver compounds. "The origin of this upturn is not known at this time."

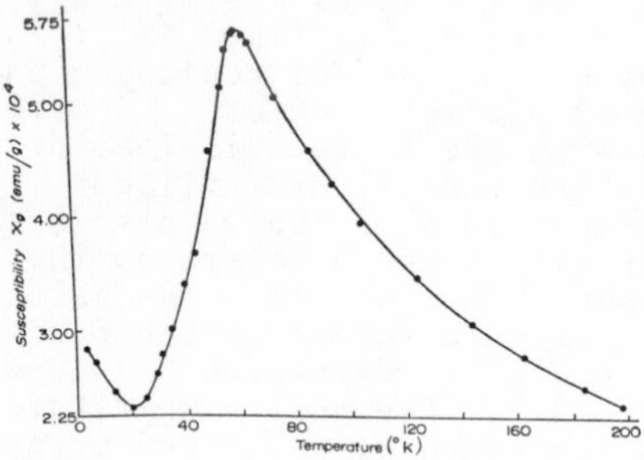

FIG. 3. Susceptibility–temperature curve for DyAg at 6.48 kOe.

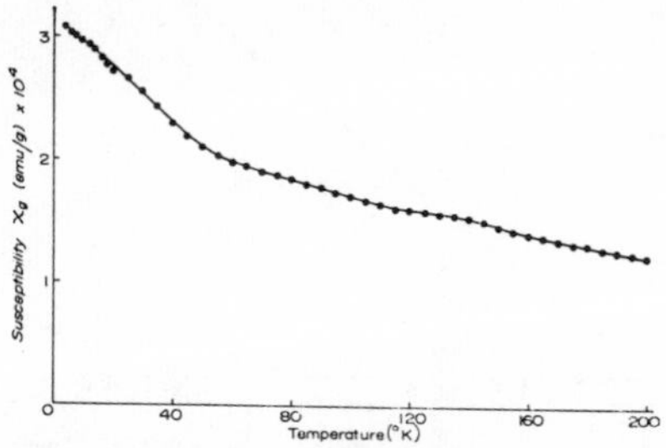

FIG. 4. Susceptibility–temperature curve for GdAg.

67. "Crystal Structure of $Li_6BeF_4ZrF_8$" D.R. Sears and J.H. Burns <u>41</u>, 3478 (1964).

In an investigation of the phase diagram of the ternary system LiF-BeF_2-ZrF_4 their ref. 6 discovered a primary phase of composition $6LiF \cdot BeF_2 \cdot ZrF_4$. Subsequently a single crystal of this composition was isolated and x-ray methods established that the unit cell was tetragonal, with a=6.57±0.02 and c= 18.62±0.06 Å. Study Fig. 2 carefully and convince yourself that there are four formula weights per unit cell. Show that the x-ray density is 3.06 g/cm^3. The original determination of four formula weights/unit cell was not trivial; an important clue was that an estimate of the density as 3.2 g/cm^3 was obtained by assuming that the molar volume of the compound was equal to the weighted sum of the molar volumes of the component salts.

Is the coexistence of two discrete complex anions, such as the octafluorozirconate and tetrafluoroberyllate, common in crystals? Do you know of any other cases?

Fig. 2 shows a perspective view of one unit cell and selected atoms of adjacent cells. The fluorine atoms surrounding one zirconium and one beryllium atom are shaded in order to emphasize their dodecahedral and tetrahedral coordinations. Lithium ions share fluorine ligands with both kinds of polyhedra in such a manner as to achieve a set of fluorine neighbors. Do

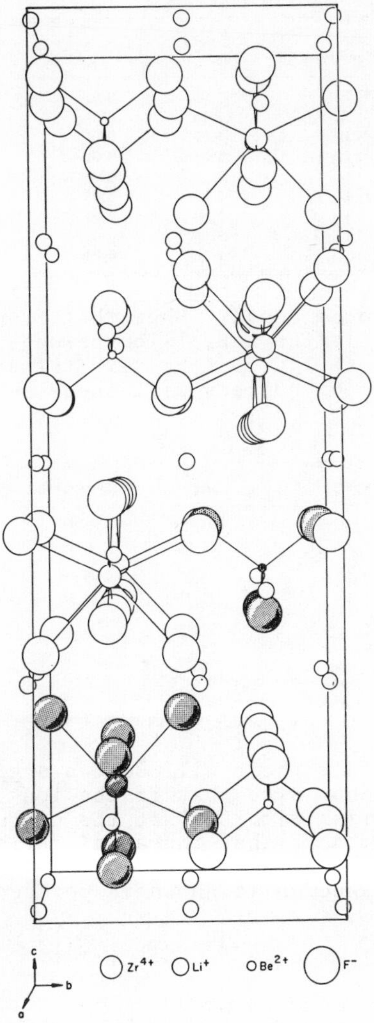

FIG. 2. Perspective view of unit cell.

you spot the lithium ion in the exact center of the perspective view of the cell? Notice that it shadows one directly behind it. How many nearest, or quasi-nearest, neighboring fluoride ions does a lithium ion have? Are there three different types of fluoride ions in this "coordination sphere"? Check your conclusions with one or two other lithium ions shown. Can then you see from Fig. 2 why the authors describe the coordination of fluorines about lithium as "distorted octahedral"? Thinking about the octafluorozirconate and tetrafluoroberyllate ions is it obvious that the form of the formula used in the title is the correct one, rather than the one expressed in terms of the three parent salts $6LiF \cdot BeF_2 \cdot ZrF_4$?

From Table III do the bond lengths of the $BeF_4{}^{2-}$ tetra-

TABLE III. Bond distances.

Bond type	Length (Å)	Standard error (Å)
(4) $Zr-F_I$	2.05	0.01
(4) $Zr-F_{II}$	2.16	0.01
(4) $Be-F_{III}$	1.57	0.01
(2) Li_I-F_{II}	1.83	0.01
(2) Li_I-F_{III}	2.10	0.03
(2) Li_I-F_I	2.32	0.03
(2) $Li_{II}-F_I$	1.89	0.03
(2) $Li_{II}-F_{III}$	2.08	0.02
(2) $Li_{II}-F_{II}$	2.16	0.03

hedral ion all seem to be equal? Show that from the F-Be-F angles given in Table IV the BeF_4^{2-} ion belongs to the point group D_{2d} rather than to T_d. Of the six "tetrahedral" angles centered on the beryllium nucleus will there be three of each

TABLE IV. Bond angles.

Type	Angle	Standard error
F_I-Zr-F_I	99.9°	0.2°
F_I-Zr-F_I $(2\theta_B)$	131.0°	0.5°
$F_{II}-Zr-F_{II}$ $(2\theta_A)$	86.0°	0.6°
$F_{II}-Zr-F_{II}$	122.3°	0.3°
F_I-Zr-F_{II}	71.5°	0.4°
F_I-Zr-F_{II}	72.3°	0.2°
$F_{III}-Be-F_{III}$	107.9°	0.7°
$F_{III}-Be-F_{III}$	110.3°	0.4°

value? Show that in fact there will be two angles of 107.9° and four of 110.3°. Show also that if the ion is placed inside a square tetragonal prism with Be^{2+} at the center and the four F^- ions at opposite corners that the axial ratio of the prism is c/a=1.028.

The ZrF_8^{4-} dodecahedron is shown in perspective in Fig. 3.

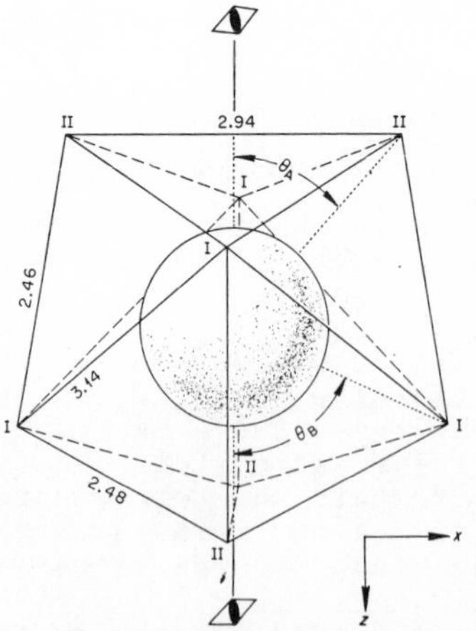

FIG. 3. Perspective view of ZrF_8^{4-} dodecahedron.

Can you determine the point group to which it belongs?

The authors conclude that the tightly coordinated BeF_4^{2-} tetrahedra and the preference of zirconium for the d^4sp^3 dodecahedral coordination seem to be the structure-determining factors for the crystal. They attribute three functions to the lithium ions; which of these create significant distortions of the ideal ZrF_8^{4-} configuration? What do they cite as evidence for their tentative conclusion that the lithium ions do not have additional stereochemical significance?

68. "Bis (3-phenyl-2,4-pentanedionato) Copper. I. Molecular and Crystal Structure" J.W. Carmichael, Jr., L.K. Steinrauf, and R.L. Belford 43, 3959 (1965).

Can this compound pictured here also be named, in increas-

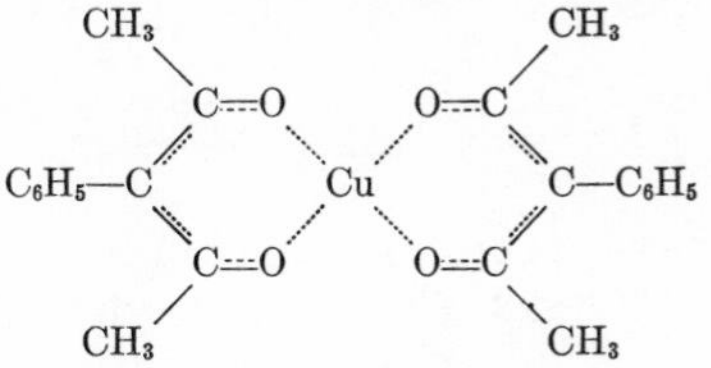

ly abbreviated forms, as copper bis(3-phenylacetylacetonate) and Cu 3ϕacac$_2$? Compute the weight percent of carbon and hydrogen in the compound and compare with their theoretical values of 63.83 and 5.35%.

Fig. 1 shows bond lengths and angles for all of the atoms save the hydrogens. Were you aware that such accuracy was

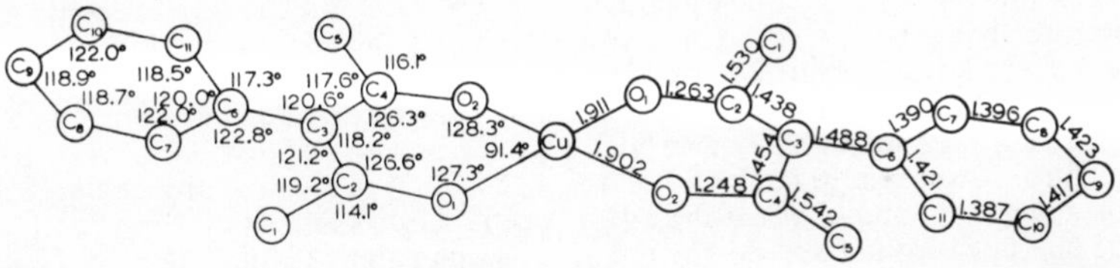

FIG. 1. Bond lengths and angles. Table III contains the listing of the estimated standard deviation for the bond lengths, Table IV those for the bond angles. Distances are in angstroms, angles in degrees.

obtainable with x-ray methods? Derive a general result for the sum of the interior angles in a planar polygon of n sides(very simple). Test your result on the angles shown in the left-hand phenyl ring. Should the residual of 0.1° be "distributed" as in a closed surveying traverse? Would this make much sense when the estimated standard deviation is the order of 1° for each angle? Would you say that a residual error had already been distributed? Table V gives angles between various mean planes in the molecule; the three angles α_1, α_2 and α_3 are shown in Fig. 2 as is the tilt of the phenyl rings from the Cu, O_1, O_2 planes. Are the six-membered rings which include the copper ion planar? Close? Do you see from what perspective Fig. 2 is drawn? The presence of the phenyl ring is thought to have little effect on the bonding in the chelate ring because of its 70° tilt. Show

TABLE V. Angles between various mean planes in bis-(3-phenyl-2,4-pentanedionato) copper. See Fig. 1 for labeling of atoms and Fig. 2 for picture.

Plane 1	Plane 2	Angle
Cu, O_1, O_2	O_1, O_2, C_3, C_4	10°24′ (α_1)
O_1, O_2, C_2,	O_2, C_3, C_4, C_6	3°29′ (α_2)
C_2, C_3, C_4, C_6	Line between C_6 and C_9	2° (α_3)
Cu, O_1, O_2	Phenyl ring	70°

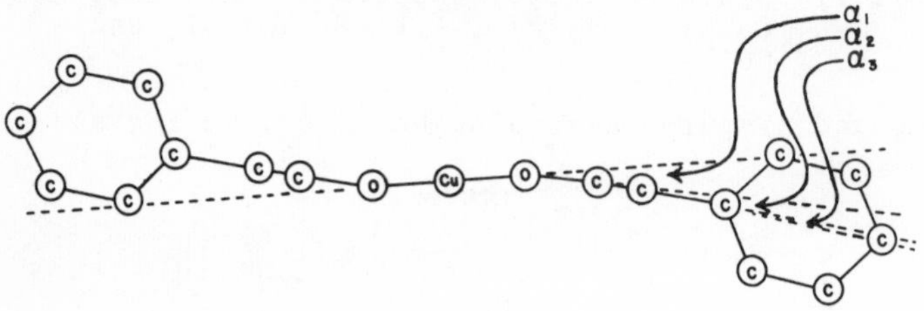

FIG. 2. The molecule is shown in projection along the bisector of the external O–Cu–O angle.

that the resonance interaction is reduced to 34 or 11% of that for coplanar rings, depending on whether one assumes a cos τ or $\cos^2 \tau$ interaction.

Show that the various angles between planes are compatible with the result that the molecule has a center of symmetry(point group ?).

Fig. 3-5 show projections of molecular skeletons onto the three principal planes [001], [010], and [100]. Is it obvious in which system the compound crystallizes? Is it also obvious how many molecules are associated with the unit cell? From the information given in the Figures you should be able to compute the x-ray density to be compared with the value of 1.442 g/cm^3 determined by flotation in KI solution; again a proof of the number of molecules/unit cell.

Let us close this exercise with a question and an activity. The question is why are two of the three projections along axes onto planes nonorthogonal? The activity is to connect each pair of atoms which are interchanged upon the symmetry operation of inversion through the center by a straight line. Must this line pass through the center? There are five perspective views in

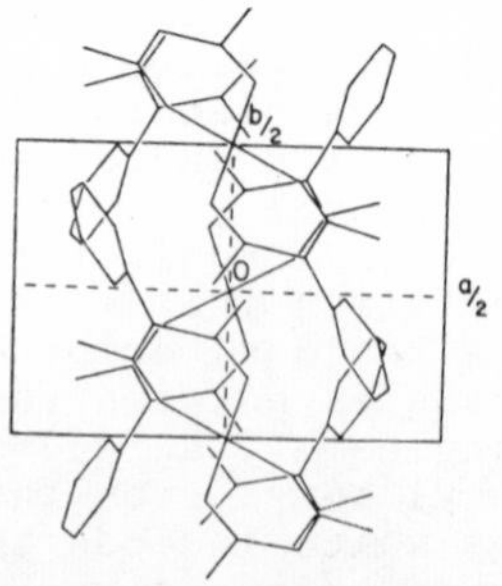

FIG. 3. Projection of molecular skeletons along [001]. This is a nonorthogonal projection along the c axis onto the ab plane. This face of the unit cell is 10.250 by 6.778 Å.

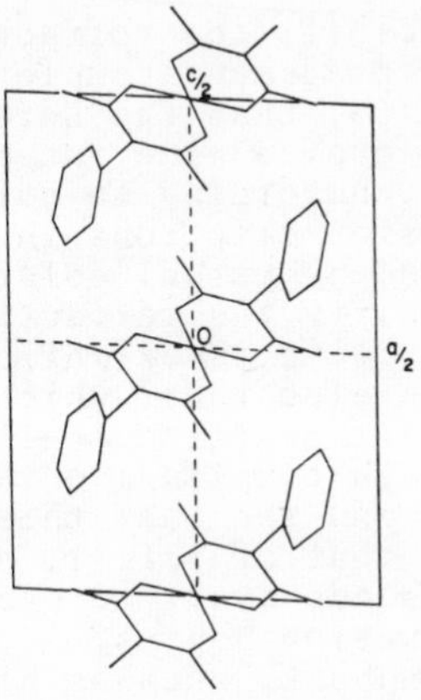

FIG. 4. Projection of molecular skeletons along [010]. This is an orthogonal projection along the *b* axis onto the *ac* plane. This face of the unit cell is 10.250 by 13.763 Å, $\beta=93°33'$.

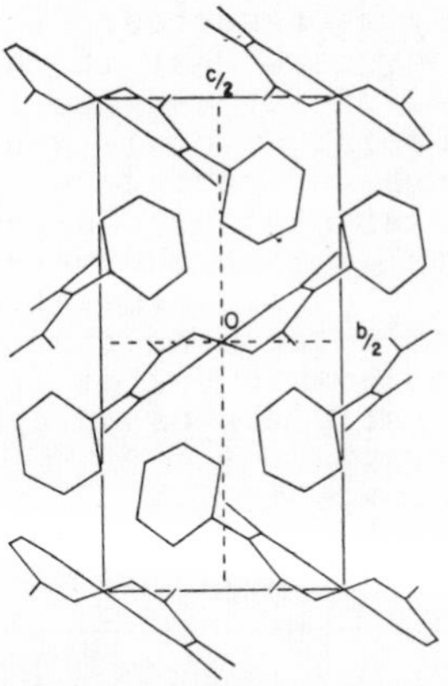

FIG. 5. Projection of molecular skeletons along [100]. This is a nonorthogonal projection along the *a* axis onto the *bc* plane. This face of the unit cell is 6.778 by 13.763 Å.

Fig. 3-5. With a sharp pencil connect as many pairs of symmetrically equivalent atoms as is possible. Should a pair of these atoms lie equidistant from and collinear with the Cu atom no matter what the perspective?

69. "Density and the Metallic State" K.F. Herzfeld 44, 429 (1966).

His ref. 1 had shown that thallous iodide became metallically conducting at 160 kbar, and ref. 2 from the same laboratory three years later had determined the volume at 160 kbar to be 0.672 that at low pressure. The author's 1927 paper(ref. 3) discussed the fact that the Clausius-Mosotti expression for the refractivity, $(n^2-1)/(n^2+2)$, is necessarily smaller than unity for a dielectric, where n is the refractive index for long waves.

According to Clausius-Mosotti the refractivity is proportional to the density. Is the statement completely equivalent to that on p. 131, exercise no. 63, that the Lorentz-Lorenz function L remains nearly constant over a wide range of temperature and pressure? This implies, according to our author, that no other coupling between dispersion electrons in different particles exists than that provided by mutual polarization. Accordingly at sufficiently high density the refractivity should then become unity and the refractive index infinite. This then would mean that the dispersion electrons, which before had been quasi-elastically bound to their own(classically) atoms or ions, have then been set free through coupling so that the solid becomes a metallic conductor. Do you see that this discussion of Herzfeld's is equivalent to that of ref. 1, according to which the valence and conduction bands start to overlap at the pressure for which metallic conduction begins?

From a standard handbook, such as his ref. 4, show that the refractive index n at normal pressures is about 2.78. Check the value of the corresponding refractivity given as 0.692. Why does he say then that the metallic state should begin at a volume ratio of 0.692 rather than 0.672 as observed? Is the dependency on density actually as asserted? To perhaps further improve the agreement he suggests that if the long-wave value for n were 2.67 rather than 2.78 for visible light the result would be exact. Check the numbers. Overall would you say the agreement is still close enough to convince us of the validity of the argument? From ref. 2 what pressure corresponds to 0.692(TlI)?

His footnote 5 warns us that the results are strictly only applicable to isotropic or cubic material. Examination of ref. 2 discourages us at first because we find that thallous iodide crystallizes in the orthorhombic system at atmospheric pressure. What happens at 5 kbar that picks us up again?

70. "Radial Distribution Function for Solid Gallium" A.F. Berndt and R. Hight <u>46</u>, 394 (1967).

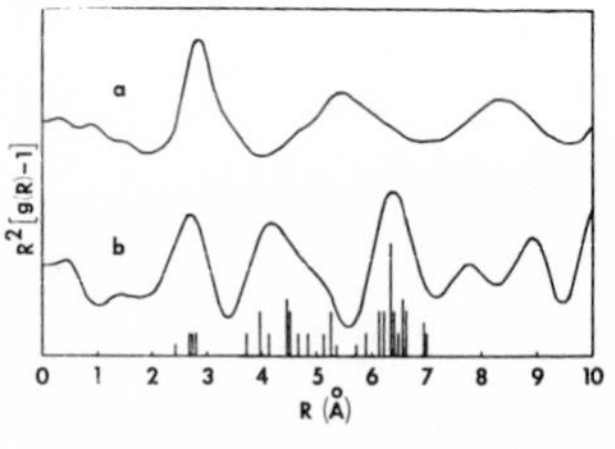

FIG. 1. Excess differential radial distribution functions for gallium. (a) Supercooled liquid gallium at 20.0°C[1], (b) stable solid gallium at 25°C.

This exercise could have been placed in the next chapter as well as not. Fig. 1 shows the excess radial distribution function $R^2[g(R)-1]$ for supercooled liquid gallium at 20.0°C (ref. 1) and for stable solid gallium at 25°C. The vertical lines represent the interatomic distances calculated from the known structure of solid gallium; their heights are proportional to the number of neighbors at distance R from a given atom. Are there no neighbors further than 7 Å away, or does the indication of neighbor density merely stop there?

The radial distribution function g(R) is defined by the statement that in a coordinate system with origin on a given particle the number of particles observed in the volume element $d\tau$ is $\rho g(R)d\tau=4\pi\rho R^2 g(R)dR$, with ρ the bulk density. Do you see that $g(R)\to 0$ as $R\to 0$ because all particles exhibit very high mutual repulsions at sufficiently small distances? How must g(R) behave as $R\to\infty$(or maybe 10-20 Å?)? Why? Notice that curves a and b of Fig. 1 are displaced vertically; probably for clarity? for any other reason? Do these excess curves behave as discussed above? The authors tell us that the appearance of detail on the curves below 1.5 Å suggests Fourier transform errors; would you physically expect any scattering centers to be present closer than the nearest neighbor? For the same reason they say that we should probably not attach significance to the fine structure beyond about 7 Å.

In solid gallium each atom has seven neighbors. One of these is extremely close at 2.44 Å, the others being in three sets of two each, at distances between 2.71 and 2.80 Å. Does this suggest dimerization in the solid? Explain in terms of these distances just given. This dimer in the solid and liquid phases is also discussed in ref. 1, and ref. 15 of that article refers us back to some 1947 work by a very famous chemist on this problem. Do the vertical lines in Fig. 1 reflect these facts for the solid at distances <3 Å?

The peaks are observed for the solid 2.70, 4.20, and 6.36 Å corresponding to the first three coordination shells. Examine critically the author's statement, with the aid of Fig. 1, that except for the first peak which is at a slightly smaller value of R in the solid than in the liquid, there is no similarity between the excess differential radial distribution curves for solid and liquid gallium. Do you see why they conclude that in the liquid there are no agglomerates of atoms, past possibly first neighbors, which retain the configuration observed in the solid?

What was the purpose of determining the distribution function for supercooled liquid gallium? You will want to look at ref. 1. Do the data in the caption of Fig. 1 allow you to bracket the melting point of gallium? Ref. 1 contains a description of the x-ray methods used.

71. "Permeation, Diffusion, and Solubility of Deuterium in Pyrex Glass" H.M. Laska, R.H. Doremus, and P.J. Jorgensen 50, 135 (1969).

We will consider the two rate processes, permeation and diffusion, in order to follow the authors' argument leading to support for earlier suggestions(ref. 4 and 5) that Pyrex glass is separated into two continous, amorphous phases on a very fine scale. One phase is almost pure silica and the other is sodium borosilicate.

A thin-walled Pyrex bulb was blown on the end of a small tube connected to the gas-handling system. For permeation deuterium gas was admitted to the inside of the bulb and the flux through it followed in time. The rate of permeation was determined at steady state, and the diffusion coefficient was calculated as in ref. 1. In both cases the volume exterior to the bulb was evacuated and the flux from the bulb was led to a mass spectrometer for analysis. How would one determine the total deuterium flux rates from the mass peak three and four heights? The peak heights were known to be proportional to the flux through calibrations.

They express the units of permeation as molecules/(cm·sec) for a 1-atm pressure difference across the bulb wall. Show that the full units of the permeation P are molecules/(sec·atm·cm). This follows when we write the expression for the total molecular flow rate as

$$\dot{n} = PA(dp/dx) \simeq PA\Delta p/\text{thickness}$$

where n is the number of molecules in the bulb, A is the area of the cross-section presented to flow, and dp/dx is the pressure gradient in the direction of flow. Notice that both sides of the equation are negative. These units of P become theirs if Δp is scaled to 1 atm and the P values adjusted accordingly. They are shown in Fig. 1, together with the diffusion coefficients D, both as functions of temperature. Rationalize the

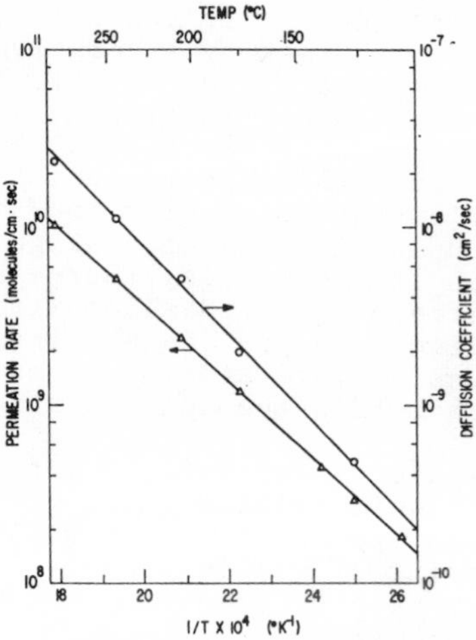

FIG. 1. The temperature dependence of diffusion (○) and permeation (△) of deuterium in Pyrex glass.

units of D, cm^2/sec, in the same manner. The equation is

$$\dot{n} = DA(dC/dx) \simeq DA\Delta C/\text{thickness}$$

with C the concentration in molecules/cm^3. Let us note quickly that these forms of the flow equations with the fluxes linear in the gradients are not the most general(see their ref. 8) but are nonetheless useful as plausibility arguments as far as the units employed are concerned.

Show that the units of P/D are molecules/cm^3. This ratio is in fact shown to be the solubility of the gas(D_2) in the glass. Show that at 200°C the solubility of deuterium in Pyrex is 5.2×10^{18} molecules/cm^3 of glass or S=0.019 cm^3(STP)/cm^3 for a gas pressure of 1 atm. The quotient P/D at 200°C is very close to that for D_2 in fused silica as Table I shows. The

TABLE I. Absolute values and activation energies for permeation and diffusion of gases in fused silica and Pyrex.

Glass	Gas	Permeation: P at 200°C (molecules/cm·sec)	Permeation: Q (kcal/mole)	Diffusion: D at 200°C (cm^2/sec)	Diffusion: Q (kcal/mole)	Ref.
Pyrex	D_2	$2.0(10)^9$	9.9	$4.1(10)^{-9}$	11.2	Present work
Fused silica	D_2	$2.5(10)^9$	9.4	$6.5(10)^{-9}$	10.5	7
Pyrex	He	$7.1(10)^{10}$	6.2	$4.7(10)^{-7}$	6.5	6
Pyrex	He	$1.2(10)^{11}$	6.5	...	5.8	Present work
Fused silica	He	$3.3(10)^{11}$	5.2	$8.0(10)^{-7}$	6.1	8

temperature dependence of solubility is examined by forming the ratio of the gas concentration in the glass phase C_i to the concentration in the gas phase C_g. This ratio is called the coefficient of solubility. Since S is determined at 1 atm, $C_i/C_g = TS/273$. Show that the coefficient of solubility is 0.033 cm^3/cm^3 for deuterium in Pyrex at 200°C, and is nearly constant in the range 100-300°C. These and other values are collected in Table II.

The constancy of solubilities with temperature shows that there is no interaction between the gas and the glass, and the agreement (better for D_2 than for He?) of solubilities of gases in fused silica and Pyrex "shows that the volume available for gas solution in these two glasses is about the same. This latter agreement is also consistent with the two-phase structure of Pyrex glass, and indicates that the silica phase is mainly responsible for the deuterium solubility in this glass."

TABLE II. Solubility of gases in Pyrex glass.

Temp. (°C)	Gas	Solubility cm^3 (STP)/ $cm^3 \times 10^3$	Solubility C_i/C_g	Ref.
23	He	5.2	0.006	12
25	He	5.5	0.006	Present work
27	He	7.8	0.009	6
200	He	5.6	0.010	6
200	He	9.6	0.017	Present work
515	He	8.4	0.014	11
100–300	D_2	19.0	0.033	Present work
465	D_2	4.4	0.012	10

72. "Interpretation of Diamond and Graphite Compressibility Data Using Molecular Force Constants" J.R. Riter, Jr. 52, 5008 (1970).

The experimental compressibility data of ref. 1 and 2 for diamond and graphite are analyzed in terms of molecular force constants, basically from ethane, benzene, and ethylene. The basic calculation was that of the quotient of potential energy with the molecular force constants and the macroscopic work of compression from $\int PdV$ using the experimental P-V relation; the quotient was evaluated at decreasing ratios of V/V_0, with V the volume at pressure P. Closeness of the quotient to 1.0 was taken as an indication that the potential energy model was at least approximately correct.

Diamond was chosen to test the idea of transferable force constants; the C-C stretching force constant in ethane was used to compute the potential energy of compression as a function of the volume ratio $\beta \equiv V/V_0$. Then three models of graphite deformation were used; in-plane ring compression, and out-of-plane buckling(boat form) and puckering(chair form).

The experimental data were cast into the form of the Murnaghan equation[ref. 2 and Proc. Nat. Acad. Sci. 30, 244 (1944)]

$$P = (B_0/B_0')[(V_0/V)^{B_0'} - 1]$$

with B_0 and B_0' the isothermal bulk modulus$(-V(\partial P/\partial V)_T)$ and its pressure derivative, both evaluated at 0-1 atmosphere. This equation follows from the assumption that the Maclaurin expansion for B can be terminated after the linear term in pressure:

$$B = -V(dP/dV) = B_0 + B_0'P, \text{ with } dT=0.$$

Show that this is the truncated Maclaurin expansion and that integration(between what limits?) does lead to the Murnaghan equation. The values of B_0 and B_0' can approximately be determined by fitting the experimental P-V data for solids to this equation, for anisotropic solids like graphite these values become just curve-fitting parameters.

Show that the work of compression, in joules/cm³ units used throughout, for diamond with B_0=5600 kbar and B_0'=4.0 is $= -\int PdV/V_0 = W(\beta) = 1.40\text{x}10^5[1/(3\beta^3)+\beta-4/3]$. This equation as it appears in the paper has suffered a transposition although the values of $W(\beta)$ in Table I are alleged to be correct. You should check one or two numerically.

TABLE I. Compressibility data,[a] work of compression,[b] and potential energy of bond compression[b] for diamond.

P(kbar)	$a/a_0=\beta^{1/3}$	$V/V_0=\beta$	$W(\beta)$	$V_{bc}(\beta)$	$V_{bc}(\beta)/W(\beta)$
0	1.0000	1.0000	0	0	...
87	0.9950	0.9851	64.4[d]	47.5[d]	0.74
180	0.9900	0.9703	260.	190.	0.73
276	0.9850	0.9557	594.	428.	0.72
(320)[c]	0.9830	0.9499	767.	549.	0.72

[a] From p. 214 of Ref. 2.
[b] From equations in the text.
[c] Computed without calibration.
[d] Units of joules per cubic centimeter.

The potential energy is given for 1 cm³ as the product of three factors

$$V_{bc}(\beta) = \rho N_0/12 \cdot \tfrac{1}{2}k_{str}(r-r_0)^2 \cdot 4/2$$

where bc stands for bond compression, ρ is the density, N_0 is Avogadro's number, k_{str} is the C-C force constant taken as 4.50 mdyn/Å(ref. 5), and r is the bond length with r_0 the equilibrium value of 1.5445 Å(ref. 2). Show that the three factors correspond to the number of atoms per cm³, the energy per bond, and the mean number of bonds per atom. Derive the numerical form $V_{bc}(\beta)=1.90\text{x}10^6(1-\beta^{1/3})^2$ and compute one or two values for comparison with Table I. Is it obvious that a/a_0 is the ratio of lattice parameters at pressure P and at 0-1 atm? The fact that the last column is reasonably close to 1.0 and shows but a slight decrease with pressure gives one confidence in the model. Is this slight function of pressure what you would expect from the neglect of anharmonicity? What about the sign? The referee had chastened the authors of the next exercise for using V for both volume and potential energy; did it probably cause you some confusion here? The remarks on p. 90, exercise no. 47, may remind us all to be careful with our notation.

The procedure for graphite is basically similar though a bit more involved, requiring one to stare at a sketch and think about the Pythagorean theorem for some time. The results are given in Table II. The last three columns may occupy our attention for a minute. The subscripts b and p stand for buckling

TABLE II. Compressibility data,[a] intraplanar work of compression,[b] and potential energies of bond compression, buckling, and puckering[b] for graphite.

P(kbar)	$a/a_0=\alpha$	$W(\alpha)$	$V_{bc}(\alpha)$	$V_b(\alpha)$	$V_p(\alpha)$	$V_{bc}(\alpha)/W(\alpha)$	$V_b(\alpha)/W(\alpha)$	$V_p(\alpha)/W(\alpha)$
0	1.0000	0	0	0	0	...	...	...
5.7	0.9987	1.36[d]	1.85[d]	131.[d]	393.[d]	1.36	96.3	289.
17	0.9977	4.57	5.79	231.	694.	1.27	50.5	152.
30	0.9966	10.7	12.7	342.	1030.	1.19	32.1	96.3
49	0.9954	21.2	23.2	463.	1390.	1.09	21.8	65.4
72	0.9943	35.2	35.6	573.	1720.	1.01	16.3	48.8
102	0.9931	56.6	52.1	693.	2080.	0.92	12.3	36.8
139	0.9919	85.7	71.8	813.	2440.	0.84	9.49	28.5
195	0.9905	132.	98.8	953.	2860.	0.75	7.23	21.7
(287)[c]	0.9890	201.	132.	1100.	3310.	0.66	5.49	16.5
(365)[c]	0.9877	282.	165.	1230.	3700.	0.59	4.37	13.1

[a] From p. 215 of Ref. 2.
[b] From equations in the text.
[c] Computed without calibration.
[d] Units of joules per cubic centimeter.

and puckering. Do you see the basis of the contention that the in-plane bond compressional mode is by far the dominant one? What about the logic here; for a fixed value of $\alpha=a/a_0$ (a is a hexagonal lattice parameter and is taken as the distance between opposite sides of the ring) would the potential energy be much larger than the work of compression if the particular mode is nothing but a minor one? Are we then forcing a minor mode to take up all of the compressional energy?

Finally a little detective work in the literature; there is a 1970 paper in J. Chem. Phys. on the Raman spectrum of graphite. See if the ring-compression force constant given there, which is obviously better than the one taken from benzene, changes $V_{bc}(\alpha)$ or the conclusion significantly.

73. "Deformation Mechanism for Ice VII" D.H. Yean and J.R. Riter, Jr. 54, 000 (1971).

Ice VII is stable at pressures above about 22 kbar(ref. 1 and 2). The oxygen atoms form a body-centered cubic unit cell as shown in Fig. 1 of ref. 3 and reproduced on the next page. The lattice parameter a was determined to be 3.30±0.01 Å at 25 kbar. This is the densest form of ice; compute the density to compare with that of ref. 2, 1.66 g/cm^3. Do you see from Fig. 1 that there will be an appreciable zero-point entropy for ice VII? Notice that each water molecule participates in four hydrogen bonds with oxygen atoms tetrahedrally arranged on the corners of the unit cube. At the other four corners of the unit cube are other oxygen atoms which do not form hydrogen bonds with the molecule in question. Do you see why ref. 2 and 3 describe the lattice as consisting of two interpenetrating but not interconnecting hydrogen-bonded frameworks? Is there enough detail shown in Fig. 1 to let you decide whether one of these networks by itself forms the wurtzite or the zincblende structure?

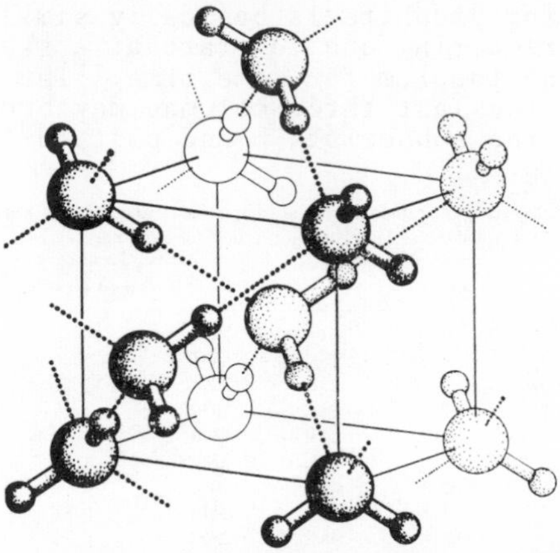

FIG. 1. The structure of ice VII. The unit cell is cubic, with $a=3.30$ Å, and the oxygen atoms are in body-centered arrangement. Hydrogen bonds are formed as shown by the dotted lines. The water molecule orientations are disordered in the unit cell, but there must be short-range order within each hydrogen-bonded framework.

The experimental P-v behavior was determined, in ref. 4, for both H_2O and D_2O ice VII. The data was again cast into the form of the Murnaghan equation(previous exercise) with the difference that $P-P_F$ is the lhs of the equation, where P_F is 22 kbar. It is the excess pressure above 22 kbar that deforms ice VII. The ratio on the rhs V_0/V becomes v_F/v, and the subscript $_0$ on B_0 and B_0' becomes $_F$. For H_2O the B_F and B_F' values are 227 kbar and 5.3. Show that the work of compression per unit volume(measured at 22 kbar) is, with $\beta \equiv v/v_F$,

$$W_{H_2O}(\beta) = 996(-5.3 + 4.3\beta + 1/\beta^{4.3})$$

in joule/cm^3.

The potential energy of compression for the harmonic model is similar to that of the previous exercise, with the differences that i) the correct molecular weight for ice should be in the denominator of the first factor, and ii) the second factor of potential energy per bond(between two oxygen atoms) should be

$$\tfrac{1}{2}[k_1 r_{1F}^2(\beta^{1/3}-1)^2 + k_2 r_{2F}^2(\beta^{1/3}-1)^2]$$

with $k_1=0.241$ mdyn/Å and $k_2=7.66$ mdyn/Å, and $r_{1F}=1.902$ Å and $r_{2F}=0.956$ Å for the hydrogen and chemical bonds respectively. Other than the obvious requirement that the two bond lengths add to give the O···O distance(neglecting subtended angles of 1° and 2°), we impose the condition that $\partial V_H/\partial r_1 = \partial V_H/\partial r_2$ so that the forces on the two types of bonds are equal. This will give us the minimum potential energy for a given β. Use this condition to determine an analytical expression for $V_H(\beta)$ and then derive the numerical form in joule/cm³, $1.058\text{x}10^5(\beta^{1/3}-1)^2$.

Compute one or two pairs of values for W(β) and $V_H(\beta)$ for H_2O and compare with those given in Table I. Do you see that the harmonic model is essentially correct, but that the potential energy drops off at higher pressures? Why is this so much more pronounced than for diamond? Is the hydrogen bond very compressible? What about the volume ratios for ice VII and diamond at comparable pressures? Is a comparison with the graphite compressibilty easy to make? Why or why not?

The Morse potential improves the picture considerably. Can you suggest an explanation to account for the difference in behavior of H_2O and D_2O? Why is it that at the same pressure V_H for the two isotopic ices differ? Is the potential energy model identical?

TABLE I. Work and potential energy of compression for H_2O ice VII.

P (kbar)	β[a]	W = work[b] (J/cm³)	V_H = harmonic[c] potential energy (J/cm³)	V_H/W	V_M = Morse[d] potential energy (J/cm³)	V_M/W
22	1.0000	0	0	···	0	···
25	0.9873	1.873	1.909	1.020	1.939	1.036
50	0.9095	114.2	102.6	0.899	121.0	1.060
100	0.8223	553.3	421.9	0.763	590.4	1.067
150	0.7703	1080.	734.7	0.680	1145.	1.060
165	0.7581	1244.	822.5	0.661	1315.	1.057

[a] From Eq. (1).
[b] From Eq. (2).
[c] From Eq. (4).
[d] From Eq. (7).

TABLE II. Work and potential energy of compression for D_2O ice VII.

P (kbar)	β[a]	W = work[b] (J/cm³)	V_H = harmonic[c] potential energy (J/cm³)	V_H/W	V_M = Morse[d] potential energy (J/cm³)	V_M/W
22	1.0000	0	0	···	0	···
25	0.9880	1.768	1.701	0.962	1.750	0.990
50	0.9114	113.3	98.09	0.866	115.2	1.017
100	0.8206	572.3	430.3	0.752	603.5	1.055
150	0.7648	1138.	773.5	0.680	1220.	1.071
200	0.7252	1739.	1091.	0.628	1874.	1.077_8
220	0.7122	1983.	1210.	0.610	2138.	1.078_3

[a] From Eq. (1).
[b] From Eq. (2).
[c] From Eq. (4).
[d] From Eq. (7).

CHAPTER VI

FLUIDS

74. "The Electrical Conductivity of Supercritical Solutions of Sodium Chloride and Water" J.K. Fogo, S.W. Benson, and C.S. Copeland 22, 212 (1954).

This exercise alone is enough to cause a title change of the present Chapter from "Liquids" to "Fluids". Is the temperature range considered of 375-393°C definitely above the critical temperature for water? Does the page number on which the article begins have any special significance?

The conductivity cell consisted of two concentric Pt-Ir cylinders which formed the electrodes and which were insulated from each other by a sapphire separator. One cylinder formed the outer wall of the cell. The cell was so constructed that internal electrodes of various sizes could be used to give a conductivity cell constant appropriate to the system investigated; about 0.01 cm^{-1} for steam measurements and about 0.09 cm^{-1} for NaCl-steam measurements. The units here really are "reciprocal centimeters" and definitely not "wavenumbers". The statement is that the conductivity cell constant is the product of specific conductance and measured resistance; show that this product has units of cm^{-1}. From plots of specific conductance against temperature for each of the eight experimental densities (0.2-0.4 g/cm^3) isotherms were obtained as shown in Fig. 1 for

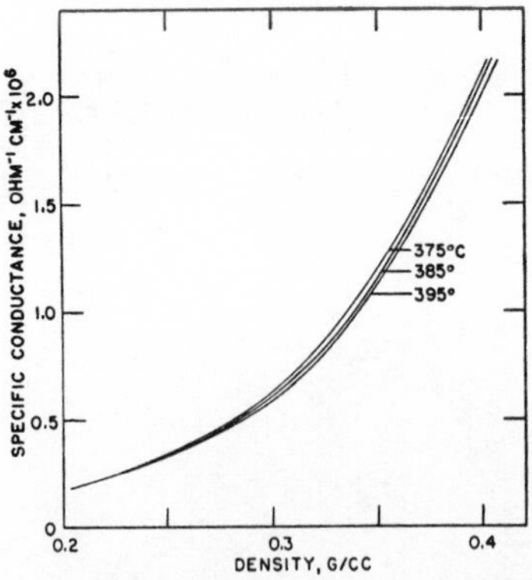

FIG. 1. The specific conductances of the steam used as solvent.

the steam used as solvent. Are the magnitude of the specific conductance shown and that for ordinary distilled water at room temperature comparable? Why or why not is this surprising?

For the solutions of NaCl in steam, a specific conductance of NaCl in the solution, designated by L, was obtained by subtraction of the specific conductance of steam from that of the solution. Is this another example of an effect discussed in Chapter 4 that the overall rate of a process is the sum of the "competing" or parallel rates? In conductivity processes what takes the role of the independent variable time of the rate processes? Does the analogy hold further when we point out that in the case of parallel rate processes the mean life or half life is the harmonic mean of the individual lives, while for resistivities in parallel the total resistivity is the harmonic mean again?

Fig. 2 presents log L vs density for various compositions, all at 388 °C. Recall molality is moles solute/kg solvent.

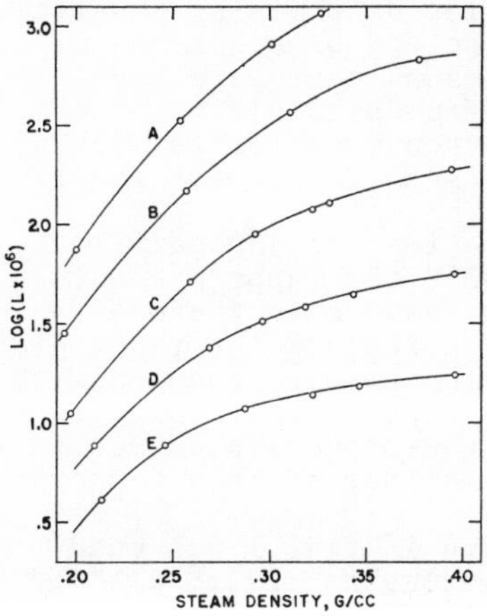

FIG. 2. The specific conductance of sodium chloride at constant composition as a function of the steam density. Molality$\times 10^5$: *A*, 1115; *B*, 246.1; *C*, 44.77; *D*, 11.96; *E*, 3.61.

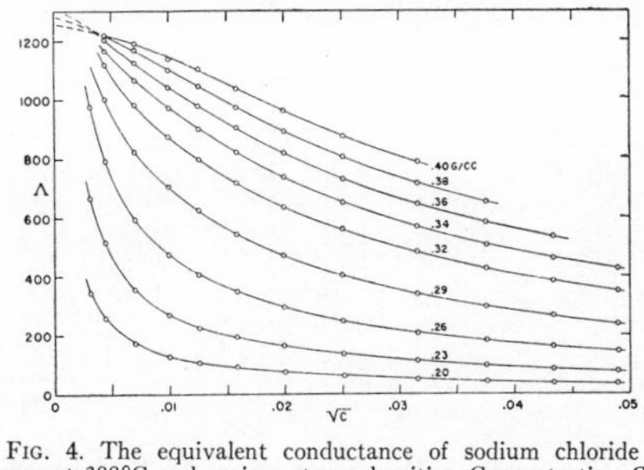

FIG. 4. The equivalent conductance of sodium chloride in steam at 388°C and various steam densities. Concentration *C* in moles NaCl/1000 cc.

At this temperature and for the average steam density what is the ratio of specific conductances for NaCl and steam itself?

Fig. 4 gives us equivalent conductance(specific conductance

divided by concentration) vs (concentration)$^{1/2}$. Show from Fig. 2 and 4 that the units of equivalent conductance Λ are $ohm^{-1}cm^2 \cdot equiv^{-1}$. Do the curves in Fig. 4 indicate that NaCl in steam behaves as a strong or weak electrolyte at these steam densities? Why is it evident that the degree of ionic association increases considerably as the steam density is lowered at constant temperature and constant molarity(moles/1000 cm^3)? What explanation do they offer for the fact that at low concentrations the curves of Fig. 4 cross, so that the limiting($c \to 0$) equivalent conductance of NaCl increases with decreasing steam density?

75. "Some Physical Properties of Pure Liquid Ozone and Ozone-Oxygen Mixtures" A.C. Jenkins and F.S. DiPaolo 25, 296 (1956).

Do you see an immediate correlation between the title of this article and the authors' affiliation at the time? This paper is part of a series determining physical and thermodynamic properties of pure ozone and ozone-oxygen mixtures. Look at ref. 1 for the method used to prepare ozone from oxygen. Have you ever smelled what you were told was ozone while using a Tesla coil to check a vacuum system for leaks? Why do these trace amounts of ozone seem to be more readily generated on a dry wintry day?

The vapor pressure test of the purified ozone is extremely sensitive since at -183°C the vapor pressure of pure ozone is 0.1 mm Hg while that of 99.9 mole % ozone is 6.5 mm Hg(ref. 3). Assume Raoult's law to calculate the vapor pressure of oxygen at this temperature. The answer, coupled with the fact that this temperature is just about the n.b.p. for oxygen, will make you aware that the oxygen-ozone system does not form an ideal liquid solution; at least not at this temperature. See ref. 3 for the phase diagram.

The apparatus shown in Fig. 1 was used to determine the melting point of pure ozone as -192.5±0.4°C. From their description of its operation, or from looking at the Figure, why

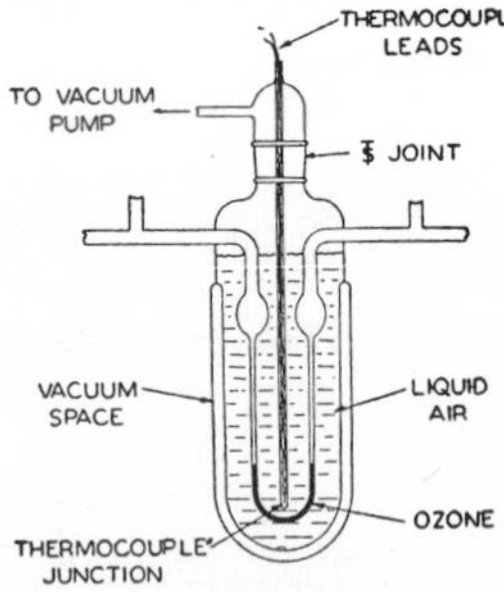

FIG. 1. U-Tube melting point apparatus.

does the helium pressure on one side of the frozen ozone(other side evacuated) drop suddenly upon melting of the ozone? The sudden drop in He pressure is easily followed on a manometer and this event is correlated in time with readings from the thermocouple as the liquid air bath is warmed. The sudden pressure drop occurs because the melted ozone flows into the enlarged portion of the U-tube. Why does it flow at all? Will a solid sustain a shear stress? Will a liquid?

The density of pure liquid ozone was determined with the apparatus of Fig. 5 by the "method of balanced columns" in which

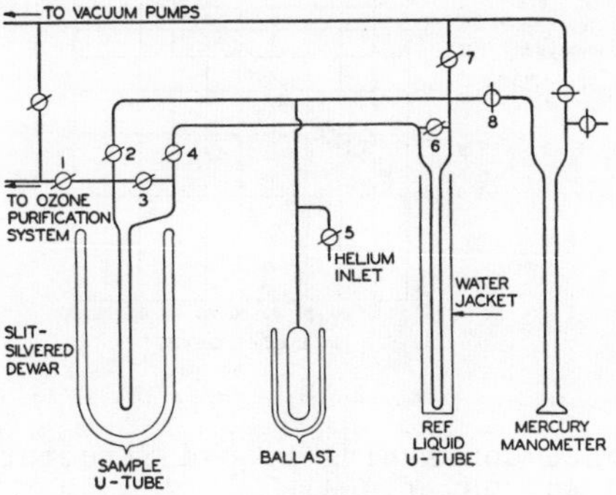

FIG. 5. Liquid density apparatus.

pressure is applied to one arm of the U-tube containing the liquid of unknown density and to one arm of a second U-tube containing a reference of known density. The liquid densities are inversely proportional to the differences in heights of the liquid columns in the two tubes(ref. 11). The first density measurements resulted in what kind of incident? A change in the reference liquid from Fluorolube FS oil to sulfuric acid resulted in two more occurrences of the same type. What did the authors suggest as the probable cause of these explosions? At what point in the experiments did they occur? The density was determined to be 1.5727±0.0004 g/cm^3 at -182.9°C.

The densities of liquid ozone-oxygen mixtures was also determined by the same method and it was found that the density of the mixture d could be accurately represented by the equation $1/d=x_1/d_1+x_2/d_2$ for weight fractions x_i and pure component densities d_i. Show that this implies the volumes are strictly additive and $\Delta V_{mixing}=0$. Does this appear to be at variance with the nonideality of the mixture(solution?) wrt Raoult's Law? Explain.

The viscosity of pure liquid ozone and its mixtures with oxygen was determined with the modified Ostwald viscometer and the results at -183°C are shown in Fig. 7. The data fit the empirical equation

$$\log_{10} \eta = x_1 \log_{10} \eta_1 + x_2 \log_{10} \eta_2$$

within 5%. Here η is the viscosity coefficient of the mixture and η_i that for component i present in mole fraction x_i. Notice that the results of Fig. 7 are interpolated through the two-phase region. It is tempting to try and make something of this equation; thinking of the reciprocal of η as the fluidity or basically the rate constant for viscous flow k you can show that $\ln k=x_1 \ln k_1 + x_2 \ln k_2$, but then what? See their ref. 18.

Compare the viscosities of liquid ozone and oxygen at -183°C with that of liquid water at 25°C.

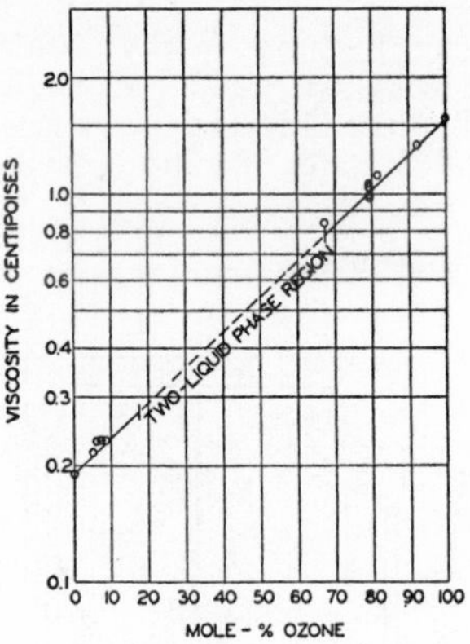

FIG. 7. Viscosities of liquid ozone-oxygen mixtures at −183.0°C.

76. "Further Evidence Concerning Liquid Structure" E.B. Smith and J.H. Hildebrand 40, 909 (1964).

Their ref. 1(1961) in which the senior author Hildebrand argues against the presence in normal liquids of "lattices", "cells", "holes", "vacancies", "dislocations", or "solid-like molecules" is said to have its claims supported by the experimental work reported in this article. Fig. 1 shows the temper-

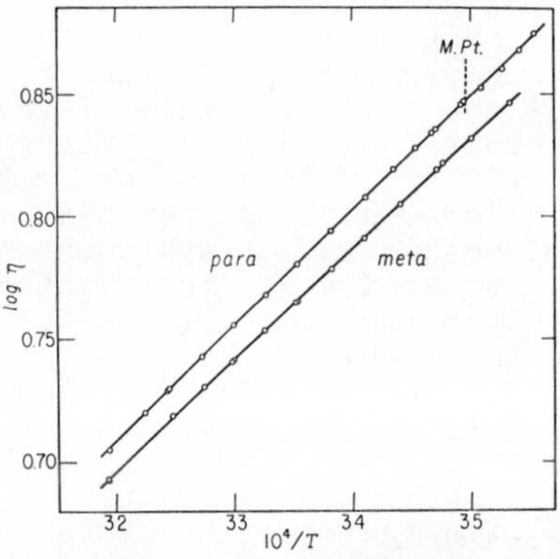

FIG. 1. Temperature dependence of the viscosities of *meta*- and *para*-xylenes.

ature dependence of the viscosities of meta- and paraxylene. Which of these two compounds has the higher melting point? Could this be rationalized in terms of which compound has the more symmetric structure and hence would "pack" more efficiently in the solid phase?

The viscosities were determined with the usual Poiseuille quotient, plus a kinetic energy correction, using water as a reference liquid. Their statement that the maximum flow velocities were approximately 20 cm/sec, well below the threshold for turbulent flow of 200 cm/sec, from the expression V_{max}= $1000\eta/r\rho$, will be recognized by students of fluid flow that a conservative estimate for the Reynolds number, at which the transition from laminar(streamline) to turbulent flow will occur, has been made at 2000. This quantity, defined as $DV\rho/\eta$ with D tube diameter, V velocity, ρ density, and η viscosity coefficient, has been shown to correlate low values with laminar and high values with turbulent flow. The transitional region is generally 1000-10 000. Show that the Reynolds number is dimensionless, then verify their expression for the threshold velocity with their radius of 0.5 mm and the viscosities of the xylenes, as well as their densities.

The authors state that the temperature-dependence of the viscosities of these two isomers is a property especially sensitive to liquid structure, and that the properties[$\eta=\eta(T)$] of liquid paraxylene are quite unaffected by nearness to its melting point. It should have been pointed out earlier that the data for paraxylene below its melting point of 13.2°C were obtained for the supercooled liquid.

The authors therefore assert again(as in ref. 1) "with even more assurance, that 'the freezing is a completely discontinous process'". Do you think the points raised in this article may still be in a state of controversy?

Does their explanation of the slightly higher viscosity of paraxylene as being due to the slightly greater angular momentum of this molecule, and its greater resistance to lamellar(laminar) flow, strike you as reasonable?

77. "Structure of Water and Hydrophobic Bonding in Proteins. IV. The Thermodynamic Properties of Liquid Deuterium Oxide" G. Némethy and H.A. Scheraga 41, 680 (1964).

Table I collects physical properties of H_2O(natural isotopic composition) and D_2O for comparison of the two. In the last column the ratio of the two values is shown for each property.

Are all of the properties intensive? Without regard to the isotopic variation of a given property, how many numerical values of the water physical properties could you have quoted, at least approximately, before having looked at the Table? As many as one-third?

Other than the first one, for the molecular weights, for how many of the properties can you ascribe the isotopic differences solely to the difference in mass? From the quotients in the last column which three properties come the closest? Do you readily understand why two of them should have the major contribution to their differences from the mass effect? What about the third?

The viscosity, the melting and normal boiling points, the temperature of maximum density, and the heat capacity are higher in liquid D_2O, as may be verified in Table I. The authors claim that this indicates that there is more structural order in liquid D_2O at a given temperature than in liquid H_2O. Do you see why? Does this imply that the degree of hydrogen bonding is higher in the former? Why? The difference in hydrogen bonding is attributed to two causes: i) intermolecular vibrational frequencies are lowered for the heavier isotope, so that the hydrogen-bonded structures contribute more heavily to the liquid partition function, and ii) hydrogen-bonded structures will also be

TABLE I. Some physical properties of H_2O and D_2O.[a]

	Unit	X_{H_2O}	Ref.[b]	X_{D_2O}	Ref.[c]	X_{D_2O}/X_{H_2O}
Molecular weight (on ^{12}C scale)		18.015	d	20.028	d	1.1117
Melting point, T_m	C°	0.00		3.81		
Triple point, T_{tr}	C°	0.01		3.82		
Temp. of max. density	C°	3.98		11.23		
Normal boiling point, T_b	C°	100.00		101.42		
Critical temperature	C°	374.15		371.2		
ΔH_{fusion} at T_m	kcal/mole	1.436	c	1.501	e	1.045
ΔH_{vap} at 3.82°C	kcal/mole	10.70	c	11.11		1.038
ΔH_{subl} at T_{tr}	kcal/mole	12.17	c	12.63		1.038
ΔE_{subl} at T_{tr}	kcal/mole	11.65		12.08		1.037
ΔS_{fusion} at T_m	eu	5.26	c	5.42		1.03
ΔS_{vap} at 3.82°C	eu	38.63	c	40.11		1.038
ΔS_{subl} at T_{tr}	eu	44.55	c	45.59		1.023
Residual entropy, S_{res}	eu	0.82		0.77	e	0.94
Entropy of the liquid at T_m	eu	14.46	f	15.80	e	1.092
Thermodynamic parameters of the liquid at 25°C:						
A	kcal/mole	−1.58	f	−1.94	g	1.23
E	kcal/mole	3.20	f	3.44	h	1.07
S	eu	16.01	f	18.08	e, h	1.12
c_p	cal/deg mole	17.99		20.16	i	1.12
c_v	cal/deg mole	17.80		20.0	i	1.12
p_{vap}	mm Hg	23.75	j	20.51	j	0.863
Molar volumes:	cm³/mole					
solid at T_m		19.65		19.679	k	1.001
liquid at T_m		18.018		18.118	l	1.0055
liquid at $T_{max.\ density}$		18.016		18.110	l	1.0052
liquid at 25°C		18.069		18.134	l	1.0036
Coefficients of thermal expansion:	$(C°)^{-1}$					
solid at T_m		1.39×10^{-4}	m	1.39×10^{-4}	m	1.000
liquid at T_m		-5.9×10^{-5}		-3.2×10^{-5}	l	0.54
liquid at 25°C		26.2×10^{-5}		21.8×10^{-5}	l	0.83
Compressibility at 25°C	atm^{-1}	4.5×10^{-5}		4.57×10^{-5}	i	0.998
Crystallographic parameters Å at 0°C:						
a		4.5239	n	4.5269	n	1.0007
c		7.3690	n	7.3706	n	1.0002
c/a		1.629	n	1.628	n	0.9994
length of the hydrogen bond		2.765	o	2.766	o	1.0003
Dipole moment	D					
measured in benzene solution at room temperature		1.76	p	1.78	p	1.01
measured in the vapor at 100° to 200°C		1.84	q	1.84	q	1.00
Dielectric constant at 25°		78.304	r	77.937	r	0.995
Refractive index, n^{20}_D		1.3326	c	1.3283		0.9968
Polarizability of vapor near 100°C	cm³/mole	58.5	q, s	61.7	q, s	1.05
Viscosity at 25°C	cP	0.895	t	1.113	t	1.243
Surface tension at 25°C	dyn/cm	71.97		71.93	u	0.9995

stabilized if the O-D···O bond is stronger than the corresponding O-H···O bond. Will both of these effects act in the direction claimed? Exactly how?

With the model developed in ref. 1, they fit many of the isotopic physical properties quite well with just one parameter, the energy of the hydrogen bond in liquid D_2O. How did this parameter turn out for the best fit to the property data? What was the difference in values between it and the energy of the hydrogen bond in liquid H_2O? Is the sign of the difference as predicted?

78. "Two-State Theory of the Structure of Water" C.M. Davis, Jr. and T.A. Litovitz 42, 2563 (1965).

The authors comment on the many theories of the structure of water that have been proposed. Their two-state model is different from the others in that it requires just 18% of the hydrogen bonds to be broken upon melting, a much smaller fraction than most other two-state moldels which required upwards of one-half of the hydrogen bonds to be broken in liquid water at 0°C. From the data on the previous page, show that about 12% of the bonds are broken upon melting from the phase-change energies and the assumption that the hydrogen bonds are the only means of holding the molecules together in the solid and liquid states. This simple calculation then assumes that the solid at 0°C has 100% unbroken hydrogen bonds, the liquid 88%, and the vapor of course 0%. Our objective in this exercise will be to examine the models for the two states proposed.

The first state is that of the puckered hexagonal rings, like those that make up the structure of ice, joined in an open packed structure as in ice. The oxygen atoms in ice form the wurtzite structure as shown in Fig. 1. The oxygen atoms are

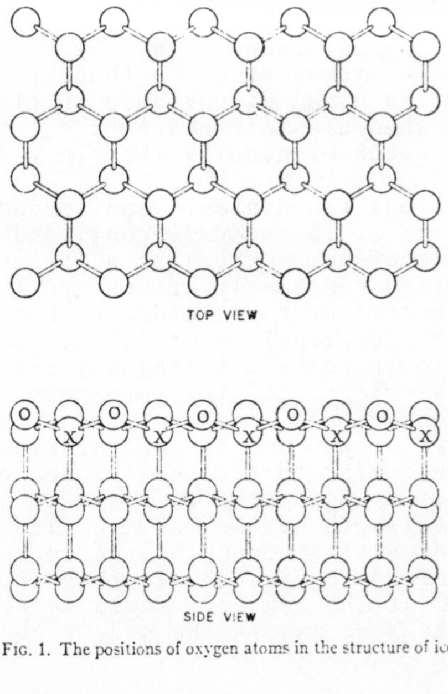

FIG. 1. The positions of oxygen atoms in the structure of ice.

connected in puckered hexagonal layers with alternate atoms raised and lowered, in the sense that the mean plane of the puckered ring is "horizontal". Can this mean plane be defined as the locus of the midpoints of the O···O lines in a given puckered ring? In a given layer? Both and/or either? Each oxygen atom is tetrahedrally coordinated to three others in the same layer and to one other in an adjacent layer. What three physical properties or characteristics of ice do they mention as being due to this layered structure of ice?

The tetrahedrally coordinated structure can be represented as body-centered cubic with the central molecule surrounded by four others situated at alternate corners of the cube as in Fig. 2. This drawing is probably familiar if you've taken a course

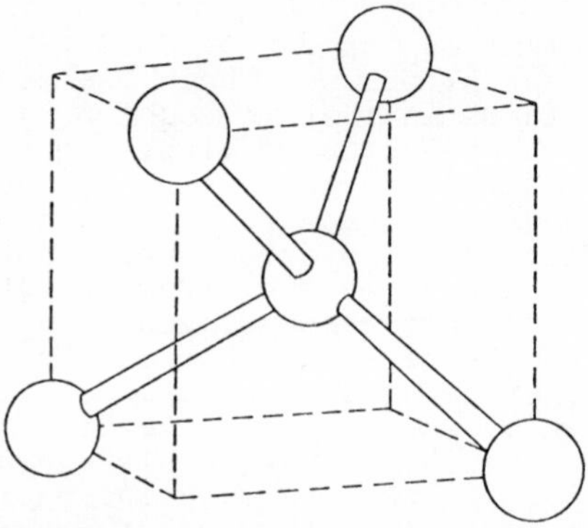

FIG. 2. The relation of the tetrahedral structure to the body-centered cubic.

in organic chemistry.

The author of this exercise book recalls a time when he was lecturing to a large group of freshmen about the tetrahedral carbon atom and had passed around a carefully-constructed cardboard tetrahedron as a visual aid. He thought at the time that there was more than the usual disturbance in class, and at the close of the period when his tetrahedron was returned he found close to a dollar's worth of pennies therein. J.J. Thomson's model of the atom?

From Fig. 2 express the nearest- and second-nearest neighbor distances in terms of the body diagonal and face diagonal.

The second state of our authors is a more closely packed structure. In order to see how it arises consider two puckered hexagonal rings connected by three hydrogen bonds as in ice, shown in Fig. 3. An additional molecule is shown dashed, in both Fig. 3 and 4, to indicate a tetragonal structure in relation to the hexagonal rings. If the three connecting hydrogen bonds between adjacent rings, e.g. the bond joining molecules A and B(Fig. 3), were broken and one of the rings rotated 60° about an axis perpendicular to the mean plane, molecule A would be above molecule C. By how much, in terms of either the O···O distance or the edge length of the cube in Fig. 2, would atom A have to be lowered until it occupies the spot of the previously vacant cube corner as in Fig. 4? Is this the body-centered cube with molecule C at the center? Could we achieve the same effect by keeping the x atoms, in the side view of Fig. 1 fixed

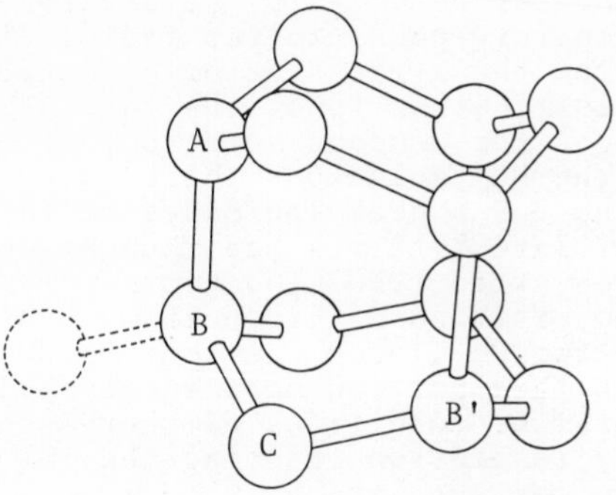

FIG. 3. Open- (icelike) packed arrangement of hexagonal rings.

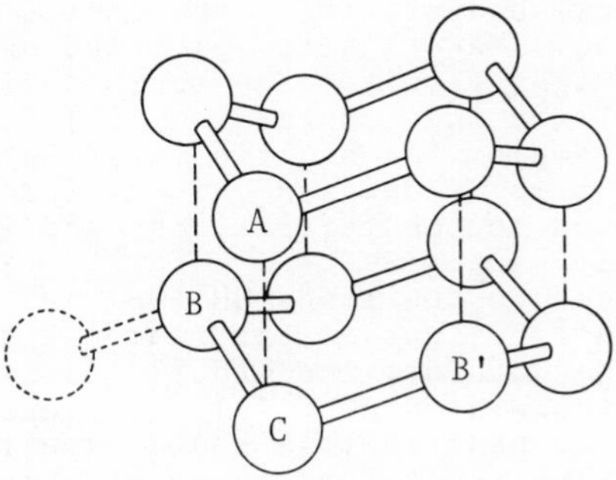

FIG. 4. Close-packed arrangement of hexagonal rings.

and lowering the o atoms (or molecules) by an amount equal to twice the separation of the x and o planes? In state or structure II, what is the distance(s) between molecules A and B and between A and B'? (Fig. 4).

The criterion for the close-packed structure II is that the maximum number of vacant corners in the bcc of Fig. 2 are filled. These two structures are discussed at some length and the agreement of the model with the radial distribution curve obtained from x-ray diffraction is noted. Thermodynamic properties are computed, including the expansivity as a function of temperature.

79. "Structure of Ferric Chloride in Aqueous Solutions" M.D. Lind 46, 2010 (1967).

Fig. 1 is a model of the $[FeCl_2(OH_2)_4]^+$ ion, thought to have D_{4h} symmetry in aqueous solution and known to have C_{2h} symmetry in the crystal. What then about the four $Fe\text{-}OH_2$ vector distances and the $H_2O\text{-}Fe\text{-}OH_2$ angles in the former case? Are the distances identical and the angles 90°? How could the ion be modified to have the symmetry it does in the crystal? Is there more than one way that this could be done?

This ion is thought to be the most predominant complex one in solutions of $FeCl_3$ in water and HCl. The vector distances for Fe-Cl=2.30 Å, and for Fe-OH_2=2.07 Å, are too nearly equal to be resolved as separate peaks in the radial distribution function obtained from the x-ray-scattering data.

These radial distribution functions (ref. 3-5) have maxima at 2.2, 3.2, and 4.6 Å for aqueous solutions of $FeCl_3$ of various compositions. For the HCl solutions there is an additional maximum at 3.6 Å and the 2.2 Å peak increases at the expense of the 3.2 Å peak. Earlier investigators had thought the first peak corresponded to three or four Fe-Cl vectors per ferric ion. Is it now reasonable to attribute this peak to overlapping contributions from the two Fe-Cl and the four Fe-OH_2 vectors per complex ion? Sketch this portion of the radial distribution curve under three or four conditions of increasing (experimentally unobtainable?) resolution, keeping the intensities of the contributions in the correct ratio.

The second experimental peak at 3.2 Å has four contributing sources. For each compute the length and number of vectors, when possible; otherwise approximate the individual contributions as best you can. Make a stick plot and show how the experimental contour might result. The four contributions are i) Cl-OH_2 vectors within the complex ion, ii) Cl-OH_2 interactions between each complex ion and the first layer of H_2O and Cl^- surrounding it, iii) Cl-OH_2 and OH_2-OH_2 interactions among Cl^- and H_2O not bonded to the ferric ions, and iv) cis OH_2-OH_2 vectors in the complex ion.

At what distances do the trans OH_2-OH_2 and trans Cl-Cl separations make contributions to the radial distribution function? Why is one seen and not the other? Is it because the number of vectors at a distance of 4.1 Å is less than those at the distance of the peak that is found experimentally? What about the relative scattering power of chlorine vs oxygen atoms?

In the solutions with HCl added, to what do the authors attribute the new maximum (3.6 Å) in the radial distribution function? If the increase in the maximum at 2.2 Å is thought to be due to an increased number of Cl-H interactions, how do the authors account for the decrease of the 3.2-Å peak in these HCl solutions?

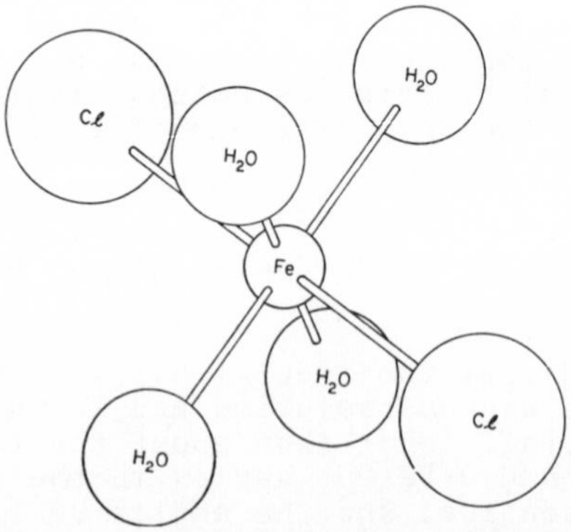

Fig. 1. Model of the $[FeCl_2(OH_2)_4]^+$ ion. In the crystals its symmetry is $C_{2h}-2/m$ but the deviations from $D_{4h}-4/mmm$ symmetry are small; therefore, the latter symmetry is likely in solutions.

80. "Compression of Mercury at High Pressure" L.A. Davis and R.B. Gordon 46, 2650 (1967).

Earlier methods of determining the equation of state of solids and liquids from high-pressure research were chiefly measurements of volume as a function of pressure. One individual and his coworkers dominated the area for several decades as the authors' references show.

The approach that is taken here, that of measuring the pressure dependence of the compressibility and then obtaining volume as a function by integration, is capable of yielding higher accuracy. The sonic velocity at temperature T is given by

$$c_T = 1/(\rho\beta_{ad})^{\frac{1}{2}}$$

with ρ the density and the adiabatic compressibility $\beta_{ad} = -(1/V)(\partial V/\partial P)_S$, the reciprocal of the adiabatic bulk modulus. This result for the sonic velocity was discussed in exercise no. 5, p.11. The pressure dependence of c_T is implicit.

The measured values for ρ as a function of temperature (21.9°, 40.5°, and 52.9°C) and pressure (0-13 kbar) are obtained from the sonic velocity measurements shown in Fig. 4 by means of

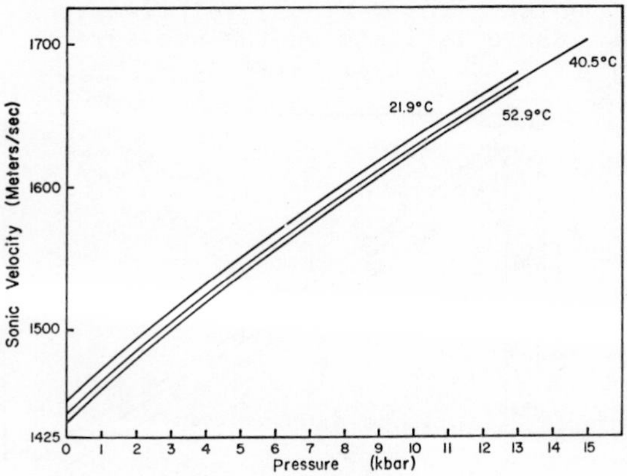

FIG. 4. The sonic velocity c_T in liquid Hg as a function of pressure at several temperatures.

thermodynamic identities. In passing, why did we pass from the adjective (adverb here?) "adiabatic" to an isentropic process? It was largely automatic of course, but wasn't the assumption of reversibility made when we replaced δq by TdS from the second law? At any rate, can you show that $\beta_T = \beta_{ad} + (T\alpha^2/\rho C_p)$? Does the intermediate result $C_p - C_V = \alpha^2 TV/\beta$ help? This last relation's derived in just about any textbook of thermodynamics. Since

$$\beta_T = -(1/V)(\partial V/\partial P)_T = (1/\rho)(\partial\rho/\partial P)_T$$

it follows that $(\partial\rho/\partial P)_T = (1/c_T{}^2) + (T\alpha^2/C_p)$. Can you show that all of these steps are consistent? On integrating wrt P at constant T show that the equation of state takes the form

$$\rho(P,T) = \rho(1\text{ atm},T) + \int_1^P dP/c_T + T\int_1^P \alpha^2 dP/C_p.$$

F*

Beginning with the sonic velocity c=c(T,P) and the one-atmosphere input data of Table IV the authors then determine all of the quantities β_{ad}, β_T, α, and ρ. C_p can be determined from the identity given earlier; it varies slightly in this temperature range as Table IV shows and its pressure dependence is negligible so that it can be taken out of the integral above.

TABLE IV. One-atmosphere input data to the compression calculation.

Temperature, t (°C)	21.9°	40.5°	52.9°
Density, ρ (g/cm³)[a]	13.54122	13.49573	13.46551
Thermal-expansion coefficient, α (°C⁻¹×10⁴)[b]	1.81069	1.80825	1.80699
Specific heat, C_P (ergs/g·deg×10⁻⁶)[c]	1.390	1.385	1.382
Sonic velocity c_T (m/sec)[d]	1450.1	1441.5	1435.8

[a] Reference 36.
[b] Reference 35.
[c] Reference 34.
[d] Reference 27.

The data are tabulated in Table V which is not reproduced. Let us instead begin with Fig. 7 which gives V=V(P,T) relative to the volume at 0°C and 1 atm. Where is 1 atm on the pressure axis? Is it distinguishable from the 0 kbar mark?

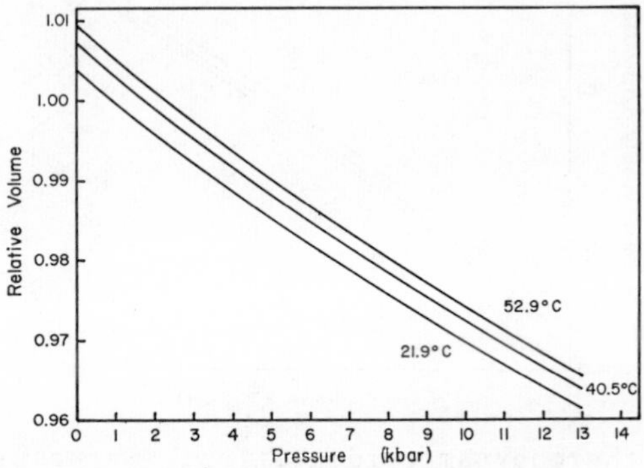

FIG. 7. The relative volume V of liquid Hg as a function of pressure at several temperatures. Volume at 1 atm and 0°C is 1.00.

From Fig. 7 compute as precisely as possible(using a good scale) values for the thermal-expansion coefficient α and the isothermal compressibility β_T at one of the three temperatures and the following pressures: 0(1 atm!), 6.5, and 13 kbar. Compare your computed results with those given in Table IV and Fig. 5 and 6. Next determine ρ for your temperature at each of the three pressures by means of Fig. 7 and the value of ρ at one atmosphere and the appropriate temperature. Will you need to make a slight extrapolation along the line P=0 to obtain the density at 0°C and 1 atm, in order to convert the relative volumes to densities?

Now with the pressure-independent value of C_p for your temperature make use of the identity to obtain β_{ad} at the three pressures. Finally with this last quantity and ρ compute the

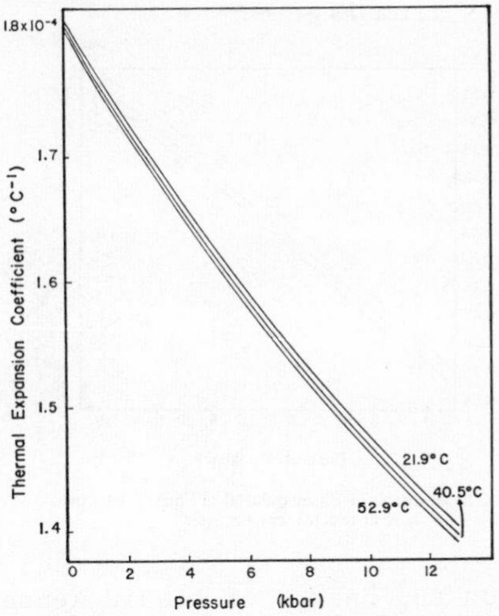

FIG. 5. The volume thermal-expansion coefficient α of liquid Hg vs pressure at several temperatures. Considering experimental error, the three curves are actually just barely distinguishable from one another.

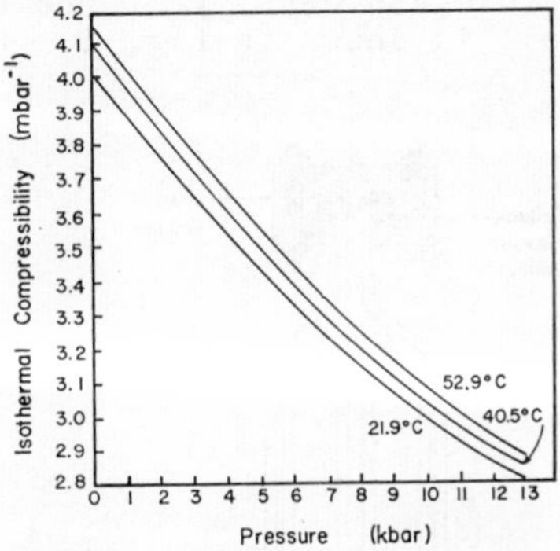

FIG. 6. The isothermal compressibility β_T of liquid Hg as a function of pressure at several temperatures.

sonic velocity, again at each pressure, and compare with those shown in Fig. 4. We have essentially reversed their path in handling the data, of course.

They find the most accurate form in which to cast the equation of state is an expansion of the pressure P in powers of $[(V_0/V)-1]$, fitting the data well within experimental error (0.01% max!) with just the linear and quadratic term. Can you see why the constant term is zero?

In exercises no. 72 and 73 we became acquainted with the Murnaghan equation; this is not as satisfactory for these high-accuracy measurements. Recall the assumption that was made in its derivation, that the isothermal bulk modulus could be expressed as a linear function of pressure. Do you see why this assumption is not valid for mercury from Fig. 9? Are the units of the ordinate millibars or megabars? Are the values the reciprocals of the β_T's from Fig. 6?

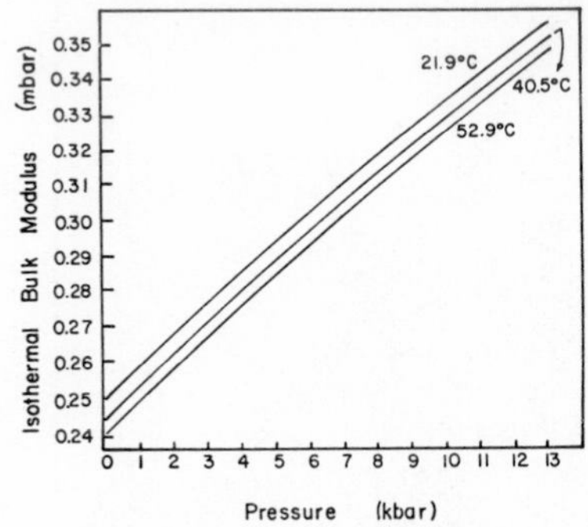

FIG. 9. The isothermal bulk modulus B of liquid Hg vs pressure at several temperatures.

81. "Surface Tension and Energy of Liquid Xenon" B.L. Smith, P.R. Gardner, and E.H.C. Parker 47, 1148 (1967).

The method used to measure surface tension was the basic one of the capillary rise, with appropriate corrections made for the radii of curvature and density of the vapor above the liquid. The sample cell is shown in Fig. 1. Precision-bore

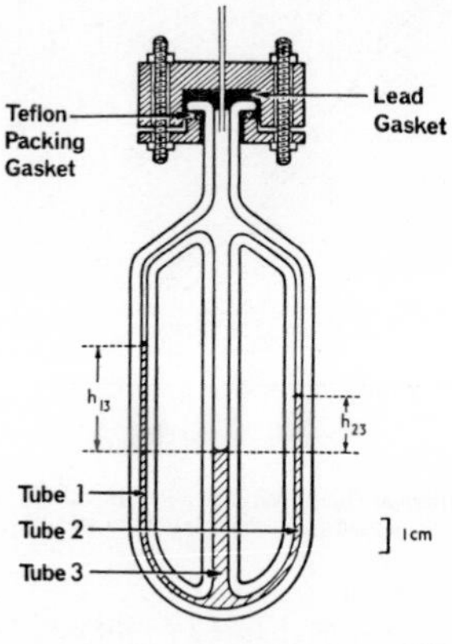

FIG. 1. Capillary tubes and pressure seal.

capillary tubing, manufactured to be of uniform cross section and with the bore advertised to be within 0.01 mm of the nominal ID, was obtained in 0.2, 0.4, and 2.0 mm ID. Are the heights of the liquid columns in Fig. 1 in the proper relationship to these diameters? After the experiment each tube was broken in several places and the internal diameter measured with a microscope. The diameters were found to vary by not more than 1% along the length of the tubes and the ellipticity was always less than 2%. Were the advertised specifications met? Why was a lead gasket used as the pressure seal?

Fig. 2 shows the overall arrangement. A sample of liquid

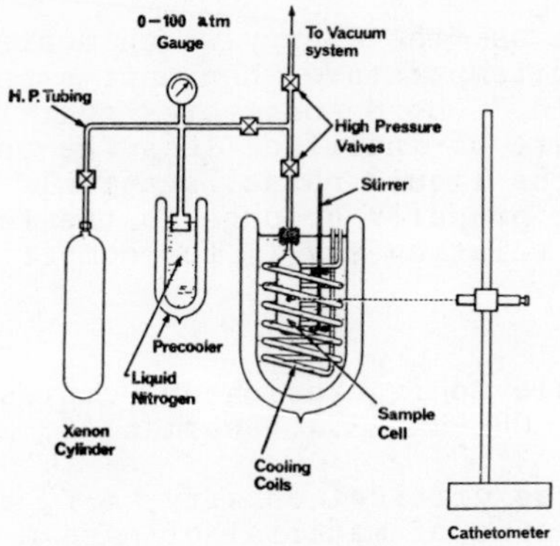

FIG. 2. General experimental arrangement.

xenon was formed in the capillary tubes and allowed to come to equilibrium. The differences between liquid levels(why three tubes rather than two?) were measured to ±0.01 cm as shown. A platinum resistance thermometer, not shown, was used to measure bath temperatures. Is this a static, rather than a dynamic measurement as described in exercise no. 75? Should the accuracy then be somewhat higher in this case?

The results are given in Table II. Their equation (6)

TABLE II. Smoothed values of the surface tension γ and surface energy E of xenon as a function of temperature.

Temp (°K)	γ (dyn cm^{-1})	E (erg cm^{-2})
165	18.46	49.88
175	16.58	49.13
185	14.74	48.25
195	12.96	47.27
205	11.23	46.17
215	9.55	44.90
225	7.94	43.44
235	6.40	41.74
245	4.94	39.71
255	3.56	37.23
265	2.30	34.04
275	1.18	29.58
285	0.28	21.53
289.74	0.00	0

$$\rho_\ell - \rho_g = 3.65[1 - (T/T_c)]^{0.345} \text{ g/cm}^3 \qquad (6)$$

with T_c the critical temperature equal to the last entry in Table II as was obvious(wasn't it?), can be used to approximately determine for which temperature(s) the relative column heights in Fig. 1 are drawn. Do it, neglecting radii of curvature corrections. Try the lowest temperature first, as they state that the maximum liquid-level difference was about 1 cm. The results for surface tension are put into the form

$$\gamma = (54.6\pm0.1)[1 - (T/T_c)]^{1.287\pm0.017} \text{ dyn/cm}$$

which has some theoretical justification as discussed.

The surface energy $E=\gamma-(Td\gamma/dT)$, whose values are also listed in Table II, is discussed by E.A. Guggenheim, "Thermodynamics" (North-Holland Publishing Co. Amsterdam 1959), p. 265, where he describes it as "the energy which must be supplied to prevent any change of temperature when unit area of surface is formed from the liquid." He demonstrates that it is equal to the energy per unit are of interface less the energy of the same material content in the liquid phase, and feels that the name "surface energy" more properly belongs to the former. Compute values of E from the relation $\gamma=\gamma(T)$ for comparison with a few entries in Table II.

82. "Effects of Sample Configuration on Compressible Fluids in the Earth's Field" B. Chu and R.J. Bearman 48, 2377 (1968).

Experiments at the critical density, m/V, are carried out by sealing a known amount of material of mass m into a sample chamber of volume V. However as the critical temperature is approached the fluid is noticeably compressed under its own weight, and the density varies markedly with height(ref. 3). It then becomes necessary to locate precisely that height in the sample chamber where the fluid is at the critical density (ref. 4,5). Usually this position is obtained from a computation of the density-height(ρ-z) profile. In this note the authors show that the density-height profile depends in general upon the shape of the container. The proof("elementary") applies to any one-component compressible isothermal one-phase fluid, not necessarily in the critical region. The effect of gravity, negligible under ordinary conditions, becomes increasingly noticeable as the critical point is approached. What is the value of the isothermal compressibility at the critical point? Compare your answer with their statement.

In the absence of convective flow the pressure P varies continously throughout the fluid, is independent of x and y, and under the assumption of local equilibrium is a function only of the density. The equation of motion becomes(see also footnote 7) $dP/dz=\rho g$. Show that upon definite integration the relation for z as a function of ρ

$$g(z - z_0) = \int_{\rho_0}^{\rho} (dP/d\rho)\rho^{-1}\, d\rho$$

is obtained. Thus given ρ_0 at any z_0 implies the ρ-z profile may be obtained without specification of shape. Show that this is so.

Experimentally, however, ρ_0 is usually not known in advance. One measures m, V, and the container shape which together determine ρ_0. The authors suggest that it might be simpler in this case to use the indefinite integral $gz=\int(dP/d\rho)\rho^{-1}d\rho+\kappa$, with the integration constant κ depending upon the sample container geometry and the nature of the fluid. It is determined by first numerically solving the integrated equation for ρ, from which $\rho=\rho(z,\kappa)$, then the result $\int\rho dV=m$. What is the region of integration for the last integral? Show that for a vessel with uniform cross-sectional area A, normal to the gravitational field, the integral reduces to $\int\rho(z,\kappa)dz=m/A$. The correct value of κ is the one for which the lhs and the rhs are equal. Thus for a fixed mass of fluid, the ρ-z profile is the same for all vessels with equal uniform cross sections, regardless of shape, so long as they are bounded horizontally by the same planes.

Show that, however, when the cross-section A varies with z the equation takes the form $\int\rho(z,\kappa)A(z)dz=m$. Then κ, or equivalently ρ_0, is affected.

Prove from the mean-value theorem for continously varying A(z) that $\int\rho(z,\kappa)dz=m/\bar{A}$, where $\bar{A}$ is some cross section in the vessel. Show that the ρ-z profile for a container of arbitrary shape with a continous A(z) is the same as that for an equivalent cylinder of cross section $\bar{A}$. Are the authors' two statements that $\bar{A}$ is usually not a simple average area and that generally $\bar{A}\times\text{height}\neq V$ equivalent? Are they if simple means most commonly taken? Were the proofs "elementary" after all?

"The effects may be secondary, but they should not be overlooked when the critical point is approached using systems supposedly at the critical density and at temperatures only a few thousandths of a degree from the critical temperature."

83. "Liquid Surface Tension near the Triple Point" P.A. Egelstaff and B. Widom 53, 2667 (1970).

It is shown in this paper that the product of the isothermal compressibility χ and surface tension σ of a liquid near its triple point may be expected to be a fundamental length which is characteristic of that liquid. The agreement with experiment is very good; they present semi-empirical relations conjectured (their verb - quite a useful one) from others with true theoretical bases which support the view of $\chi\sigma$ as a fundamental characteristic length.

First, show that the product does indeed have the dimension of length. Next review the derivation of the excess pressure δp under which a liquid droplet of radius r finds itself when suspended in a vapor with which it is in equilibrium. The derivation is found in physical chemistry texts among other sources and yields the result $\delta p=2\sigma/r$ where σ is the surface tension. Because the vapor pressure is low at the triple point the pressure exerted on a small drop is primarily this surface pressure $2\sigma/r$. Check the authors' statement that this quantity in argon does not become as small as the vapor pressure until r becomes as large as 5000 Å.

Compared to a spherical portion of the bulk liquid phase containing an equal number of molecules, show that in response to this excess pressure the (compressible)drop suffers a fractional decrease in its volume $v^{-1}\delta v=\chi\delta p$ and a fractional decrease in its radius $r^{-1}\delta r=(1/3)v^{-1}\delta v=(1/3)\chi\delta p$. The radius decrement(another good word; look it up if unfamiliar) δr is then $(1/3)r\chi\delta p$, or

$$\delta r = (2/3)\chi\sigma.$$

So long as $r\gg\delta r$, which was implicitly assumed, the radius decrement is seen from this equation to be independent of r, and therefore a fundamental length characteristic of the substance. The authors argue that because it has the dimension of length and is an intensive property it may be expected to be of the order of an angstrom unit. Table II gives σ, χ, and the product for several types of substances.

The comparison of liquid metals with nonmetals is said to be significant because interionic forces in metals are density

TABLE II. Values of $\chi\sigma$ for liquids at or near their triple points.

	σ (dyn/cm)	χ ($cm^2/dyn \times 10^{12}$)	$\chi\sigma$ (Å)
Alkali metals			
Sodium	194[a]	21[b]	0.40
Potassium	113[a]	40[b]	0.45
Rubidium	95[a]	49[b]	0.46
Cesium	71[a]	67[b]	0.47
Other metals			
Iron	1790[c]	1.43[d]	0.25
Copper	1280[c]	1.45[d]	0.19
Silver	940[e]	1.86[d]	0.18
Zinc	785[e]	2.4[b]	0.19
Cadmium	666[e]	3.2[b]	0.21
Lead	470[e]	3.5[b]	0.17
Bismuth	395[f]	4.3[b]	0.17
Molten salts[g]			
Sodium chloride	116	29	0.34
Potassium chloride	97	38	0.37
Sodium bromide	99	34	0.34
Potassium bromide	90	44	0.39
Sodium iodide	88	40	0.35
Potassium iodide	78	50	0.39
Water[h]	76	50	0.38
Simple nonmetallic liquids			
Bromine	41.5[f]	63[f]	0.26
Xenon	18.7[i]	166[j]	0.31
Oxygen	18.4[k]	120[k]	0.22
Krypton	16.3[k]	172[j]	0.28
Argon	13.1[k]	212[k]	0.28
Nitrogen	11.8[k]	211[k]	0.25
Other liquids[f]			
Ethylene glycol	48	37	0.18
Nitrobenzene	44	49	0.22
Aniline	43	45	0.19
Chlorobenzene	33	74	0.24
Benzene	29	94	0.27
Ethanol	23	112	0.26

[a] D. Germer and H. Mayer, Z. Physik **210**, 391 (1968).
[b] O. J. Kleppa, J. Chem. Phys. **18**, 1331 (1950).
[c] P. Kozakevitch, in *Liquids: Structure, Properties and Solid Interactions* (Elsevier, Amsterdam, 1965), p. 243.
[d] S. I. Filippov, N. B. Kazakov, and L. A. Pronin, Chernaya Met. (USSR) **9**, 8 (1966).
[e] J. W. Taylor, Rept. A.E.R.E. M/TN 24 (1954).
[f] S. W. Mayer, J. Phys. Chem. **67**, 2160 (1963).
[g] G. J. Janz, *Molten Salts Handbook* (Academic, New York, 1967).
[h] *Handbook of Chemistry and Physics*, edited by R. C. Weast (Chemical Rubber, Cleveland, 1969).
[i] A. J. Leadbetter and H. E. Thomas, Trans. Faraday Soc. **461**, 10 (1965).
[j] Yu. P. Blagoi, A. E. Butko, S. A. Mikhailenko, and V. V. Yakuba, Russ. J. Phy. Chem. (Eng. Transl.) **41**, 908 (1967).
[k] *Simple Dense Fluids*, edited by H. L. Frisch and Z. W. Salsburg (Academic, New York, 1968).

dependent. Why should this be so? If the density gradient had not been sharp at the interface, it might show up through a difference in the product $\chi\sigma$ for metals and insulators. Is this effect seen?

CHAPTER VII

MASS SPECTROMETRY AND IMPACT PHENOMENA

84. "Ionization and Dissociation of Nitrogen Trifluoride by Electron Impact" R.M. Reese and V.H. Dibeler 24, 1175 (1956).

Fig. 1 shows the ionization efficiency curves for three of the positive ions formed when NF_3 is bombarded with 70-v electrons and the products analyzed in a mass spectrometer.

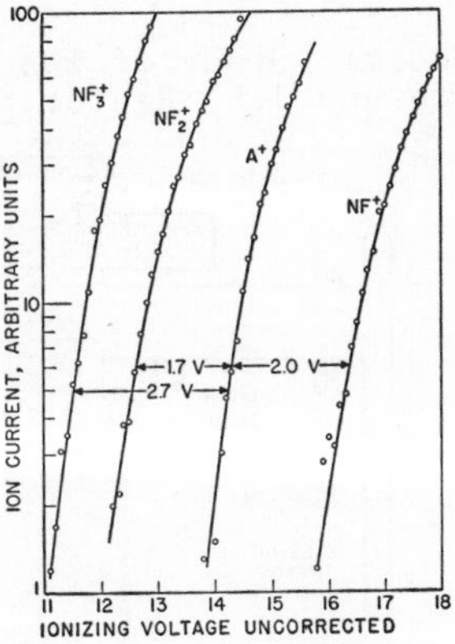

FIG. 1. Initial portions of the ionization efficiency curves for several positive ions in NF_3 and the calibrating gas, argon.

The calibrating gas argon is included in with the nitrogen trifluoride; its ionization potential for the process

$$Ar\ 3p^6\ {}^1S_0 \rightarrow Ar^+\ 3p^5\ {}^2P_{3/2}$$

is known(ref. 5) to be 15.76 ev. This ionization potential is known from spectroscopy and can therefore be used to calibrate the instrumental ionizing voltages obtained. When corrected in the manner indicated these potentials are called appearance potentials, as they are minimum voltages for which the respective ions appear at the detector. They must be at least as large as the energy required to break the appropriate bond in the parent molecule plus that necessary to ionize the resulting fragment. Draw an energy diagram indicating these processes with the energies of NF_3, NF_2, and NF_2^+ in the proper relationship. In addition there may be kinetic energy of the neutral fragment, some vibrational excitation of the products, and finally the electronic state reached for the ion may not be its lowest one. May these uncertainties be eliminated one by one in the case of the calibrating gas argon?

Show that this calibration has been used to obtain the AP's in Table I for the three positive ions of Fig. 1.

TABLE I. Mass spectral data and appearance potentials of NF_3.

Ion	Relative abundance[a]	Appearance potential (ev)	Probable products
NF_3^+	59.7	13.2 ±0.2	NF_3^+
NF_2^+	100.0	14.2 ±0.3	$NF_2^+ + F$
NF^+	39.1	17.9 ±0.3	$NF^+ + 2F$
N^+	5.4	22.2 ±0.2	$N^+ + 3F$
F^+	4.8	25±1	$F^+ + 2F + N$
F_2^-	0.07 (at 3 v)	Approx. 0	?
F^-	20 (at 3 v)	Approx. 0	$F^- + NF_2$
	1	Approx. 22	$N + F^+ + F^- + F$

[a] Abundance for 70-v electrons except as noted.

Notice that the older symbol, A rather than Ar, is used for argon in Fig. 1. The father of a seventh grade science student was amused not long ago when his son reported to him that the atmosphere contained 1% "airgon".

85. "Mass-Spectrometric Study of Ion Profiles in Low-Pressure Flames" H.F. Calcote and J.L. Reuter 38, 310 (1963).

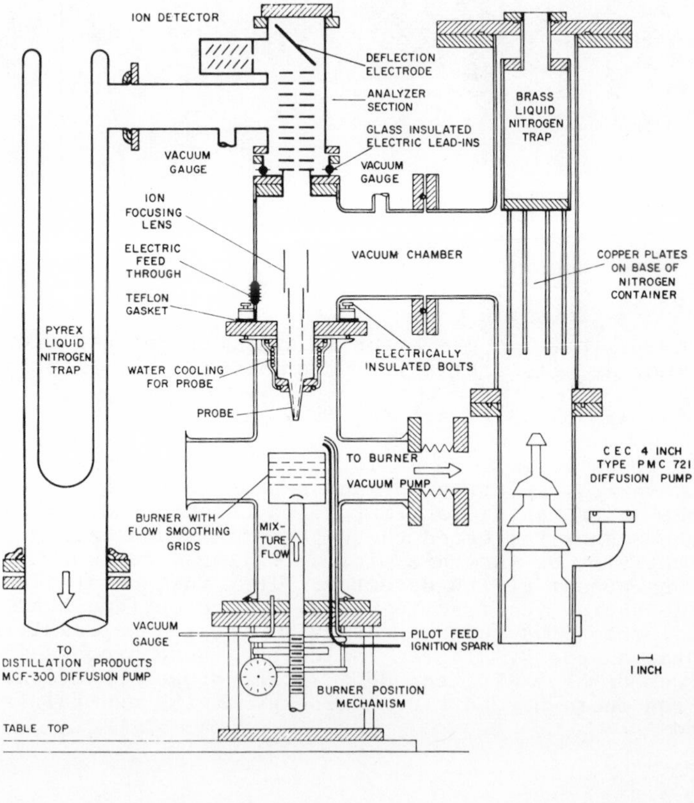

FIG. 1. Analyzer and vacuum system.

Fig. 1 shows the experimental setup. The pressure was 1-6 mm Hg and a flat flame several cm thick was produced on the burner. The flame was moved vertically by suitable gearing to obtain ion profiles, such as those shown in Fig. 6 and 7.

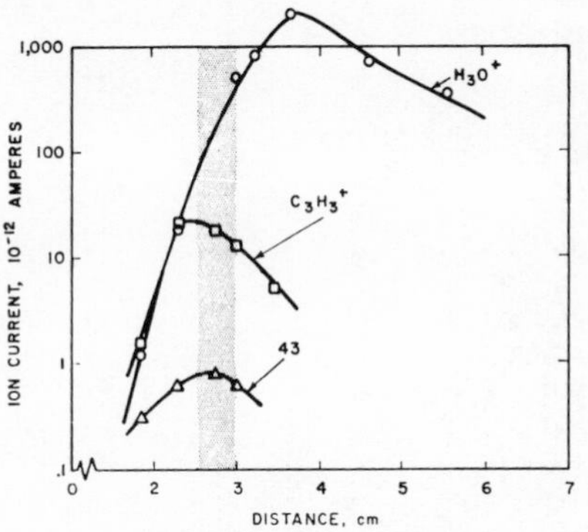

FIG. 6. Ion profiles in ethylene-oxygen flame. Equivalence ratio = 0.78, p = 2.5 mm Hg.

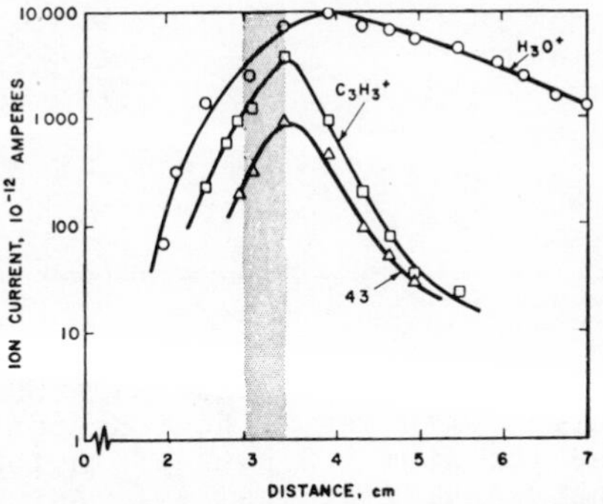

FIG. 7. Ion profiles in ethylene-oxygen flame. Equivalence ratio = 1.0, p = 2.5 mm Hg.

The equivalence ratio is defined in footnote 9 as the stoichiometric oxygen/fuel ratio divided by the actual oxygen/fuel ratio. Can you rationalize the relative abundances of the H_3O^+ and $C_3H_3^+$ ions at the two equivalence ratios shown? From Fig. 1 does it appear as though the burner could travel the distances required to sample the flame gases at the positions shown in Fig. 6 and 7? The shaded area in these two figures denotes the downstream edge of the luminous combustion zone. Does it strike you as curious that the $C_3H_3^+$ ion is so prominent? Can it be a parent ion from one of the two fuel molecules used, ethylene or acetylene? What mechanisms do the authors suggest for its formation?

The other ion shown, at mass 43, is thought to be due to C_2H_3O or C_3H_7 as Table I reveals. From a study of the trends in this Table, could you suggest another possible ion at mass 37? By the addition of deuterated acetylene to an acetylene-oxygen flame, it was demonstrated that mass 19 had three hydrogens and

TABLE I. Summary of ions observed by mass spectrometry in hydrocarbon flames.

Mass	Probable ion	Observed by Sugden	Observed by van Tiggelen	Observed by AeroChem
13	CH	x	x	a
14	CH_2	x	x	a
15	CH_3	x	x	a
18	H_2O	x	x	a
19	H_3O	x	x	x
24	C_2	x	x	...
25	C_2H	...	...	x
26	C_2H_2	x	x	x
27	C_2H_3	x	...	x
28	C_2H_4 or CO	x	x	x
29	C_2H_5 or CHO	x	x	x
30	CH_2O	...	...	x
31	CH_3O	x	...	x
32	CH_4O or O_2	x	...	x
33	HO_2	x	b	x
35	H_3O_2	...	b	x
37	H_5O_2	x	x	x
39	C_3H_3	x	x	x
41	C_2HO or C_3H_5	x	...	x
42	C_2H_2O or C_3H_6	...	...	x
43	C_2H_3O or C_3H_7	x	x	x
45	CHO_2	...	b	...
47	CH_3O_2	...	b	...
49	CH_5O_2	...	b	...
57	C_2HO_2	x	x	...
59	$C_2H_3O_2$	x	b	a

[a] Same magnitude as noise level.
[b] Only with halogen present.

was therefore due to H_3O^+, and also that mass 39 contained hydrogens. These spectra, with and without deuterium, are shown in Fig. 9. Determine the atom % D present in the mixture for the solid line in the vicinity of mass 19. Assume that the C-H

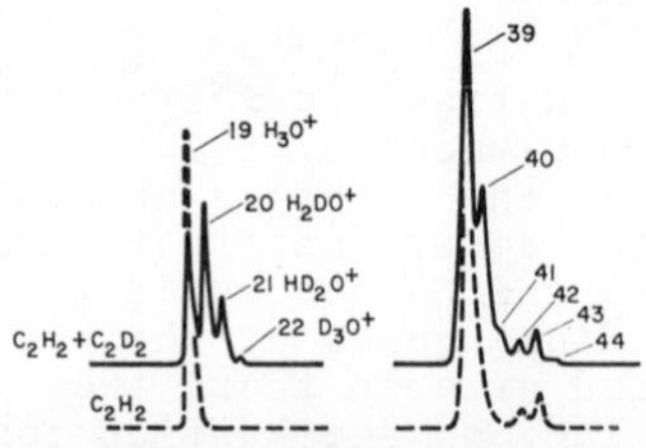

FIG. 9. Deuterated acetylene addition to acetylene-oxygen flame. $p=2.2$ mm Hg, 2.54 cm above burner.

and C-D bonds are equivalent in the ion(isoelectronic with ammonia?). This should be about 30 atom % D. Will this type of analysis yield the same result in the mass 39 region shown? If not why not? Could it be that the hydrogen atoms in C_3H_3 are not equivalent? Draw one or two reasonable structures with non-equivalent hydrogen atoms.

86. "Investigation of a Fragmentation Model for n-Paraffins" J.C. Schug 38, 2610 (1963).

A fragmentation model is postulated for the interpretation of n-paraffin mass spectra. Considering only the breaking of C-C bonds and ignoring the possible losses of H atoms, several predictions are made regarding the mass spectra: i) At low ionizing voltages the spectra will be dominated by C_n^+ ions, where $N/2 \leq n \leq N-2$ (beginning with C_NH_{2N+2}). ii) At all ionizing voltages the smaller ions in this range should be more abundant than the larger ones. iii) Fragment ions having N-1 C atoms should at all voltages constitute a negligible part of the spectrum. iv) It is predicted that the larger the molecule the greater the ionizing voltage required before the fragmentation pattern becomes independent of voltage. v) The presence of numerous metastable ion peaks is predicted; e.g. in n-decane insofar as activation energies and frequency distributions applicable to the primary formation of C_6^+, C_7^+, and C_8^+ are equivalent as was assumed here, a metastable ion for each of these primary reactions is expected.

Test his first three predictions with the fragmentation pattern of Fig. 2. He claims that all three points are satisfied. Do you agree? You will need to extrapolate to lower

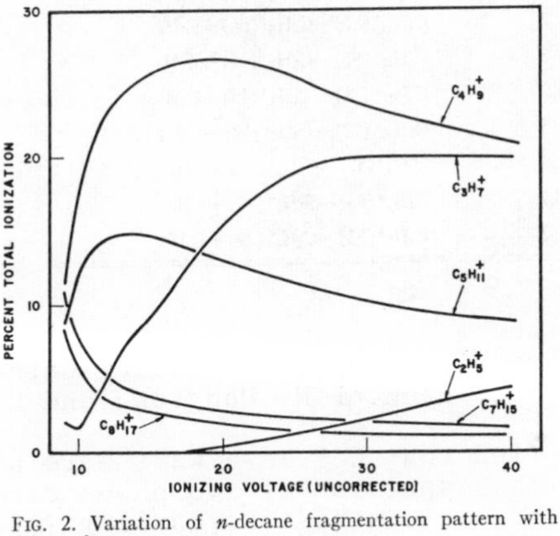

FIG. 2. Variation of *n*-decane fragmentation pattern with ionizing voltage.

voltages to check i). Prediction iv) may be verified by examin-

TABLE I. Comparison of relative abundance curves for three *n*-paraffins.

Feature	Ionizing voltage *n*-Decane	*n*-Pentadecane	*n*-Eicosane
$C_2H_5^+$ Appearance	~17 V	~20 V	~24 V
$C_5H_{11}^+$ Maximum	13	17	22
$C_4H_9^+$ Maximum	18	27	32
$C_3H_7^+$ Plateau	28	46	>60

ation of Table I, as you should do.

For the schematic dissociation reaction $m_1^+ \rightarrow m_2^+ + (m_1 - m_2)^0$ for mass m, it is shown in his ref. 12 that when the half-life for this reaction is comparable with the time-of-transit that a metastable peak at mass $m^* = m_2^2/m_1$ may appear. Subject this result to a test with two or three of the metastable peaks shown in Table III. Is prediction v) borne out as well? In the second row of the Table occurs the most trivial of misprints.

TABLE III. Some observed metastable peaks in mass spectrum of *n*-decane.

m^*	Dissociation
89.9	$C_{10}H_{22}^+(142) \rightarrow C_8H_{17}^+(113) + C_2H_5$
70.4	$C_{10}H_{22}(142) \rightarrow C_7H_{16}(100) + C_3H_6$
63.4	$C_8H_{18}^+(114) \rightarrow C_6H_{13}^+(85) + C_2H_5$
61.6	$C_8H_{16}^+(112) \rightarrow C_6H_{11}^+(83) + C_2H_5$
44.7	$C_8H_{17}^+(113) \rightarrow C_5H_{11}^+(71) + C_3H_6$
50.4	$C_7H_{16}^+(100) \rightarrow C_5H_{11}^+(71) + C_2H_5$
48.6	$C_7H_{14}^+(98) \rightarrow C_5H_9^+(69) + C_2H_5$
33.6	$C_7H_{13}^+(97) \rightarrow C_4H_9^+(57) + C_3H_4$
32.8	$C_7H_{15}^+(99) \rightarrow C_4H_9^+(57) + C_3H_6$
38.3	$C_6H_{13}^+(85) \rightarrow C_4H_9^+(57) + C_2H_4$
36.5	$C_6H_{14}^+(86) \rightarrow C_4H_8^+(56) + C_2H_6$
21.8	$C_6H_{13}^+(85) \rightarrow C_3H_7^+(43) + C_3H_6$
26.0	$C_5H_{11}^+(71) \rightarrow C_3H_7^+(43) + C_2H_4$
24.5	$C_5H_{12}^+(72) \rightarrow C_3H_6^+(42) + C_2H_6$
29.5	$C_4H_9^+(57) \rightarrow C_3H_5^+(41) + CH_4$
27.7	$C_4H_7^+(55) \rightarrow C_3H_3^+(39) + CH_4$
15.3	$C_4H_7^+(55) \rightarrow C_2H_5^+(29) + C_2H_2$

87. "Mass Spectrographic Study of Molecular Carbon Ions Formed in a Spark Source" W.L. Baun, F.N. Hodgson, and M. Desjardins <u>38</u>, 2787 (1963).

These authors find that C_2^+ is always more abundant than C_3^+ in the sparked mass spectrum of amorphous carbon, graphite, and synthetic diamond, in contrast to earlier work. Show that separation of the $^{24}Mg^+ - ^{12}C_2^+$ doublet requires a resolution $(m/\Delta m)$ of 1600 or better; they do claim this resolution for their instrument. Would this then meet possible objections that the electrodes may have contained magnesium as an impurity?

In the catenation or association curve of Fig. 1, is it

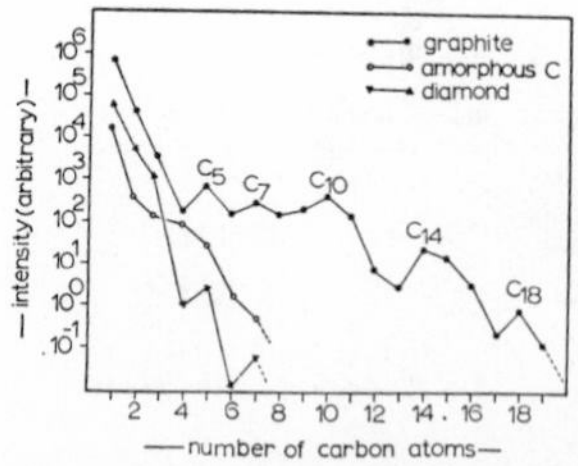

FIG. 1. Catenation curves for graphite, diamond, and amorphous carbon, showing abundance of positive ions in the radio-frequency spark source. Diamond and amorphous carbon were sparked in Al electrodes.

at all unexpected that the graphite curve shows maxima at C_5 and C_7 rather than at C_6? If we attribute the C_{10} maximum to a naphthalene-like skeleton, to what could the C_{14} and C_{18} maxima be due? Notice especially that the C_{18} intensity is about an order of magnitude greater than the C_{17} intensity; is this surprising in view of the authors' statement that catenation curve of coronene(look up the structure?) very closely approximates that of graphite?

88. "Formation of Fragment Ions from CH_3Te^{125} and $C_2H_5Te^{125}$ Following the Nuclear Decays of CH_3I^{125} and $C_2H_5I^{125}$" T.A. Carlson and R.M. White <u>38</u>, 2930 (1963).

TABLE I. Relative abundances of the observed fragment ion resulting from the nuclear decay (1) of CH_3I^{125} and (2) of $C_2H_5I^{125}$.

CH_3I^{125}		$C_2H_5I^{125}$	
Ion	% Abundance	Ion	% Abundance
CH_3Te^+	1.0±0.3	$CH_3CH_2Te^+$	1.2±0.5
CH_3Te^{2+}	0.3±0.4	CH_2Te^+	0.0±0.2
CH_3^+	0.8±0.2	$C_2H_5^+$	1.3±0.3
CH_2^+	0.0±1.0	$C_2H_4^+$	0.0±0.3
CH^+	0.2±0.2	$C_2H_3^+$	0.4±0.2
C^+	1.7±0.2	$C_2H_2^+$	0.3±0.4
CH_3^{2+}	0.0±0.1	C_2^+	0.8±0.3
CH_2^{2+}	0.1±0.1	CH_3^+	0.3±0.1
CH_2^+	0.05±0.1	CH_2^+, $C_2H_4^{2+}$	0.0±1.5
C^{2+}	1.7±0.1	CH^+, $C_2H_2^{2+}$	0.0±0.2
C^{3+}	0.6±0.1	C^+, C_2^{2+}	3.2±0.3
C^{4+}	0.0±0.1	$C_2H_5^{2+}$	−0.1±0.3
H_2^+	0.0±0.2	$C_2H_3^{2+}$	−0.2±0.3
H^+	5.5±0.3	CH_3^{2+}	0.3±0.3
Te^+	0.2±0.3	CH_2^{2+}	−0.3±0.3
Te^{2+}	0.5±0.2	CH^{2+}	−0.2±0.3
Te^{3+}	1.2±0.3	C^{2+}	1.4±0.2
Te^{4+}	2.1±0.2	C^{3+}	0.4±0.1
Te^{5+}	2.6±0.2	C^{4+}	0.0±0.1
Te^{6+}	5.1±0.3	H_2^+	0.0±0.2
Te^{7+}	9.1±0.5	H^+	6.4±0.5
Te^{8+}	12.9	Te^+	0.5±0.3
Te^{9+}	13.5±0.8	Te^{2+}	0.5±0.3
Te^{10+}	15.0±0.5	Te^{3+}	1.1±0.3
Te^{11+}	11.4±0.6	Te^{4+}	2.3±0.5
Te^{12+}	5.2±0.2	Te^{5+}, C_2H^+	3.3±0.3
Te^{13+}	2.8±0.4	Te^{6+}	5.4±0.8
Te^{14+}	2.8±0.2	Te^{7+}	13.6±0.8
Te^{15+}	1.8±0.4	Te^{8+}	13.3
Te^{16+}	1.2±0.2	Te^{9+}	12.3±0.8
Te^{17+}	0.5±0.1	Te^{10+}, C_2H^{2+}	12.2±0.8

The radioactive methyl and ethyl iodides were separately allowed to leak into a source volume of a specially designed mass spectrometer; as the gases underwent nuclear decay in the source volume, the ions that were formed as a result were collected and analyzed for their ratios of mass to charge. An abbreviated Table I shows most of the results; the remainder are given in Fig. 2 with the exceptions of $Te^{19,20+}$.

Check and see that both ions, where pairs are given, in Table I have their ratios of mass to charge within the error limits given.

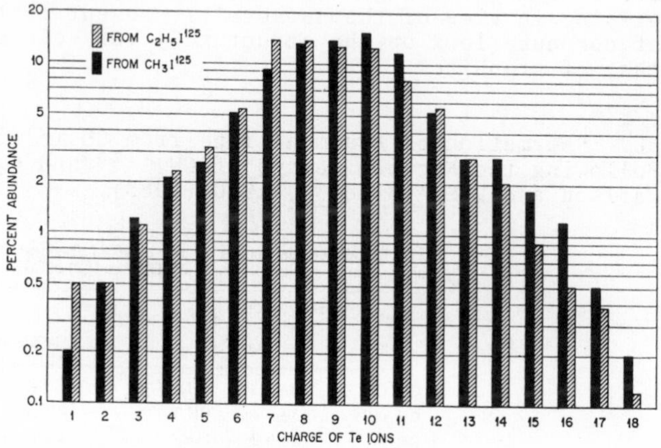

FIG. 2. Charge spectra of the tellurium ions formed following the decay of (1) $C_2H_5I^{125}$ and (2) CH_3I^{125}.

Why is it that charged fragments of the methyl or ethyl radicals are formed at all?

Draw a schematic mass spectrum for those species in Table I and in Fig. 2; one each for methyl and ethyl iodide.

Is the amount of H^+ recorded surprising in view of its light weight which would lead it to receive a rather large share of kinetic energy, and thus be collected with poor efficiency?

89. "Mechanism of Formation of the Helium Molecule Ion Under Electron Impact" J.L. Franklin and F.A. Matsen <u>41</u>, 2948 (1964).

The appearance potential of $He_2{}^+$ had been determined by electron impact to be 23.1 eV. These molecular ions arise from the process

$$He^* + He \rightarrow He_2{}^+ + e$$

where He^* is any excited helium atom of energy equal to or greater than(why?) the AP. From a quantum-mechanical calculation in ref. 4 a rigorous lower bound of 2.2 eV for the dissociation energy of $He_2{}^+$ was obtained. Is the <u>lower</u> bound to be expected from the variation theorem?

Rationalize their equation $AP(He_2{}^+) \geq IP(He) - D_0(He_2{}^+)$ and numerical result $AP(He_2{}^+) \geq 22.4$ eV. Perhaps the discussion of exercise no. 84 will be of benefit.

The excitation energies, in eV, of the relevant helium states are taken from ref. 5, the volume by C.E. Moore referenced on p. 70 of this book. You might verify the values given from that source. The states and energies are $2^1P(21.22)$, $3^3S(22.72)$, $3^1S(22.92)$, $3^3P(23.01)$, and $3^1P(23.09)$. Which state is prevented from functioning as He^* by the quantum-mechanical calculation? Which state actually serves as He^*? What explanation is given for the inability of the other three states to function as the appropriate excited helium atom? Would you now regard the value of the AP of He_2^+ as understood?

90. "Vibrational Intensities of the $A^1\Pi \leftarrow X^1\Sigma^+$ Transition in Carbon Monoxide" A. Skerbele, V.D. Meyer, and E.N. Lassettre 44, 4069 (1966).

Fig. 1 shows the high-resolution electron-impact spectrum

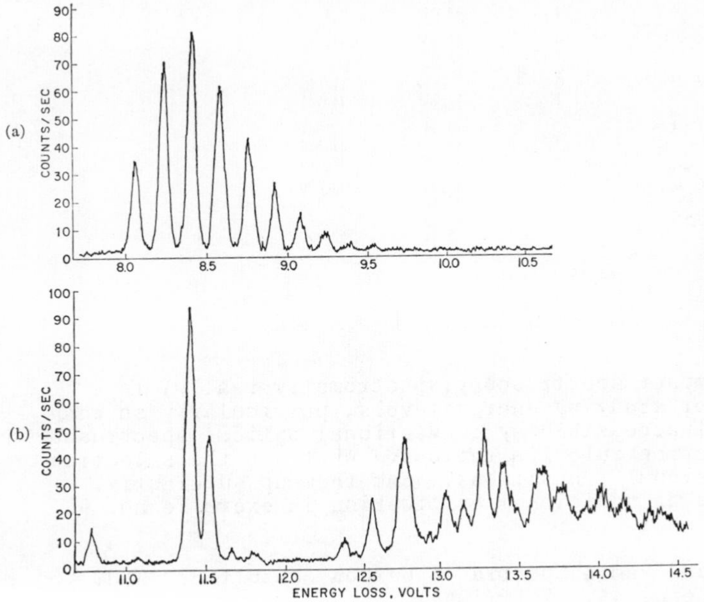

FIG. 1. High-resolution electron-impact spectrum of carbon monoxide. (a) Fourth positive bands, (b) remainder of the spectrum. Collision energy is 200.0 V, velocity analysis at 100.0 V, and $\theta \sim 0°$.

of CO obtained under the conditions stated in the caption. Were you aware that the individual vibrational transitions could be obtained so clearly? The fourth positive bands(what is a band by the way?) are the vibrational transitions all associated with the electronic transition(absorption or emission here?) $A^1\Pi \leftrightarrow X^1\Sigma^+$ as discussed in their ref. 3, the volume by Herzberg referenced on p. 93 herein. From that source we find ν_{00} for this electronic transition to be 64 746.5 cm^{-1}. See if this energy coincides with the peak just beyond 8 eV. Do you suspect this band may have been used to help calibrate the energy loss scale? We see that a progression is present, $v''=0 \rightarrow v'=0,1,2,...,10$, with the last band observed in several spectra using a more sensitive scale on the count-rate meter. Using the values, from ref. 3, of $\omega_e'=1515.61$ cm^{-1} and $\omega_e x_e'=17.250$ cm^{-1}, and making measurements from Fig. 1(a) as precisely as possible, draw the predicted spectrum and compare with the observed one. Is it possible to detect the anharmonicity in the latter?

The additional transitions shown in Fig. 1(b) are assigned in Table I. What factor is it that causes fewer members of these progressions to be observed? From ref. 3-5 and this Table check the value of ω_e' for each excited electronic state. The

E state had been assigned as $^1\Sigma^+$ in ref. 3. The two entries in the third from last row should be increased by 1.0 as was probably obvious.

TABLE I. Transitions observed in high-resolution spectra of carbon monoxide.

State	v'	Excitation potential (volts)[a] Electron impact	Ultraviolet
$B\ ^1\Sigma^+$	0	10.777*	10.777[b]
	1	11.051	
$C\ ^1\Sigma^+$	0	11.400	11.395
	1(?)	11.659	
$E\ ^1\Pi$	0	11.519	11.520[c]
	1	11.785	11.765
$F\ ^1\Pi$	0	12.367	12.367[d]
		12.577	12.580
		12.789	12.796
$G\ ^1\Pi$	0	13.050	13.049
	1 (?)	12.17	12.172
		13.29	13.285
		13.41	13.390

[a] Reference peak indicated by asterisk.
[b] Reference 5.
[c] Reference 4.
[d] R. E. Huffman, T. C. Larrabee, and Y. Tanaka, J. Chem. Phys. **40**, 2261 (1964).

Electron-impact spectroscopy(spectrometry really) is a valuable tool for studying energy levels, particularly so when the states are inaccessible by conventional optical spectroscopy because of selection rule limitations. What are the selection rules for collisional transitions? Sum them up succinctly.

We will see an additional application in exercise no. 92.

91. "Infrared and Raman Spectra of Carbon Subsulfide" W.H. Smith and G.E. Leroi <u>45</u>, 1778 (1966).

This exercise is not misplaced from the next chapter, but rather is included here because of Table I, the analysis of

TABLE I. Major peaks in the mass spectrum of carbon subsulfide.[a]

m/e	Ion	Peak height	Intensity relative to $C_3S_2^+$	Intensity of corresponding peak[b] in C_3O_2
100	$C_3S_2^+$	980	100	94
88	$C_2S_2^+$	12	1.2	...
76	CS_2^+	46	Impurity	...
68	$CCCS^+$	116	12	2.4
64	S_2^+	52	5.3	...
56	CCS^+	266	27	100
50	$C_3S_2^{++}$	124	13	9
44	CS^+	134	14	26
36	CCC^+	34	3.5	1.2
32	S^+	220	23	0.7
24	CC^+	29	3.0	12.6

[a] Internal standard: *S*.
[b] T. J. Hirt and J. P. Wightman, J. Phys. Chem. **66**, 1756 (1962).

C_3S_2. This material is a "brilliant orange-red, nauseous, mobile liquid freezing at around 5°C." A small amount of CS_2 was the only detectable impurity as Table I shows. How were the identifications confirmed? Sketch the pattern, including the correct intensities, to be expected for the isotopes of S^+.

The cracking patterns of C_3O_2 and C_3S_2 are compared in the last two columns. The authors surmise that the C-S bond may be more easily broken than the C-O bond, as expected, and therefore that the C-C bonds in the latter compound are broken relatively less often than those in C_3O_2. Why would this conclusion follow? Check both the premise and the conclusion at as many pairs of peaks as possible in Table I. Do you find any contradictions?

92. "Electron-Impact Spectrum of Ethane" E.N. Lassettre, A. Skerbele, and M.A. Dillon <u>46</u>, 4536 (1967) and <u>48</u>, 539 (1968).

The second reference is to an Erratum containing a corrected Table I which we have included here.

A vibrational progression is shown in the spectrum of Fig. 1 with excitation potentials to the various values of v' in

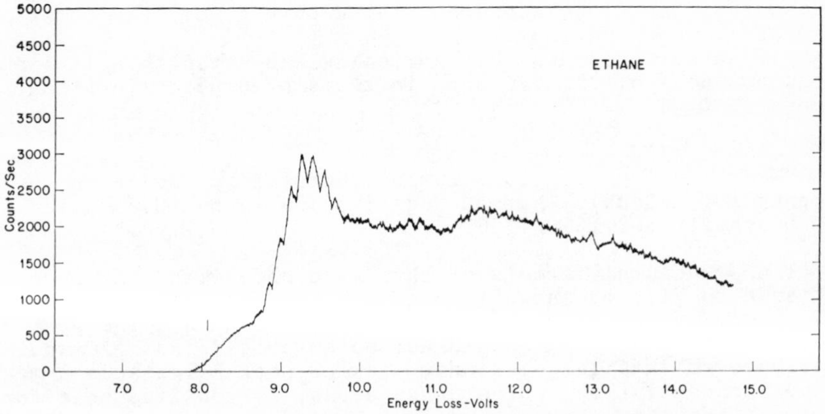

FIG. 1. Electron-impact spectrum of ethane. Accelerating voltage 50 (subtract 2.0±0.3 to obtain kinetic energy of incidence). Scattering angle 2°. Energy-loss scale calibrated by mixing mercury vapor with ethane. Spectra obtained at accelerating voltages of 100 are almost identical in shape to the above except that the intensity maximum is (by a very narrow margin) at $v'=5$ due (probably) to a small change in the underlying continuum.

TABLE I. Ethane excitation potentials.

v'	Excitation potential (eV)	Spacing
0	8.73	
1	8.87	0.14
2	9.00	0.13
3	9.13	0.13
4	9.26	0.13
5	9.40	0.14
6	9.54	0.14
7	9.69	0.15
8	9.85	0.16
9	(9.97)	(0.12)

Table I. Unlike the usual diatomic molecule, very little is known about the excited electronic states of most polyatomics. This is particularly so for a molecule like ethane where all valence electrons are used in single bonds, with neither lone pair nor π electrons; cf. Herzberg, pp. 545-6(referenced p. 109 herein).

The authors relate the lack of any perceptible convergence in the first 10 members of the progression to very strong binding in the excited electronic state; why can they make this inference? The vibrational mode under study is ν_3, basically a C-C stretch with frequency 994.8 cm^{-1} in the ground electronic state. This paper then reports the first information about this frequency in the excited electronic state. If it arises from a C-C stretch in the excited electronic state, would that imply that the excited state is more tightly bound? Why? From the experimental result that excitation to the $v'=4$ or $v'=5$ state is the most intense, and that the Franck-Condon principle therefore tells us that the outer turning point in those upper states is directly above(in what sense?) the point of maximum probability density in the state $v''=0$, see if you can duplicate their results for the decrease of 0.19 and 0.21 Å, for $v'=4$ and 5 respectively. Since the spacing is approximately constant, the potential will be parabolic or harmonic. In passing why was it assumed that the Franck-Condon turning points were the outer ones rather than the inner?

The authors discuss the shortening and strengthing(two phenomena or one?) of the C-C bond in terms of hyperconjugation arguments.

93. "Negative-Ion Mass Spectrum of Benzo[cd]pyrene-6-one. Evidence for a Doubly Charged Negative Ion in the Gas Phase" R.C. Dougherty 50, 1896 (1969).

The structural formula of this compound, with name somewhat shortened to "I", is shown in Fig. 1. Verify that the parent peak should appear at mass/charge of 254 as shown. The peak at 255 is due to what? We should not be surprised that ^{13}C would contribute to this large extent. Taking peak heights as a measure of concentration, rather than areas, do you find that for the least sensitive(lower) positive ion spectra that the 254/255 height ratio is $\sim 5/1$? This can be understood in terms of the binomial expansion $(a+b)^n$, with a the fraction of ^{12}C, b the fraction of ^{13}C, and n the number of C atoms in the molecule. The various terms in the expansion, whose sum is unity, then represent the fraction of molecules with a fixed $^{12}C/^{13}C$ ratio. Show that the first two terms of the expansion, representing the fraction of molecules with no ^{13}C and one ^{13}C atom, are in the ratio $a/(nb) \sim 5$. Upon what basic assumption does this treatment(in common with most simple statistical approaches) depend?

Surely you noticed the peak at 254.5; the author attributes it to the doubly charged dimer of I with a single ^{13}C atom as the most likely candidate. Can you think of another one? The half-integer would rule out any singly-charged ions, while the proximity to the parent peak would call for an n-charged n-mer. What would be the m/e ratio of the doubly charged dimer with no ^{13}C atoms? Does the 1% of this isotope, naturally abundant, come in very handy here?

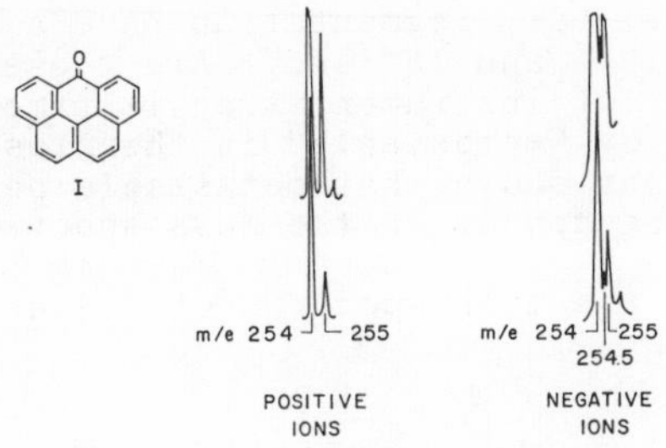

FIG. 1. Molecule-ion region in the positive and negative-ion mass spectra of I.

This may well be the first doubly charged negative ion found in the gas phase(ref. 2 and 3). He points out that these ions should be stable in the gas phase if the additional bond energy due to the second electron matches or exceeds the Coulomb repulsion in the ion. Why should this be so?

Assuming the same C-C bond length as in benzene, and a typical C=O distance, compute the two longest perpendicular in-plane distances. If a doubly charged dimer of I were to be formed by collision between two singly charged monomers, could the two extra electrons reasonably be expected to be ∿12.4 Å apart as he suggests is possible? Show that the extra Coulomb repulsion at that distance corresponds to 27 kcal/mole. He quotes a value of near 31 kcal/mole for the bond energy of an oxygen-oxygen single bond which could be formed in this geometry (ref. 12). Do you think this consideration led him to suggest 12.4 Å? Assume that the extra electron is localized on the carbon atom directly above the label "I" in Fig. 1. Then assuming that the dimer has D_{2h} symmetry, how long would the resultant O-O single bond become to conform to his suggested distance of charge separation? Is this unusually long? Impossibly long? Show that the argument is still very reasonable by demonstrating that only 29 kcal/mole of repulsion energy is acquired with the C=O bonds and the O-O bond all just 1 Å in length.

In addition to the negative ion peaks shown, others were observed at 226 and 224(presumed singly-charged); these would correspond to the loss of what two stable small molecules, with one C atom apiece, from the parent monomer I?

94. "Metastable Peaks in the Mass Spectra of N_2 and NO" A.S. Newton and A.F. Sciamanna <u>50</u>, 4868 (1969).

The metastable peaks discussed very briefly in exercise no. 86 will be examined in a bit more detail here. Show that for the process $N_2^+ \rightarrow N^+ + N$ a metastable peak may occur at M/q=7.0.

Earlier results(ref. 8) had shown that a delayed unimolecular dissociation of the diatomic ion HS^+ into the fragments H and S^+(lower energy than H^+ and S? by how much?) takes place. These results showed that there is no theoretical prohibition to the existence of excited states of diatomic ions which are long lived wrt fragmentation. The only criteria are that i) the state be one that can be excited from the ground-state diatomic by electron impact, and ii) the rate-determining step for depopulating the excited state by either radiation, crossovers (what might these be?), or fragmentation have a half-life between approximately 10^{-7} and 10^{-5}sec. Are these times then necessarily the order of instrument transit times? Is this really what is meant by "metastable" in the present context? In this research it was shown that metastable peaks arising by unimolecular processes appear in the mass spectra of both N_2 and NO.

Fig. 1 displays the mass spectrum of N_2 in the relevant

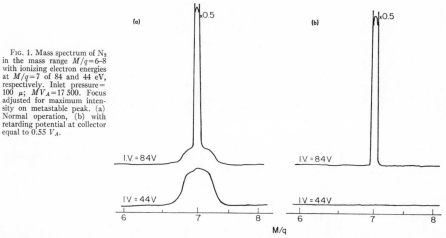

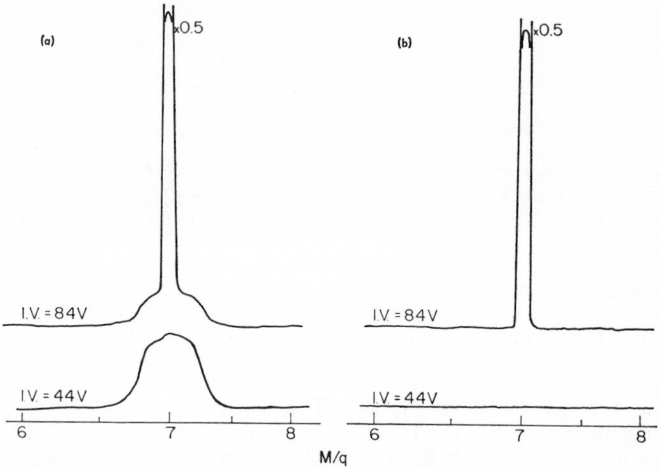

FIG. 1. Mass spectrum of N_2 in the mass range $M/q=6$–8 with ionizing electron energies at $M/q=7$ of 84 and 44 eV, respectively. Inlet pressure = 100 μ; $MV_A=17\,500$. Focus adjusted for maximum intensity on metastable peak. (a) Normal operation, (b) with retarding potential at collector equal to 0.55 V_A.

mass range. The two lower curves were obtained at an ionizing voltage of 44 V, said to be below the appearance potential of N^{2+}. It is obvious this peak at M/q of 14/2 is absent. It would be comforting to check the first two ionization potentials of the nitrogen atom, for instance from Moore's work referenced on p. 70 herein, and see if their sum exceeds 44 eV. Why don't you do this? Fig. 1(a) shows the metastable peak recognized by its broad diffuse nature and its occurrence(though not in this particular case) at nonintegral M/q in general. As pointed out by F.W. McLafferty in his "Interpretation of Mass Spectra" (W.A. Benjamin Inc. New York 1966), pp 64-68, variation in both decompositional kinetic energies and lifetimes of the parent ion contribute to the diffuse nature of the metastable peak. Is it really the decomposition fragment itself that is metastable (wrt instrument transit time)? Or is it the parent ion that is?

In Fig. 1(b) we are told that the retarding potential at the collector is 55% that of the accelerating voltage V_A. This retarding potential is also called the metastable suppressor voltage V_{mss}; do you see why?

In N_2 the transition associated with the metastable peak is shown to occur from a state of the diatomic ion with an appearance potential of 24.9±0.3 eV. The kinetic-energy release (meaning exactly what?) accompanying the metastable decomposition was found to be 0.55±0.10 eV. Show that this proves that the dissociation of the diatomic ion is to the products $N^+(^3P)$ and $N(^4S)$. Finally show that this energy of the products, 24.30 eV in the absence of kinetic energy, is just what one obtains for the energy of the atom and the ion(ground electronic states) wrt the ground state of N_2. Will the dissociation energy of N_2 plus the first ionization potential of the atom do it? These last two calculations are really one and the same.

A discussion of several states that might be the originating one of the diatomic ion for the metastable transition is given; it was concluded that none of the states were well enough defined to allow selection of the most probable one involved.

95. "Mass Spectrometric Detection of Cs_2^+ Produced by Associative Ionization" Y. Ono, I.Koyano, and I. Tanaka <u>52</u>, 5969 (1970)

The authors report the first mass spectrometric observation of Cs_2^+ produced by the reaction

$$Cs^*(8P_{1/2}) + Cs \rightarrow Cs_2^+ + e.$$

We will check some of the points involved in their experiment.

Earlier measurements of the photoionization spectra of cesium had shown that the monatomic vapor could be photoionized by light of energy less than its ionization potential. The ionization had been known to take place by absorption of light at the principal series line of Cs(6S→nP), $n \geq 8$. The mechanism had been explained in terms of the formation of molecular ions.

The light source, a helium discharge lamp, emitted two intense lines in the 3000-4000 Å region at 3188 and 3888 Å. Identify the transitions responsible. The first line was removed by a filter which cut off completely light shorter in wavelength than 3200 Å. Look up the ionization potential of atomic cesium to compare with their quoted value of 3.89 eV, corresponding to 3184 Å. Thus there is definitely not enough energy available to ionize a cesium atom by absorption of a single photon from the filtered helium-lamp output. How much energy is required to excite a cesium atom to the $8P_{1/2}$ state? Is the resonance between this particular transition and the one in helium producing the λ3188 line especially close?

The energies of formation of the various gaseous species are Cs(0), $Cs^*8P_{1/2}$(25 700), Cs^+(31 400), Cs_2(-3600), and Cs_2^+ (25 200 cm^{-1}). The last two may be obtained from the last sentence of the article under discussion and their ref. 3. Now show that the reaction given above is exothermic; by how much? Show also that the reaction of the molecular ion with Cs^* to produce a Cs ion is exothermic whether the other products of the reaction are the neutral diatomic or two ground state atoms with the change in energy computed for each case.

Is the role of the Cs_2^+ ion in the photoionization now obvious? What about the possibility of the formation of the diatomic ion by photoionization of the neutral diatomic with the filtered helium-lamp output? Is it possible to definitely rule this out? Why?

96. "Dissociation Energy of F_2" J.J. DeCorpo, R.P. Steiger, J.L. Franklin, and J.L. Margrave <u>53</u>, 936 (1970).

In this final exercise of this Chapter we will take a closer look at some of the energetics of three of the processes taking place under the category of electron impact phenomena.

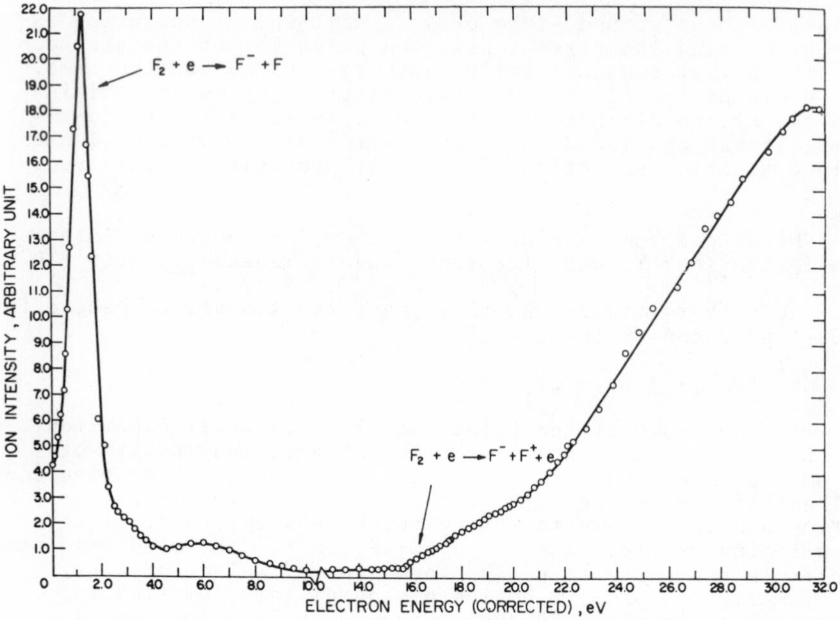

FIG. 1. Ionization efficiency curve of F^- from F_2.

Fig. 1 shows the ionization efficiency curve of F^- from F_2. Molecular fluorine can dissociate upon electron impact in the following ways:

$F_2 + e \rightarrow F^- + F$ dissociative electron attachment (1)

$F_2 + e \rightarrow F^- + F^+ + e$ ion-pair formation (2)

$F_2 + e \rightarrow F^+ + F + 2e$ dissociative ionization (3)

Are the names of the processes sufficiently descriptive? There is of course another possibility $F_2+e \rightarrow 2F+e$; is it possible to gain any information about the products from mass spectrometry? Process (1) is shown in Fig. 1 by the typical sharp resonance peak. Notice that the appearance potential AP for process (2) is obtained at 15.8±0.1 eV in Fig. 1; this from monitoring the negative F^- ion intensity. In the retarding potential difference(RPD) curve of F^+ from F_2 presented in Fig. 2, notice that the initial onset or AP is also at 15.8±0.1 eV, exactly the same value as for F^-. Assign the state of F^+ responsible for the break in the curve of Fig. 2 giving an AP of about 18.4 eV from the following energy level data taken from Moore(referenced p. 70 herein):

Term	Energy(cm^{-1})
$2p^4\ ^3P_2$	0.0
$2p^4\ ^3P_1$	341.8
$2p^4\ ^3P_0$	490.6
$2p^4\ ^1D_2$	20 873
$2p^4\ ^1S_0$	44 919

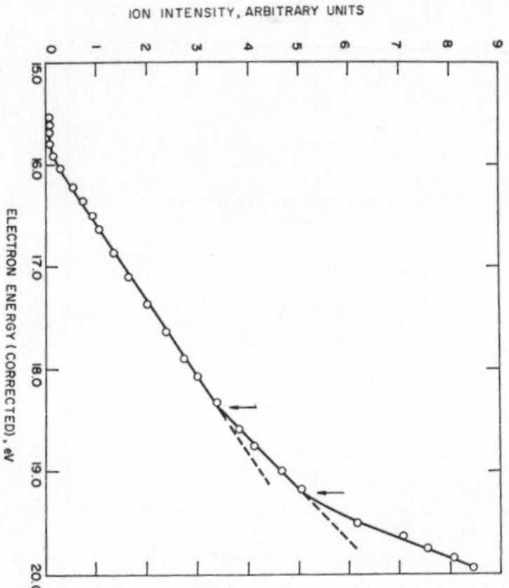

FIG. 2. Retarding potential difference curve of F^+ from F_2.

The additional break in the curve at 19.2±0.2 is attributed to process (3); is it obvious that it requires more energy than (2)? Is the capture of an electron by a fluorine atom an exothermic process?

$E^*>0$ is the excess translational energy of the fragmentation products, and was determined for each process at the threshold by the method of ref. 7 and 8. The results are given in Table I along with the AP's and the values of the dissociation energy obtained from each process in a manner to be discussed shortly. Fig. 3 gives E^* for the dissociative ionization process (3). It is seen to be zero within experimental error.

TABLE I. Results obtained from each dissociation process.

Process	Ion mode	A.P. (eV)	E^* (kcal/mol)	$D°(F_2)$ (kcal/mol)
(1) $F_2+e\rightarrow F^-+F$	negative	0.0±0.1	41.2±2.0	38.3±3.0
(2) $F_2+e\rightarrow F^++F^-+e$	negative	15.8±0.1	6.0±0.5	36.1±2.4
(3) $F_2+e\rightarrow F^++F^-+e$	positive	15.8±0.1	5.0±0.5	37.1±2.4
(4) $F_2+e\rightarrow F^++F+2e$	positive	19.2±0.2	0.0±1.0	38.6±4.8

The labeling of the processes in Table I should be in the order (1), (2), (2'), and (3), with the difference in (2) and (2') being in the ion monitored. The value of $D^0(F_2)$ obtained for each process will be illustrated by the result for process (1). The energy change for this reaction is equal to the AP for the F^- ion minus the excess translational energy E^*. This might be called the instrumental energy change. The thermodynamic energy change is the dissociation energy of F_2 minus the electron affinity of the fluorine atom taken as 3.448 eV(ref. 17). On this basis, what substance is defined to have an energy of formation of zero? What basis was used in the previous exercise for the cesium atoms and diatomics, neutral and charged? Are the bases the same? Convenient? The two energy changes here must be equal of course giving us in kcal/mole

$$0.0 - 41.2 = D^0(F_2) - 3.448 \times 23.06$$

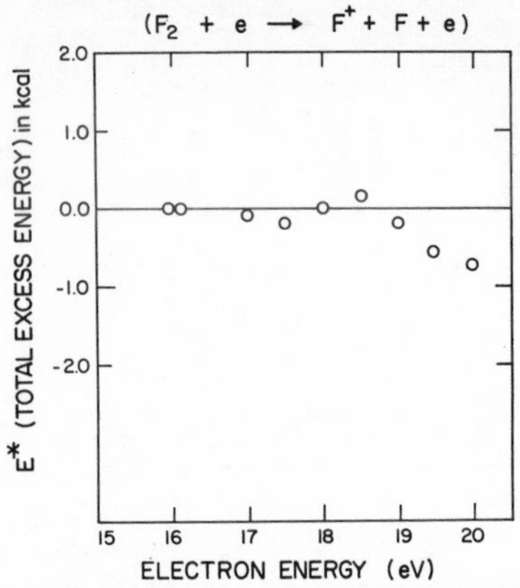

FIG. 3. Excess energy for the dissociative ionization process.

with $D^0(F_2)$ then equal to 38.3 kcal/mole. Continue in a similar fashion to evaluate the dissociation energies energies from the other three processes.

From this work then, the average of the values in Table I is 37.5±2.3 kcal/mole as you may easily check; does the value determined in exercise no. 1 fall within these authors' error limits?

CHAPTER VIII

INFRARED SPECTROSCOPY

97. "Substituted Methanes. XVIII. Vibrational Spectra of $C^{13}H_3I$" R.B. Bernstein, F.F. Cleveland, and F.L. Voelz 22, 193 (1954).

In this chapter we will expect to refer to the volume by Herzberg referenced on p. 58 of this exercise book; let us agree to refer to it as IRRS for "Infrared and Raman Spectra(of Polyatomic Molecules)".

For instance on p. 314 of IRRS are shown the normal modes of the methyl halides CH_3X and CD_3X. Should they be similar for isotopic substitution at the carbon atom? Of course. On the opposite page of the reference are given the fundamental frequencies of that epoch; they have changed a bit for $C^{12}H_3I$ at the time of this research as can be seen in Table VI, and have

TABLE VI. Fundamentals for gaseous $C^{12}H_3I$ and the isotopic shifts for $C^{13}H_3I$.

Fundamental	$\nu_{vac}C^{12}H_3I$	$\nu(C^{12}H_3I)-\nu(C^{13}H_3I)$
$\nu_1(a_1)$	2969.0[a]	10 ± 3 cm^{-1}
$\nu_2(a_1)$	1251.4[a]	5.5 ± 1.0
$\nu_3(a_1)$	{533.4	16.1 ± 1.0
	{525.1[b]	17.7 ± 0.5
$\nu_4(e)$	3062.2[c]	9.4 ± 2.3[d]
$\nu_5(e)$	1437.8[c]	4.4 ± 0.8
$\nu_6(e)$	882.5[c]	3.7 ± 0.3

[a] Lagemann and Nielsen value, corrected to vacuum.
[b] Raman value for the liquid state.
[c] These are the ν_0 values from Table V.
[d] Calculated from the Redlich-Teller product rule.

surely changed a bit since then. That's really not the way to state things, is it? We really mean that our accuracy of measurement continues to improve with time. This article was not based upon a mere refinement of frequency measurements but rather was apparently the first spectrum of a C^{13} molecule with more than three atoms to have been studied in any great detail.

Follow through their chemistry used in preparing the C^{13}-enriched methyl iodide starting with barium carbonate. What is the one great rule to be followed religiously when preparing a compound that is to be isotopically enriched? Something about checking the procedure out first with the normally isotopic chemicals? Why, most basically, are infrared(and occasionally other) spectroscopists so interested in isotopically substituted compounds?

In Fig. 2A and Fig. 4A and C are classic examples of parallel and perpendicular bands of a symmetric top or rotor. What

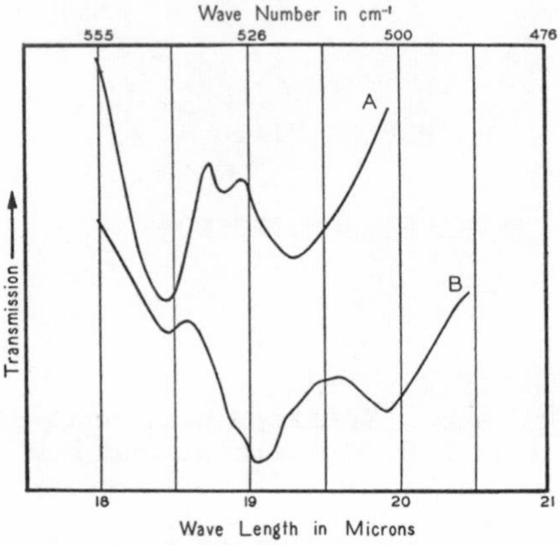

FIG. 2. Infrared absorption of gaseous CH_3I in the region of the $\nu_3(a_1)$ fundamental: (A) normal CH_3I, (B) 61 percent C^{13} enriched sample.

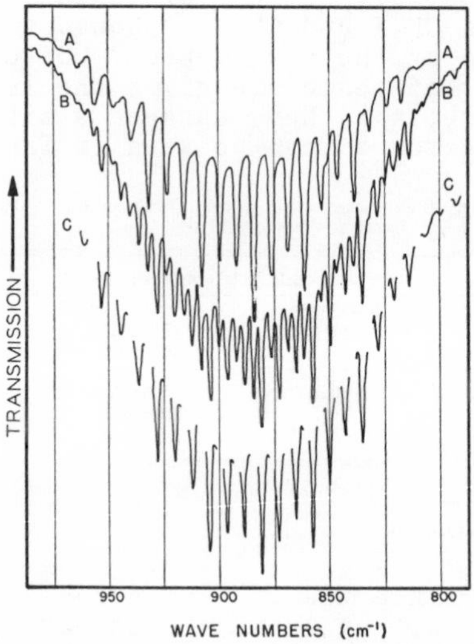

FIG. 4. Infrared absorption of gaseous CH_3I in the region of the $\nu^8(e)$ fundamental: (A) normal CH_3I, (B) 61 percent C^{13} enriched sample, (C) the deduced absorption for $C^{13}H_3I$ alone.

connection do these two adjectives have to the dipole moment change during vibration and the threefold symmetry axis? From the resolution shown can you determine if a prism or grating instrument was used? Fig. 4's caption should read $\nu_6(e)$.

In Fig. 4A and C are shown quite clearly the intensity alternation strong, weak, weak, strong,... characteristic of molecules with three-fold axes with identical nuclei having non-zero spin. See IRRS p. 432. Table II shows the K assignments as you should check against Fig. 4; ΔK=+1,K=0 is a strong line.

TABLE II. Positions of maximum absorption in the Q branches of the $\nu_6(e)$ band.[a]

Negative branch ($\Delta K=-1$)			Positive branch ($\Delta K=+1$)		
	ν_{vac} (cm^{-1})			ν_{vac} (cm^{-1})	
K	$C^{12}H_3I$	$C^{13}H_3I$	K	$C^{12}H_3I$	$C^{13}H_3I$
9	(818.2)[b]	814.3	0	884.9	881.2
8	(825.4)	821.3	1	892.6	889.0
7	832.3	828.7	2	900.5	896.5
6	839.8	836.0	3	908.3	904.7
5	847.4	843.4	4	916.1	912.8
4	854.7	851.0	5	924.1	920.4
3	862.2	858.4	6	932.3	928.2
2	869.5	866.2	7	940.3	936.7
1	877.3	873.7	8	948.7	945.0
			9	(956.8)	953.2

[a] ν_{vac} = wave number corrected to vacuum; K = quantum number; ΔK = change in the quantum number during the transition.
[b] Values in parentheses were calculated from the formula of BM, reference 4; all other $C^{12}H_3I$ values are from their data.

One last little exercise if you happen to have this issue of the journal handy. Look on p. 208, the first paragraph under section 3. In it you will see a marking made on a galley proof that has been innocently(not maliciously?) incorporated into the text. A caveat to you as a prospective, if not current, author and contributor to the scientific literature!

98. "Pure Rotational Spectrum of H_2Se between 50 and 250 cm^{-1}" E.D. Palik 23, 980 (1955).

Fig. 1 shows the far infrared spectrum of this molecule.

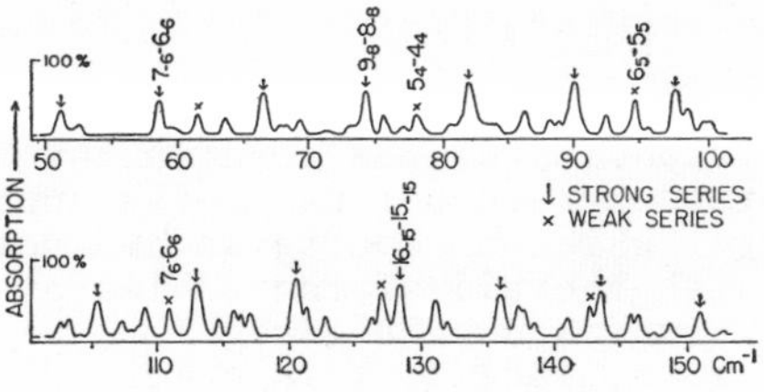

FIG. 1. Pure rotational spectrum of H_2Se.

It turns out that while H_2Se is an asymmetric rotor, with the three principal moments of inertia all different, the two smaller moments are not greatly different from one another while being about half of the value of the third. The molecule is thus approximately an oblate(coin, planar BCl_3) symmetric top. Methyl iodide, from the previous exercise, is an example of a prolate symmetric top. His approximate formula for the energy levels of this quasioblate top

$$F(J,K)_{oblate} = \tfrac{1}{2}(a+b)J(J+1) + [c-\tfrac{1}{2}(a+b)]K^2$$

is taken from IRRS p. 48. The rotational constants a, b and c are conventionally taken in the order $a \geq b \geq c$. If the first equality sign holds the molecule is an oblate symmetric top or rotor, and the second equality would classify it as prolate. What type of rotor has all three rotational constants equal? Give two or three molecular examples. These rotational constants, in cm^{-1}, are basically the reciprocal of the moments of inertia, e.g. $a=h/(8\pi^2 cI_A)$. J is the quantum number of the total angular momentum and K is the angular momentum <u>component</u> quantum number, where the component is that along the symmetry axis. K is not an exact quantum number in this case, but still it is useful for the low J transitions observed here.

Let us make a rough measurement on the spectrum. The transitions marked with a cross form part of the "weak series" with the rotational transitions at $\sim$78, 95,111 cm^{-1} labeled. The large numbers are the J values; in this absorption experiment is $\Delta J \pm 1$? The smaller subscripted number are the τ values where τ is an index on the 2J+1 different levels belonging to a given J value. Notice that for the three weak series transitions labeled that $\Delta\tau$ is zero. For our purpose we will take that to mean that ΔK is also zero. Now show from the approximate formula above that the spacing between adjacent members of this series is a+b. Measure the two spacings from the three labeled lines and compare the averaged answer with the results of Table I. Notice the regularity in the line assignments of

TABLE I. Molecular constants of H_2Se.

$a=8.16\ cm^{-1}$	$I_a=3.42\times10^{-40}\ g\ cm^2$
$b=7.69$	$I_b=3.64$
$c=3.91$	$I_c=7.16$
$r(H-Se)=1.47\ A$	$2\alpha=91°$

the weak series; J" and τ increase by one in regular fashion. We have made the process of analysis of the asymmetric top appear deceptively simple by introducing one or two inaccuracies. Note that the regular spacing of the members of the strong series is about half that of the weak series, yet the progression of the labeling of the transitions looks similar. Let us close this discussion of the spectrum by saying that the analysis was first carried out using the approximate formula to full advantage, which of course we did not do, then refining the results with the aid of the tables of ref. 5 and 6 to end up with the values shown in Table I.

For a planar figure, including this molecule, the sum of the two smaller moments of inertia is equal to the larger. Do the data of Table I bear this out? If not you might check the calculation of the moments from the rotational constants shown. Finally you may wish to read about the inertial defect Δ, 0.10 $\times10^{-40}\ g\cdot cm^2$ for H_2Se, on p. 461 of IRRS. Results for H_2O and H_2S are given on pp. 488-9 as well.

From the moments of inertia determine the bond length and angle for comparison with those shown. You will need to determine the center of mass, through which the principal inertial axes pass.

The trend in the bond angle when moving through the series from H_2O through H_2Te is an interesting one to follow. One or two standard explanations in terms of decreasing hybridization and/or decreasing bonding electrons/protons repulsion. The structural information which is the basis of these arguments came from infrared spectroscopy, and is contained in Herzberg's volume on electronic spectra of polyatomics, pp. 585-7. This book is referenced on p. 109 herein. You may wish to make your own correlation of this bond angle, perhaps with electronegativity or atomic number.

Are the values from this article for the bond length and angle so very different from more recent determinations?

99. "Infrared Spectrum of Iridium Hexafluoride" H.C. Mattraw, N.J. Hawkins, D.R. Carpenter, and W.W. Sabol <u>23</u>, 985 (1955).

The compound, a golden yellow gas, was prepared by direct combination of the elements. How was purification accomplished?

The observed bands, intensities, and assignments are given in Table I. From IRRS p. 122 we see that only ν_3 and ν_4 have changes in the dipole moment(zero in the static nonvibrating state) during vibration and hence are infrared active. It was

TABLE I. Infrared spectrum of IrF_6.

Observed band, cm^{-1}	Intensity[a]	Assignment
1414	M	$\nu_1+\nu_3$
1361	M	$\nu_2+\nu_3$
972	W	$\nu_1+\nu_4$
		$\nu_3+\nu_5$
918	W	$\nu_2+\nu_4$
850	MW	$\nu_2+\nu_6$
785	VW	$3\nu_4$
718	VS	ν_3
465	VW	$\nu_5+\nu_6$
450	VW	$\nu_3-\nu_5$
440	VW	$\nu_2-\nu_6$
324	W	$2\nu_5-\nu_6$
276	VS	ν_4

[a] Symbols: M = medium, W = weak, MW = medium weak, VW = very weak, VS = very strong.

then most reasonable to assign the two strongest bands(by far) as these fundamentals. The infrared inactive fundamentals were then determined from the combination(sum) and difference bands which were found. If the assignment of the band at 785 to $3\nu_4$ is correct, is this normal mode <u>strongly</u> harmonic? Would we expect to see $2\nu_4$ with an intensity of MW or W? At about what position in the spectrum? Is there a band in this vicinity? Can you make an alternate assignment of the 785 band using the fundamentals determined in Table II? Does the rest of the assignment of the spectrum depend on this particular one? In the same vein, for every difference band assigned we expect to see the corresponding sum band at an intensity at least as great. Is this condition fulfilled for their three assigned difference bands? For the one at 324, is the corresponding sum band liable to be hidden in the shoulder of the VS fundamental nearby?

Given the assignment of Table I, which was surely aided by the results for the similar molecules of Table II, show how you would determine the other four infrared inactive fundamentals.

What is the total number of fundamentals including degeneracies? Given that there are only six unique ones, what is the degeneracy pattern(i.e. how many singly, doubly and triply degenerate fundamentals)? Recall that the breathing mode, with the metal atom stationary(in rotating translating coordinate system!) and the six fluorine atoms moving out and in in phase, is nondegenerate.

One of the ReF_6 fundamentals appears to be off the trend.

TABLE II. Fundamental vibrations.

	Species	WF_6[a]	ReF_6[b]	IrF_6
ν_1	A_{1g}	769	753	696
ν_2	E_g	670	600	643
ν_3	F_{1u}	712	716	718
ν_4	F_{1u}	256	393	276
ν_5	F_{2g}	322	246	260
ν_6	F_{2u}	216	170	205

[a] Burke, Smith, and Nielsen, J. Chem. Phys. **20**, 447 (1952).
[b] J. Gaunt, Trans. Faraday Soc. **50**, 209 (1954).

100. "Infrared Spectra of Liquid Anhydrous Hydrogen Fluoride, Liquid Sulfur Dioxide, and Hydrogen Fluoride-Sulfur Dioxide Solutions" R.H. Maybury, S. Gordon, and J.J. Katz 23, 1277 (1955).

Are the freezing points of these two materials similar? What about their normal boiling points? To what would you attribute a large share of the responsibility for this last difference? The measurements were made at room temperature; does this say something about the construction of the cell? Other than being strong enough to withstand the vapor pressure, why was the cell constructed of nickel? Of what material were the windows? What agent was used to fluorinate traces of water present in the HF?

Fig. 1 presents the infrared spectrum of liquid HF. The very strong band centered at about 3450 cm^{-1} is attributed to absorption by polymers of HF. The spectrum of the gas phase monomer has P and R branch maxima at 3760 and 4100 cm^{-1} from

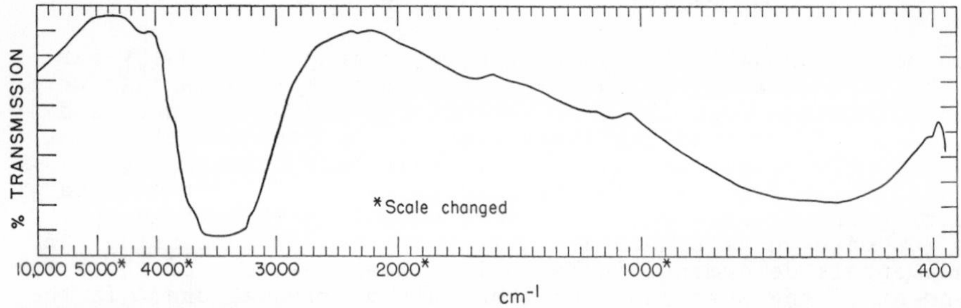

FIG. 1. Infrared spectrum of liquid hydrogen fluoride from 10 000 cm^{-1} to 400 cm^{-1}. Cell thickness 6μ. * indicates scale change.

their ref. 2 and 4. Is it an equivalent, if not preferable, viewpoint to state that the transition in the liquid is the same as in the vapor, with the positions of the levels perturbed somewhat by the hydrogen bond interactions? Fig. 2 shows the spectrum of liquid SO_2. It is quite a bit more complicated. Most fundamentally, why should this be so? Is the reason something other than a larger number of atoms giving a larger number of fundamental vibrational modes? Neglecting the very broad absorption of liquid HF between ~400-1000 cm^{-1}, could the reason be that the lowest fundamental of HF is toward the high end of the wavenumber scale and thus the opportunity for overtone and sum and difference bands is cut down? Of course there can be no sum and difference bands without more than one fundamental! Determine the three fundamental frequencies of SO_2 in the liquid

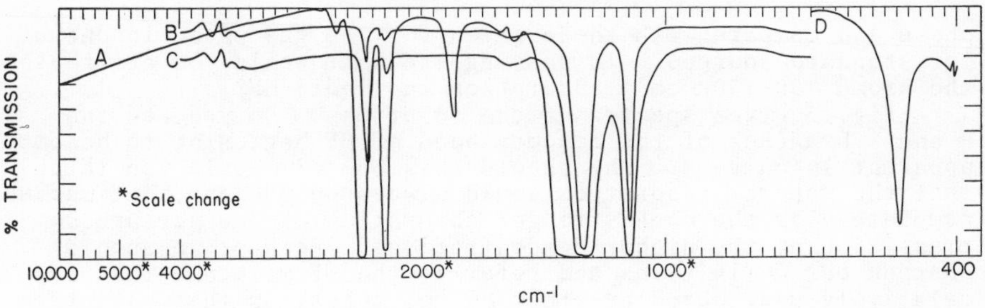

Fig. 2. Infrared spectrum of liquid SO_2 from 10 000 cm^{-1} to 400 cm^{-1}. *A*, 1 mm cell; *B*, 6μ cell; *C*, 39μ cell; *D*, 17.8μ cell.

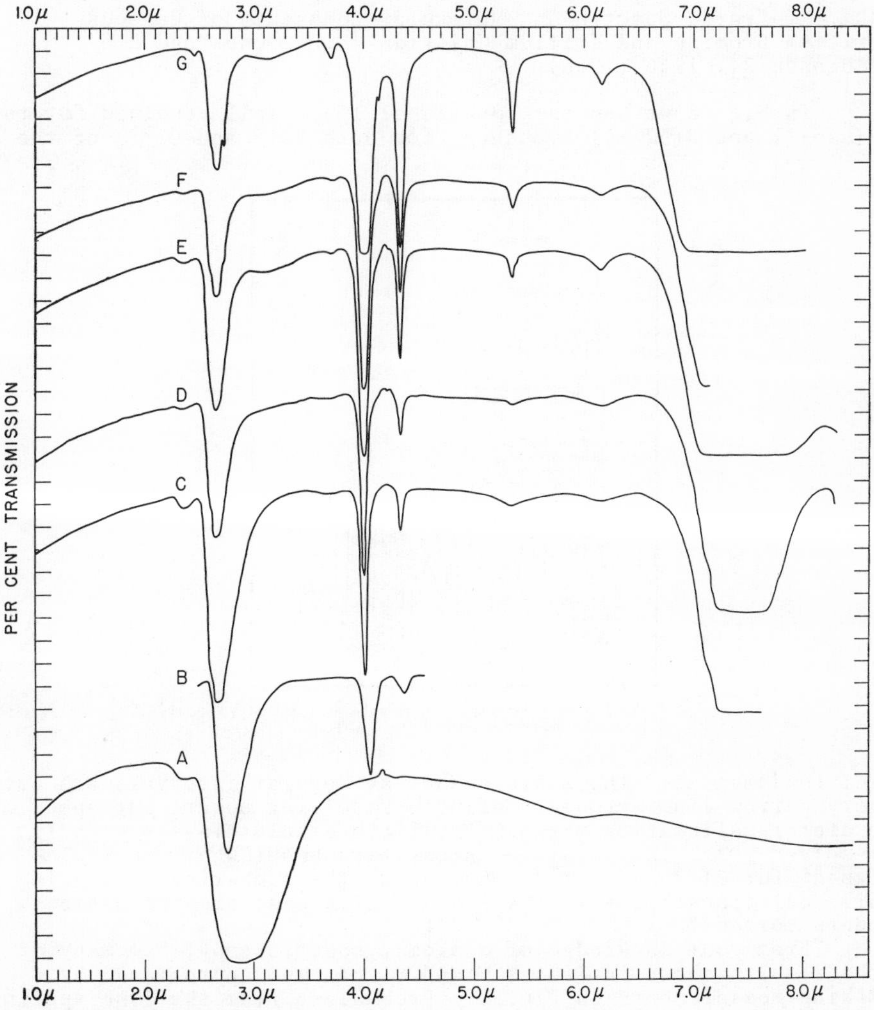

Fig. 3. Infrared spectra of SO_2—HF solutions. *A*, pure liquid HF, 6μ cell; *B*, 3.9 *f* HF, 39μ cell; *C*, 1.9 *f* HF, 73μ cell; *D*, 0.64 *f* HF, 73μ cell; *E*. 0.27 *f* HF, 255μ cell; *F*, 0.14 *f* HF, 255μ cell; *G*, 0.034 *f* HF, 1 mm cell.

phase and compare with those given for the gas phase in one of our standard sources. In passing, to what would you attribute the broad 400-1000 cm^{-1} absorption in liquid HF?

Fig. 3 gives spectra of the solutions. Do you see the P and R branches of the monomer band of HF beginning to become apparent in curve G? Why should this be so? Would you think that the monomer absorption would always be present, but masked completely by the much stronger absorption of the perturbed transition at the higher concentrations? Perhaps we should sharpen our definitions and refer to the HF molecule that is relatively unaffected by other HF molecules, or that is not participating in hydrogen bonding, as the monomer.

101. "Infrared Spectra and Molecular Constants of Gaseous Tritium Bromide and Tritium Chloride" L.H. Jones and E.S. Robinson 24, 1246 (1956).

In Fig. 2 we see the rotational lines well resolved for the $P(\Delta J=-1)$ and $R(\Delta J=+1)$ branches, for both Cl^{35} and Cl^{37}, of the

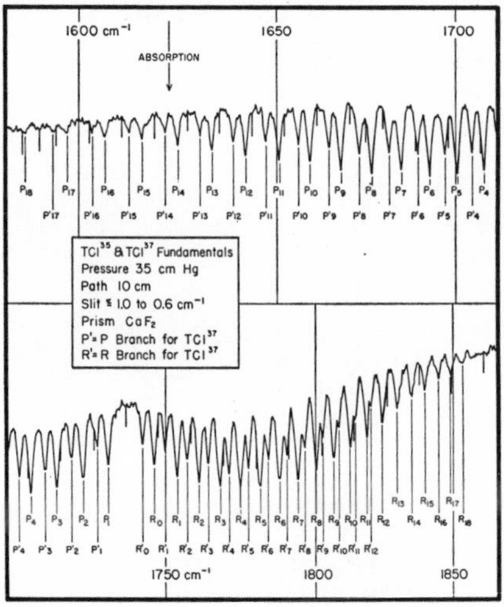

FIG. 2. Infrared absorption spectrum of the fundamental vibration-rotation bands of TCl^{35} and TCl^{37}.

TCl fundamental. The short dashes at regular intervals are not very narrow lines at near infinite resolving power, but spectrometer calibration marks. The isotope splitting is clearly resolved. Are the relative intensities about what you would expect for Cl^{35} and Cl^{37}? What about the numbering of the individual transitions in the P and R branches; are the assignments correct?

From your knowledge of diatomic spectroscopy, how many of the results shown in Tables V and VI can you approximate by making measurements on Fig. 2? Probably B from the line spacing and hence r, and of course ν_0 for the two isotopes. The other quantities, including the dependence of B and r on v, were

TABLE V. Molecular constants for TCl^{35}.

	Observed	Calc from HCl^{35}	Calc from DCl^{35}
B_e, cm^{-1}	3.7458	3.7378	3.7367
α_e, cm^{-1}	0.0611	0.0637	0.0638
D_e, cm^{-1}	7.7×10^{-5}	6.602×10^{-5}	6.464×10^{-5}
β_e, cm^{-1}	0.84×10^{-5}	-5.9×10^{-7}	-2.4×10^{-7}
r_e, A	1.273	1.274_4	1.274_6
ν_0, cm^{-1}	1739.10	1739.09	...
$\omega_e x_e$, cm^{-1}	...	18.36	...
$\omega_e y_e$, cm^{-1}	...	0.012	...
ω_e, cm^{-1}	1775.86	1775.85	...

TABLE VI. Molecular constants for TCl^{37}.

	Observed	Calc from HCl^{37}	Calc from DCl^{37}
B_e, cm^{-1}	3.7226	3.7217	3.7204
α_e, cm^{-1}	0.0655	0.0633	0.0634
D_e, cm^{-1}	6.75×10^{-5}	6.56×10^{-5}	6.41×10^{-5}
β_e, cm^{-1}	-1.15×10^{-5}	-5.9×10^{-7}	-2.4×10^{-7}
r_e, A	1.274	1.274_4	1.274_6
ν_0, cm^{-1}	1735.51	1735.39	...
$\omega_e x_e$, cm^{-1}	...	18.28	...
$\omega_e y_e$, cm^{-1}	...	0.012	...
ω_e, cm^{-1}	1772.11	1771.99	...

obtained by smoothing the measured values and using the method of combination-differences as described below.

Obtain the expression for the vibration-rotation term values including the centrifugal distortion constant $D>0$. It is of course $F_V(J)=B_VJ(J+1) -D_VJ^2(J+1)^2+\delta_{V,1}\nu_0$ with $v=0,1$. ν_0 is the forbidden($J''=0 \rightarrow J'=0$) transition of the band center. Now show that the combination-difference relations

$$R_{J-1}-P_{J+1} = (4B_0-6D_0)(J+\tfrac{1}{2})-8D_0(J+\tfrac{1}{2})^3 \quad (2)$$

$$R_J-P_J = (4B_1-6D_1)(J+\tfrac{1}{2})-8D_1(J+\tfrac{1}{2})^3 \quad (3)$$

$$R_{J-1}+P_J = 2\nu_0-2\alpha_eJ^2-2\beta_eJ^2(J+1)^2, \quad (4)$$

with $B_V=B_e-\alpha_e(v+\tfrac{1}{2})$ and $D_V=D_e+\beta_e(v+\tfrac{1}{2})$, can all be derived from the expression for the term values. The J values in the above equations are all those of the ground vibrational state, J". Hence $R_J=F_1(J+1)-F_0(J)$ and $P_J=F_1(J-1)-F_0(J)$. The constants α_e and β_e merely reflect the slight dependence of B and D on the vibrational quantum number. Prove that $\alpha_e>0$ from consideration of the shape of the potential energy curve.

With the calculated(smooth) values of the transitions in the R and P branches given in Table II for TCl^{35}, see if you can match all of the numbers in the second column of Table V save the last one(ref. 13). For example with equation (2) you should divide both sides by the quantity $J+\tfrac{1}{2}$, then a plot of $(R_{J-1}-P_{J+1})/(J+\tfrac{1}{2})$ vs $(J+\tfrac{1}{2})^2$ should yield a straight line with slope $-8D_0$ and intercept $(4B_0-6D_0)$.

Why are there no entries for the anharmonicity constants ω_ex_e and ω_ey_e in the second column of Tables V and VI? We will gain a bit of experience with the first of these quantities in exercise no. 103.

Could we make a slight, almost trivial, correlation between the authors' affiliation and the use of the tritium isotope?

TABLE II. Observed and calculated frequencies for TCl^{35}.

	$\nu_{cm^{-1}}$ observed	$\nu_{cm^{-1}}$ calculated	$\nu_{obs}-\nu_{calc}$
R 18	1853.75	1853.63	+0.12 cm^{-1}
17	1849.07	1849.08	−0.01
16	1844.27	1844.34	−0.07
15	1839.29	1839.41	−0.12
14	1834.23	1834.31	−0.08
13	1828.84	1829.04	−0.20
12	1823.51	1823.58	−0.07
11	1817.98	1817.97	+0.01
10	1812.16	1812.19	−0.03
9	1806.30	1806.26	+0.04
8	1800.24	1800.23	+0.01
7	1793.83	1793.97	−0.14
6	1787.56	1787.56	0.00
5	1780.95	1781.04	−0.09
4	1774.17	1774.37	−0.20
3	1767.59	1767.58	+0.01
2	1760.66	1760.65	+0.01
1	1753.60	1753.59	+0.01
0	1746.43	1746.41	+0.02
P 1	1731.93	1731.67	+0.26
2	1724.12	1724.12	0.00
3	1716.42	1716.45	−0.03
4	1708.53	1708.67	−0.14
5	1700.76	1700.77	−0.01
6	1692.71	1692.76	−0.05
7	1684.60	1684.62	−0.02
8	1676.43	1676.41	+0.02
9	1668.01	1668.08	−0.07
10	1659.54	1659.56	−0.02
11	1650.86	1650.98	−0.12
12	1642.17	1642.28	−0.11
13	1633.40	1633.48	−0.08
14	1624.80	1624.57	+0.23
15	1615.53	1615.54	−0.01
16	1606.32	1606.39	−0.07
17	1596.96	1597.14	−0.18

102. "Xenon Difluoride" D.F. Smith 38, 270 (1963).

The infrared spectrum was obtained and used to make a rough estimate of the bond length(s) in this molecule. A mass spectrometer scan of the products of the preparative reaction showed peaks of Xe^+, XeF^+, and $XeF_2{}^+$, without any $XeF_3{}^+$ which is the most common fragment from the decomposition of XeF_4 in the mass spectrometer(ref. 1b). Is this convincing proof of the absence of XeF_4? What about the interpretation of the two bands found at ∿558 and 1070 cm^{-1}, with unresolved P and R branches, as suggesting that the molecule is linear? The bands are suggested to be due to ν_3 and $\nu_1+\nu_3$ transitions. If linear, in what region of the spectrum should the bending vibration ν_2 lie? What about the symmetric stretch or breathing mode ν_1? Would it be allowed if XeF_2 is linear? If nonlinear could ν_2 still occur in the low-frequency region? Would ν_3 then be expected at a frequency somewhat greater than ν_1? Is the lack of a discernible Q branch proof that none exists? Of course not; it could be quite weak and/or not resolved from the P or R branch. If a Q branch were found would it prove nonlinearity? See IRRS p.380 for a discussion. Would it prove linearity? Would a bent molecule exhibit a complicated spectrum, similar to that of H_2Se in exercise no. 98, in the vibration region? Suggest another form of spectroscopy whose application could conclusive prove nonlinearity or strongly suggest linearity. Do you think the molecule is linear?

For the band at 558 cm^{-1} the two maxima are at 550 and 566

cm^{-1}, with the latter about 20% more intense. For the purpose of this exercise we will define them as the P and R branches. From the author's ref. 2(IRRS p. 391) show that this leads to an internuclear distance of 1.7 Å by assuming that the spectrum was taken at room temperature. Should this assumption be made automatically from the description of the experimental conditions?

Try out this formula for the case of TCl in the last exercise. Take the most prominent transition in each branch, then move one to either side to gain a feeling for the accuracy. Would these lines be much closer together in XeF_2? Can you derive this formula(IRRS pp. 18-19), ignoring the alternation of statistical weights for the presumably linear symmetric case?

103. "Ground State of the HeH^+ Molecule Ion" B.G. Anex 38, 1651 (1963).

Is this species purely fictional? What about the β decay product of TH? The author quotes results of two independent investigations(ref. 2 and 3; the latter contains an informative footnote) by mass spectrometry indicating that about 90% of the TH disintegrations result in He^3H^+ ions stable for at least the mass-spectrometer transit time. A Morse function was fitted to four points calculated on the potential energy curve and the energies of the four vibrational states v=0-3 are shown. From these determine ω_e and $\omega_e x_e$ with two checks. It is a property of the Morse potential that no higher anharmonic terms are present from the energy eigenvalue as you may know. Draw a spectrum of the fundamental and first two overtones that one would expect in the infrared region. Should a Q branch be exhibited? Why or why not? Can a diatomic give rise to a Q branch? No? NO(nitric oxide)? Does the same situation arise here? See Fig. 1 of the next exercise.

This calculation goes well beyond the Hartree-Fock limit discussed on p. 49 herein, obtaining a very large portion of the correlation energy through configuration interaction. The bond has been described as the prototype of the two-electron heteropolar bond. Do you agree with the description?

TABLE XII. Physical constants of He^4H^+.

	Evett	This calculation
Equilibrium internuclear separation (bohrs)	1.432	1.446
Uncorrected dissociation energy (eV)	1.90	1.931
Zero point energy (eV)	0.22	0.204
Corrected dissociation energy (eV)	1.68	1.727
Vibrational energies (cm^{-1})		
ν_0	1750	1643
ν_1	4948	4654
ν_2	7744	7299
ν_3	10138	9577

104. "Absolute Intensity Measurement for the NO Fundamental at 5.3 Microns" B. Schurin and S.A. Clough <u>38</u>, 1855 (1963).

The integrated band intensity A is defined in terms of the spectral absorption coefficient α by

$$A = \int_{\text{band}} \alpha(\nu)d\nu = (1/p_0\ell) \int \ln (I_0/I)\ d\nu$$

with p_0 the equivalent pressure of an ideal gas, ℓ the sample path length, and (I_0/I) the ratio of incident to transmitted light intensity as a function of wavenumber ν. The spectral absorption coefficient α is closely related to the molar extinction coefficient ε; can you write the equation connecting the two? What is the reason for using such high pressures?

The measured value of A is $111 \pm 7\ cm^{-1}/(cm \cdot atm)$ at 273°K. See if you can match this value by approximating the area under the absorption curve for each of the three pressures in Fig. 1. Neglect corrections for gas imperfections (4% maximum), but assume that the temperature is somewhat higher than room temperature, perhaps 310-320°K. Correct your results to 273°K. Does the Q branch shown depend upon high pressure for its existence?

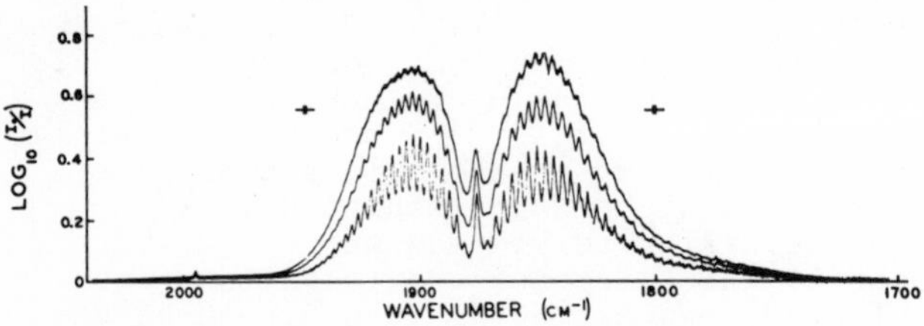

FIG. 1. Spectra of the NO fundamental taken with a 0.5 mm path length and pressures of 500, 365, and 210 psia.

Note the linearity of the plot in Fig. 2; determine the value of A at 273°K. This of course is just what the authors did. Our measurement suggested above, from Fig. 1, will be only a rough approximation.

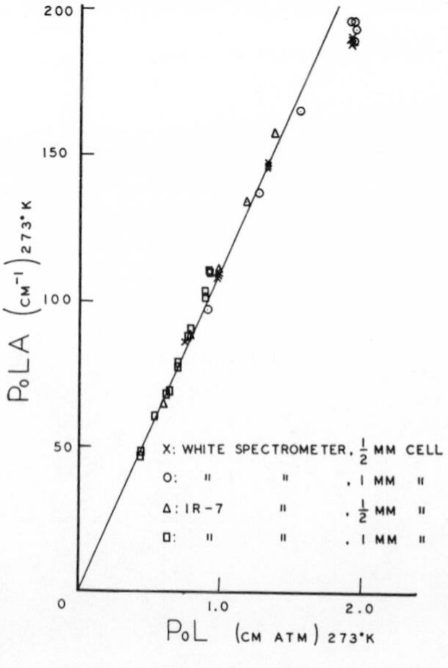

FIG. 2. Beer's law plot for the NO fundamental.

105. "Infrared Spectroscopic Evidence for the Species HO_2"
D.E. Milligan and M.E. Jacox 38, 2627 (1963).

Mixtures of HI and O_2, in an Ar matrix at 4°K, were photolyzed and infrared absorption spectra of the products were obtained. A spectrum is shown in Fig. 1. The band at 1399 cm^{-1}

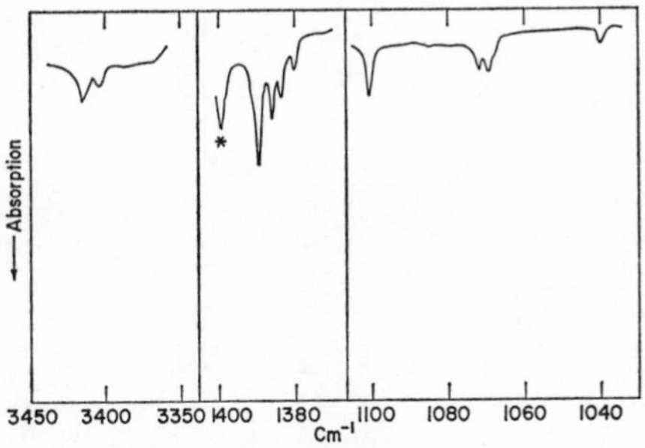

FIG. 1. Ar:O_2(27.7%O^{18}):HI = 200:4:1. Spectrum following photolysis. Present in initial sample.

marked with an asterisk is the only one appearing in these regions of the spectrum before photolysis, and is presumed to be due to HDO remaining in the deposition line from a previous experiment. Is this a reasonable supposition? See e.g. IRRS p. 282.

Table I gives the results including those for isotopically

TABLE I. Photolysis product frequencies (cm^{-1}) observed in Ar:O_2:HI experiments employing isotopic substitution.

Ar:O_2^{16}:HI	Ar:O_2^{18}:HI	Ar:O_2(27.7% O^{18}):HI	Ar:O_2^{16}:DI	Ar:O_2^{18}:DI
1101	1040	1040	1020	982 (w)
		1069.5		998
		1072		
		1101		
1389.4	1380.3	1380.3		
		1383.6		
		1386.0		
		1389.4		
3414	3402	3402	2530	2499
		3414		2515

substituted parent molecules. The frequencies in the first column are assigned as the O-O stretch, the bending mode, and the H-O stretch respectively. The bond angle is unknown though thought to lie between 90° and 120°. For all seven of the O-O stretching frequencies observed, assign each to a given isotopic species making use of the diatomic approximation HO-O.

Note that structurally the molecule may be thought of as a fragment of hydrogen peroxide.

Have you seen HO_2 postulated as an intermediate in the H_2+O_2 reaction mechanism?

106. "Infrared Spectra of Methylgermane, Methyl-d_3-Germane, and Methylgermane-d_3" J.E. Griffiths 38, 2879 (1963).

TABLE I. Infrared spectrum of methylgermane (cm^{-1}, vac).

Observed	Calculated	Intensity	Assignment	
3775	3779.8	w	$\nu_1+\nu_4$	(A_1)
3508 *R*	3503.1	w	$\nu_7+\nu_{12}$	(A_1+A_2+E)
3486 *P*				
2997.2	···	w	ν_7	(e)
2947.5 *R*	···	s		
2937.8 *Q*	···	s	ν_1	(a_1)
2924.8 *P*	···	s		
2868 *R*	···			
2856 *Q*	2855.4	m	$2\nu_9$	(A_1+E)
2846 *P*	···			
2588	2590.7	w	$\nu_2+\nu_{12}$	(E)?
2569				
2512.4 *R*	···			
2501.2 *Q*	2508.2	w	$2\nu_3$	(A_1)
2487.6 *P*	···			
2386	2386	vw	$\nu_2+2\nu_6$	(A_1)
2240	2240	w	$\nu_6+\nu_8$	(E)
2097.2 *R*	···			
2084.8 *Q*	···	vs	ν_2	(a_1)
2073.8 *P*	···			
2084.3	···	s	ν_8	(e)
1867.3 *R*				
1854.5 *Q*	1855.7	w	$\nu_3+\nu_5$	(A_1)
1842.5 *P*				
1783	1783	vw	$\nu_2-2\nu_6$	(A_1)
1738	1742.9	vw	$\nu_4+\nu_{10}$	(E)
1686	1685	vw	$2\nu_4$	(A_1)
1575	1583	vw	$\nu_6+\nu_9$	(E)
1427.7	···	w	ν_9	(e)
1266.4 *R*	···			
1254.1 *Q*	···	s	ν_3	(a_1)
1241.4 *P*	···			
1028.6	1027.9	vw	$2\nu_{12}$	(A_1+E)
900.4	···	s	ν_{10}	(e)
853.5 *R*	···			
842.5 *Q*	···	vs	ν_4	(a_1)
834.5 *P*	···			
847.5	···	vs	ν_{11}	(e)
613.9 *R*	···			
601.6 *Q*	···	vs	ν_5	(a_1)
589.9 *P*	···			
522	···	s	$\nu_{12}+\nu_{12}-\nu_{12}$	
505.9	···	s	ν_{12}	(e)

These eight-atom molecules all have the same symmetry distribution of the fundamental frequencies; ν_1 through ν_5 are parallel vibrations of species a_1, ν_6 is the torsional vibration of species a_2, and ν_7 through ν_{12} are perpendicular vibrations of species e. Is this the correct number of frequencies? All of the fundamentals except ν_6 are allowed in the infrared and observed as in Table I for CH_3GeH_3. Neglecting anharmonicity, what value(s) do you deduce for the torsional vibration from sum and difference bands in Table I? Discuss the 522 cm^{-1} band.

Fig. 2 shows the P,Q and R branches of ν_5 for each of the

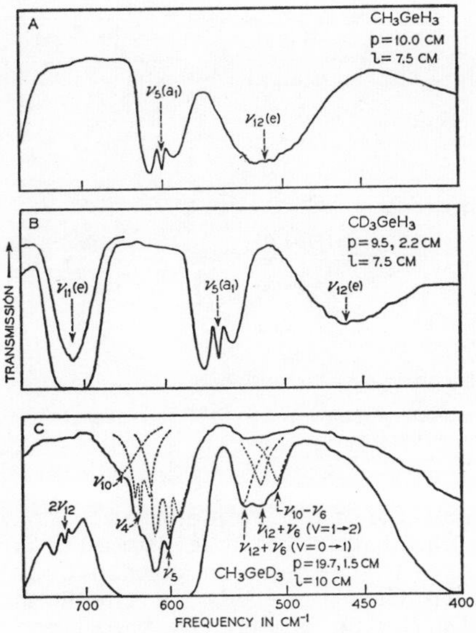

FIG. 2. Infrared spectra of (a) CH_3GeH_3, (b) CD_3GeH_3, and (c) CH_3GeD_3 from 800 to 400 cm^{-1}.

three molecules. The fundamentals are 549.6 and 598 cm^{-1} for CD_3GeH_3 and CH_3GeD_3 respectively. How would you approximately describe the normal mode corresponding to this fundamental? From the evidence available, can you muster it into a form that will support your statement?

107. "Vanadium Tetrachloride Vibrational Spectrum" E.L. Grubb and R.L. Belford 39, 244 (1963).

What is the Jahn-Teller effect? May it occur in the VCl_4 molecule and why? These authors as well as those that they reference believe that the spontaneous distortion, if it does indeed occur, will be in the doubly-degenerate ν_2 mode. From IRRS p. 100 this is seen to correspond to a "scissors", or better perhaps a "grass clippers", motion(in one of its degenerate components) such that the molecular distortion and subsequent removing of the orbital degeneracy is accomplished. Make a sketch to show that the molecule is then inscribed in a tetragonal prism rather than inside of a cube as is a regular tetrahedron.

This work presents a further study of the infrared spectrum in order to help clarify the situation. Three points mentioned specifically are: i) all bands were observed in the vapor phase; why? ii) great care was taken to detect and exclude impurities which interfere and are thus extraordinarily troublesome, iii) as a result some suggested assignments of both the infrared and Raman bands are made as shown in Table I.

TABLE I. Infrared and Raman bands in VCl_4. The observations and assignments of Dove *et al.* (Ref. 4) are listed in the first three columns. Those of the present study are listed in the last two columns. All frequencies are in kaysers.

Raman[a] (Ref. 4)	Infrared[b] (Ref. 4)	Preferred assignment[c] (Ref. 4)	Infrared[d] (this work)	Proposed assignment
128 m		$\nu_2(\epsilon)=\nu_4(\tau_2)$		$\nu_4(\tau_2)\approx 128$
383 vs		$\nu_1(\alpha_1)$		$\nu_1(\alpha_1)\approx 383$
475 w	490 vs (475, 482)	$\nu_3(\tau_2)$	487 vs	$\nu_3(\tau_2)\approx 487$
	509 s (504, 505 sh)	$\nu_1+\nu_4=511$	510 vw, sh	$VOCl_3$, *maybe* $\nu_1+\nu_4$
	620 vs (610 m)	$\nu_3+\nu_2=603$	570 vw	Impurity or $\nu_3+\nu_2(\nu_2<100K)$
	640 m (626 w)	$\nu_3+\nu_4=603$	613 w	$\nu_3+\nu_4=615$
	(855 vw)	$\nu_1+\nu_3=858$	861 w	$\nu_1+\nu_3=870$
	(955 w)	$2\nu_3=950$	962 w	$2\nu_3=974$

[a] CCl_4 solution.
[b] Infrared bands from reference 4 are those of saturated vapor at 25°C; those of pure liquid or CCl_4 solution are indicated in parentheses.
[c] Another set based on $\nu_2=128$, $\nu_4=150$ is shown in reference 4.
[d] Vapor, saturated, 80°C.

What unit is used for the frequencies? To what is it equivalent? The designations for the symmetry species α, ε and τ correspond to what other set(s) in common use? Which fundamentals are allowed in the infrared? in the Raman? The scissors mode ν_2 is accompanied by no change in the dipole moment as you can see. The authors suggest that ν_2 has not been observed at all in the Raman spectrum and only in the very weak 570-K sum band in the infrared, if at all. In passing, if you had to assign one of the 510, 570K pair to VCl_4 and one to an impurity which choice would you make based on their discussion? They say that their suggestion was frankly colored by the expectation that ν_2 would be strongly involved in a Jahn-Teller interaction; what would this do to its vibrational level spacing and why? What experiment do they suggest to observe ν_2? To what do they attribute their suggested failure of ref. 4 to observe this fundamental in the Raman studies? Speaking of their suggested experiment, the present authors say "In the absence of that experiment, we can infer confidently neither the influence nor lack of influence of a possible Jahn-Teller effect in the vibrational spectrum of VCl_4."

Were the three points mentioned above a concrete step in the search for a possible Jahn-Teller effect? Make a literature search from the time of this article forward to see if the suggested experiment was performed. For how many molecular or ionic substances has a definite Jahn-Teller distortion been established?

108. "Effect of Pressure on the Infrared Spectra of Some Hydrogen-Bonded Solids" J. Reynolds and S.S. Sternstein 41, 47 (1964).

Fig. 4 shows the pressure dependence of the hydrogen-bonded

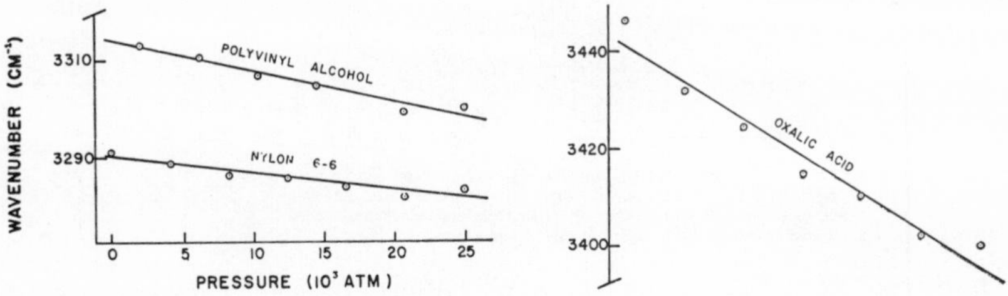

FIG. 4. Average frequency of the –AH stretching band as a function of pressure for the hydrogen-bonded system A–H···B.

O-H···O stretching frequencies in oxalic acid and polyvinyl alcohol, and the N-H···O stretching frequency in nylon 6-6. How is this explained in terms of the electrostatic attraction between the proton and atom B(here oxygen)? Will increased pressure force A and B atoms more closely together? What then happens to the H···B attraction? the A-H distance? the A-H bond strength? and finally, the A-H stretching frequency? Do you have a suggestion as to why this effect is more pronounced in oxalic acid than in the other two?

The second aspect of the study is the effect of pressure on the positions and intensities of both the antisymmetric and symmetric $-CH_2$ stretching frequencies. The spectra are shown in Fig. 5. Is it a general rule that the antisymmetric bands

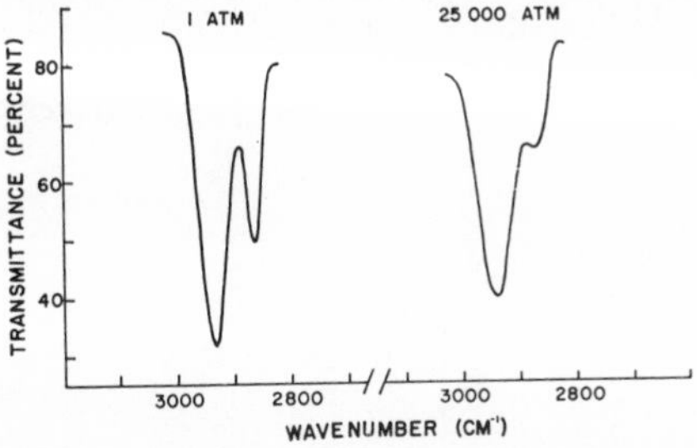

FIG. 5. $-CH_2$ stretching bands of nylon 6–6 at two pressures.

occur at somewhat higher frequencies than the symmetric ones? Why is this so? What is the sign of $d\nu/dP$ for these bands? Is it opposite to that for the other bands? Does it imply a shortening of the C-H bonds with increasing pressure as expected? Their possible explanation of these effects, including the striking decrease in the intensity of the symmetric stretching band, is tied to the change in the magnitude of the dipole moment changes themselves as shown in Fig. 6. For the symmetric stretch the change is along the dipole(2-fold symmetry) axis itself; it is perpendicular to the dipole axis for the antisymmetric stretch. Summarize their arguments and explanations for

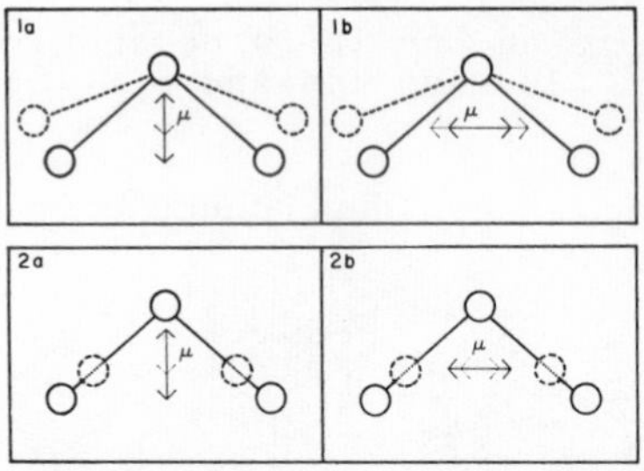

FIG. 6. The effect of (1) an H–C–H angle increase and (2) a C–H bond shortening on the dipole moment changes for the symmetric (a) and asymmetric (b) stretching vibrations. The circles represent the mean positions of the atoms.

these observed effects in nylon 6-6. What is their essential assumption relating dipole moment change to bond length? Is this closely akin to assuming a point charge model? How is this dipole moment change expressed quantum mechanically(that is to say correctly!)? What is the relation between this last quantity and the observed intensity?

109. "Infrared Absorption of Ce^{3+} in LaF_3 and of CeF_3" R.A. Buchanan, H.E. Rast, and H.H. Caspers 44, 4063 (1966).

The trivalent cerium ion contains one 4f electron outside of closed shells. The resultant ^2F term has states with J=5/2 and 7/2; which of these would lie the lowest? Would the situation be reversed for Yb^{3+} ions? The states belonging to a given J value are further split into three and four Stark levels respectively by the action of the crystalline field; why isn't the degeneracy completely removed for the 2J+1 components? The residual degeneracy is called the Kramers' degeneracy; to what is it due? Fig. 3 presents the energy level scheme of the ground ^2F term as determined from the present work. The upper

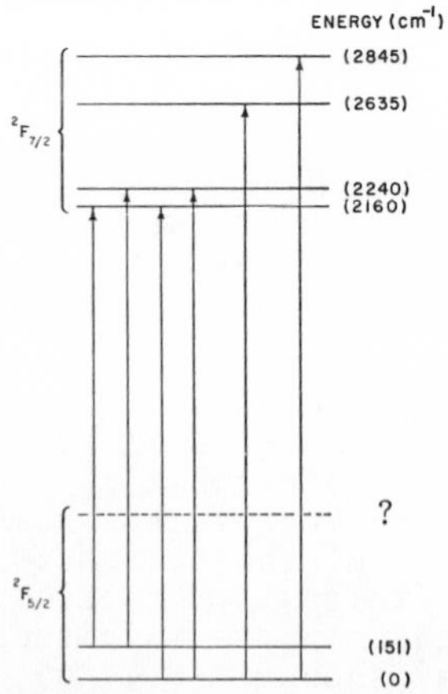

FIG. 3. Energy-level scheme for 1% Ce^{3+} in LaF_3. The transitions indicated correspond to the major electronic-absorption lines observed in the spectra of Fig. 1.

level belonging to the J=5/2 state could not be located experimentally.

Table I shows the selection rules for the three proposed site symmetries. The Ce^{3+} concentration in LaF_3 was 1%. The

TABLE I. Summary of electronic selection rules for Ce^{3+} in LaF_3 and CeF_3.

La site symmetry	Electric and magnetic dipole transitions
D_{3h} (bimolecular unit cell)	Some transitions forbidden Some transitions polarized either σ or π Some transitions unpolarized
C_{2v} (hexamolecular unit cell)	All transitions allowed no polarization
C_2 (hexamolecular unit cell)	All transitions allowed no polarization

spectra of the cerium ion in both the doped LaF_3 and CeF_3 is nearly identical as Fig. 1 and 2 show. The observed lines are

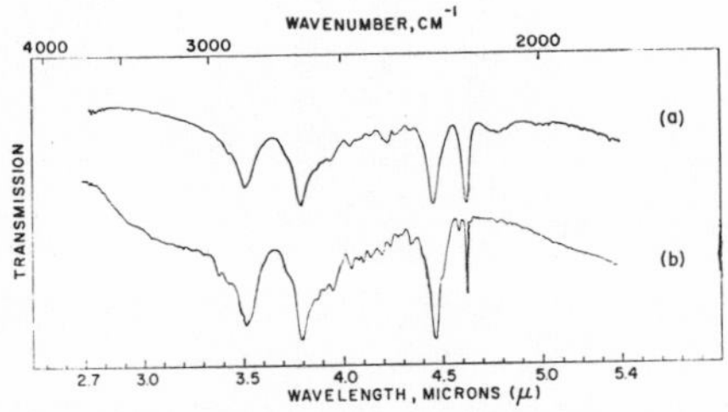

FIG. 1. Spectra of 1% Ce^{3+} in LaF_3 sample 6.76 mm thick. (a) Sample cooled with liquid nitrogen; (b) sample cooled with liquid helium.

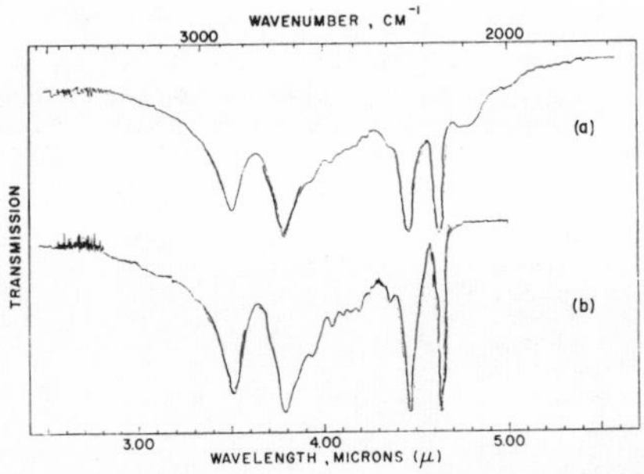

FIG. 2. Spectra of CeF_3 crystal 0.125 mm thick. (a) Sample cooled with liquid nitrogen; (b) sample cooled with liquid helium.

listed in Table II. Check the conversion from wavelength in air to wavenumber in a vacuum for one or two cases. You will need the refractive index of air at those wavelengths. Is the correction significant for five significant-figure wavelengths? Identify the two "hot" lines from Table II; is their intensity temperature dependent? From the article we see that study of

TABLE II. Absorption spectra of CeF_3 and of Ce^{3+} in LaF_3.

Ce^{3+} in LaF_3		CeF_3	
μ(in air)	cm^{-1}(in vac)	μ(in air)	cm^{-1}(in vac)
3.5140*[a]	2845.0	3.510*	2848
3.7936*	2635.3	3.789*	2638
4.4640*	2239.5	4.468*	2238
4.6272*	2160.5	4.633*	2158
4.785	2089	4.780	2092
4.975	2010	4.980	2007

[a] Asterisks indicate lines that persist near the boiling point of liquid helium.

the intensity of the 2092 cm^{-1} line as a function of temperature yielded and activation energy of 141±13 cm^{-1} for this line. Explain. Why would it be desireable to search for an additional line in the range 66-71 cm^{-1}? No transitions were found in the range of 25 to 100 cm^{-1}. Can we then regard the transitions shown in Fig. 3 as established?

110. "Infrared and Raman Spectra of Vinylene Carbonate" K.L. Dorris, J.E. Boggs, A. Danti, and L.L. Altpeter, Jr. 46, 1191 (1967).

The molecule

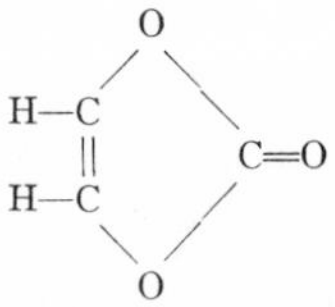

was shown to be planar with C_{2v} symmetry. The fundamental frequencies, where determined, and the approximate description of the corresponding normal modes are contained in Table II. By

TABLE II. Fundamental frequencies of vinylene carbonate (cm^{-1}).

Vibration number	Approximate description	Raman (liquid)	Infrared (liquid)	Infrared (vapor)
	Species A_1			
1	Sym C–H stretch	3171	3170	3179
2	C=O stretch	1835	1831	1867
3	Skeletal stretch	1627	1620	...
4	C–H in-plane bend	1165	1160	1170
5	Skeletal stretch	...	1100	1097
6	Skeletal breathing	907	905	883
7	Skeletal bend	744	740	735
	Species A_2			
8	C–H out-of-plane	?	...	...
9	Skeletal out-of-plane	?	...	...
	Species B_1			
10	Sym C–H stretch	3171	3170	3179
11	C–H in-plane bend	...	1347	1347
12	Skeletal stretch	1085	1081	1086
13	Skeletal stretch	1040	1035	1030
14	Skeletal bend	?	?	...
15	C=O in-plane bend	565	564	...
	Species B_2			
16	C-H out-of-plane	...	711	701
17	C=O out-of-plane	532	532	...
18	Skeletal out-of-plane	...	258	...

the standard methods of group theory verify that each of the irreducible representations shown contains the number of normal modes given. For as many of the 18 modes as possible sketch the form of the vibration and demonstrate that it does indeed belong to a given symmetry species. The standard reference work is the book "Molecular Vibrations", by E.B. Wilson, Jr., J.C. Decius, and P.C. Cross (McGraw-Hill Book Co. Inc. New York 1955), Chapter 5.

Fig. 1 shows a portion of the vapor spectrum with frequency increasing right to left. The 1867 cm^{-1} band is seen to be the C=O stretching fundamental of species A_1. Discuss their suggestion that the weaker bands are hot bands arising from the $v=0 \rightarrow 1$ transition of the C=O stretch in successively higher excited states of the ring-puckering mode. The B_2 ring-puckering motion involves only bending around single bonds, and they suggest should appear at a lower frequency than the A_2 ring-puckering vibration which requires bending of the double bond.

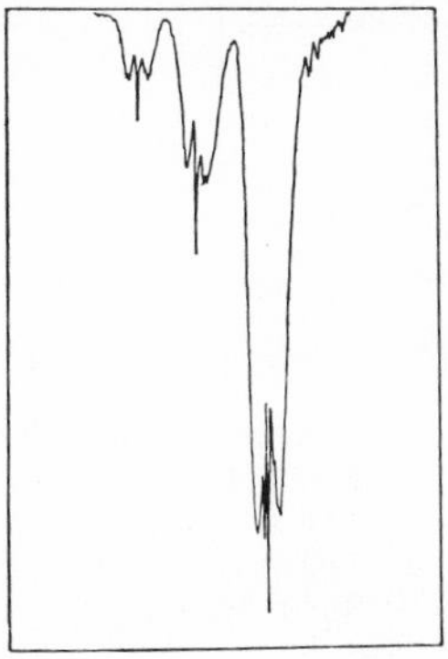

FIG. 1. The absorption bands of vinylene carbonate at 1973, 1924, and 1867 cm^{-1}.

The unusually short C=O bond length required to fit the moments of inertia from microwave studies correlates well with the somewhat higher C=O stretching frequency. What are the more usual values for these two quantities?

111. "Spectroscopic Observation of Nuclear Spin Conversion in Methane" F.H. Frayer and G.E. Ewing 46, 1994 (1967).

In methane the arrangement of nuclear spin with rotational levels is more involved than in hydrogen. In CH_4 the state J=0 contains a single nuclear spin isomer, the quintet(meta-methane). Likewise the J=1 state is a nuclear spin triplet(ortho-methane). The J=2 state contains two spin isomers, a triplet and a singlet(para-methane). Higher rotational levels also contain a mixture of spin states. See ref. 1 and IRRS p. 38 and 449. At the lowest temperatures, certainly those obtainable with liquid helium, all of the molecules would tend to be in their lowest rotational states, J=0. It had been shown earlier (ref. 4) that the rotational motion of methane in rare-gas matrices is only slightly hindered and that infrared vibration-rotation features could be clearly distinguished. The ortho$\rightarrow$meta(triplet$\rightarrow$quintet) nuclear spin process might then be observed. The absorption spectra after bringing the CH_4/Ar solid film from liquid nitrogen to liquid helium temperatures is shown after two time intervals in Fig. 1. Discuss the four

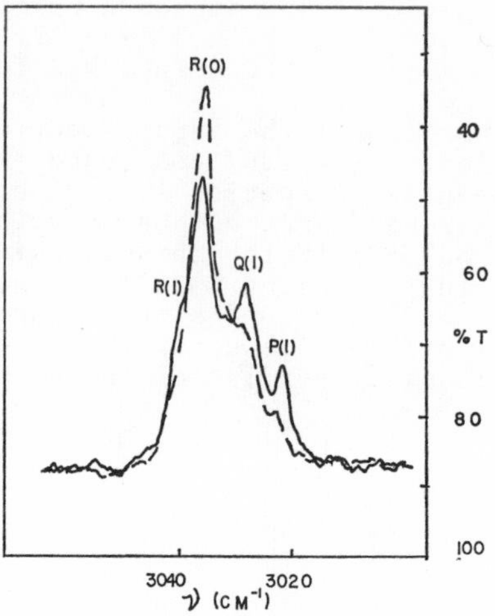

FIG. 1. The absorption spectra of a solution of 0.2% methane in argon in the ν_3 region. The solid curve is after 8 min at helium temperatures, the dashed curve is after 500 min.

labeled transitions. What about the P(0) and Q(0) lines? Do these two spectra prove the presence of the ortho→meta transition? Why or why not? It is suggested that the weak absorption on the high-frequency side of R(1) may be R(2). Rationalize its possible appearance early after the film is brought to ~4°K. Would its presence change the argument? What do they suggest as a possible external perturbation which may mix the spin states and thus provide the mechanism for the triplet→ quintet transition?

112. "Pure Rotational Spectrum and Electric Dipole Moment of CH_3D" I. Ozier, W. Ho, and G. Birnbaum <u>51</u>, 4873 (1969).

The effective electric dipole moment of this molecule was measured and found to be $(5.68\pm0.30)\times10^{-3}$ D, independent of J for $5\leq J\leq12$. Is this down several orders of magnitude from the usual values of dipole moments in molecules? Is it remarkable that the value is something other than zero? To what effect is the nonvanishing dipole moment attributed?

The value of the electric dipole moment was determined by measuring the absolute intensity of the individual rotational transitions. We shall be content with examining the pure rotational spectrum and making approximate measurements of the rotational constant B for this symmetric top molecule. Would we expect the intensities of these transitions to be comparable to say those of the methyl halides? Why or why not?

Fig. 1 shows the spectrum. Is the path length unusually long? Physically how was this accomplished? What exactly is meant by a density of 3.55 amagat? Why are the data presented

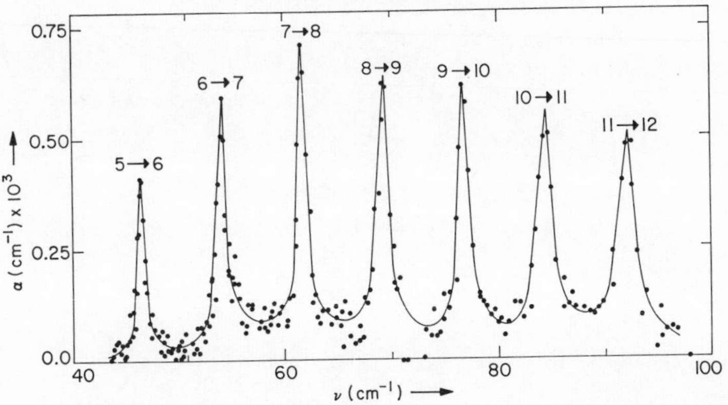

FIG. 1. Preliminary scan of the pure rotational spectrum of CH_3D taken in a 5.0-m cell at a density of 3.55 amagat and a temperature of 296.5°K. A representative smooth curve has been drawn in by hand. The numbers above each line specify the initial and final rotational levels involved in the corresponding transition.

as dots rather than the usual recorder-drawn curves?

Make measurements to determine the approximate value of the rotational constant B. By assuming tetrahedral angles you will be able to compute one bond length. Will it be for the C-H bond(s) or for the C-D bond? Can you detect any centrifugal distortion from your measurements?

It should be pointed out that ref. 5(1963) had calculated a value of $4\text{x}10^{-3}$ D for μ on the basis of a theoretical model.

CHAPTER IX

RAMAN SPECTROSCOPY

113. "Raman Spectra of Chlorofluoromethanes in the Gaseous State" H.H. Claassen 22, 50 (1954).

What advantage is there to studying the Raman spectra in the gaseous state rather than in the liquid? What is the chief experimental disadvantage? Why was it necessary to heat the Raman sample tube to 150-175°C when determining the spectrum of CCl_4?

Fig. 1 and 2 present the spectra of four of the five molecules. The frequency shifts, in wavenumbers, are from the Hg λ4358 exciting line. Four components of this line, designated

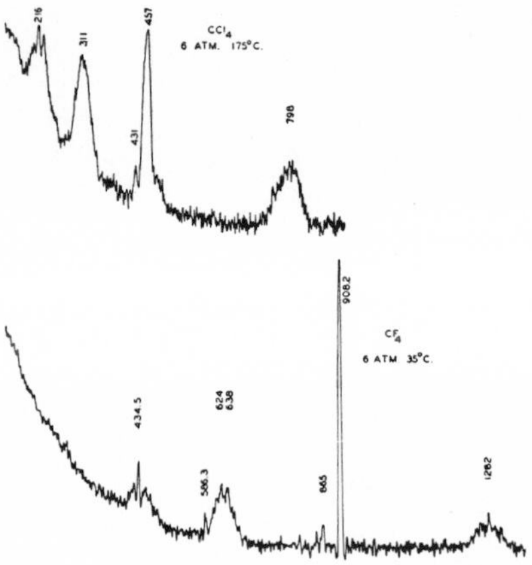

FIG. 1. Raman spectra of gaseous CCl_4 and CF_4.

e,f,k and i are shown in Table I. From this Table find the lines listed for all of the molecules CCl_xF_y, except CCl_2F_2, in the spectra. Do the overtone and combination assignments seem reasonable? The two ÷ signs should of course be +.

For the five molecules all of the fundamentals were observed save one of those of the b_2 species in CCl_2F_2. This one was observed however, in the infrared by previous workers. From Table I do you see that a fundamental at 446 cm^{-1} might be hidden by the neighboring strong and medium bands?

TABLE I. Raman and infrared spectral data for CCl_4, CCl_3F, CCl_2F_2, $CClF_3$, and CF_4.

Present Raman Data, Gaseous State					
Wave number	Description	Exciting Hg lines	Previous Raman data	Previous infrared data gaseous state	Interpretation
			CCl_4, Symmetry T_d		
			Gas (ref. 3)	(ref. 1)	
216	m[a]	e	221	···	e fundamental
311	m,d	e	310.0	···	f_2 fundamental
431	w, sh	e	434	···	$2\times216 = 432(A_1+E)$
457	s	e, k	459.0	···	a_1 fundamental
···	···	···	756	778[b]	$311+457=768\ F_2$
798	m, d	e	794.3	795[b]	f_2 fundamental
			CCl_3F, Symmetry C_{3v}		
			Liquid (ref. 5)	(ref. 1)	
241	s, d	e	244.6	···	e fundamental
349.5	s, sh	e, k	350.5	350	a_1 fundamental
394	m, d	e	397.8	401	e fundamental
535.2	vs, sh	e, k	535.8	535	a_1 fundamental
847	m, d	e	835.4	847	e fundamental
1090	vw	e	1068.1	1085	a_1 fundamental
			CCl_2F_2, Symmetry C_{2v}		
			Liquid (ref. 6)	(ref. 1)	
261.5	m, sh[a]	e, k	260	···	a_1 fundamental
322	w, d	e	320	···	a_2 fundamental
433	m	e, k	433	437	b_1 fundamental
···	···	···	···	446	b_2 fundamental
457.5	s, sh	e, k, i	455	···	a_1 fundamental
640	w	e	···	···	$2\times322=644\ A_1$
667.2	s, sh	e, f, k, i	664	667	a_1 fundamental
923	w, d	e	919	922	b_1 fundamental
1098	m	e	1082	1101	a_1 fundamental
1167	w, d	e	1147	1159	b_2 fundamental
			$CClF_3$, Symmetry C_{3v}		
			Liquid (ref. 7)	(ref. 1)	
350	m, vd	e	356	···	e fundamental
468.2	m, sh	e, k	···	···	a_1 fundamental Cl^{37}
475.8	s, sh	e, k, i	478	···	a_1 fundamental Cl^{35}
560	w	e	560	563	e fundamental
781.7	vs, sh	e, k, i	784	781	a_1 fundamental
1106	w	e	1092	1105	a_1 fundamental
1217	w, d	e	1205	1212	e fundamental
			CF_4, Symmetry T_d		
			Liquid (ref. 4)	(ref. 1)	
434.5	m	e	437	···	e fundamental
586.3	w	e	···	···	Impurity?
624	m	e }	635	630	f_2 fundamental
638	m	e }			
865	w	e	···	···	$2\times434.5 = 870(A_1+E)$
908.2	vs, sh	e, k, i	904	···	a_1 fundamental
1282	w, d	e	···	1277	f_2 fundamental

[a] Apparently a sharp Q branch and rotational branches.
[b] These values are interpolated from Fig. 12 of reference 1.

Are the number of fundamentals, including degeneracies, correct for each molecule? Check their assignment of the lines at 468.2 and 475.8 cm^{-1}, for $CClF_3$, as being due to the ^{35}Cl and ^{37}Cl isotopes both in the C-Cl stretching mode. Do this, of course, by means of the diatomic approximation as used in exercise no. 105 and 106. In passing, attack or defend the description of this normal mode as a C-Cl stretch.

An interesting bit of empiricism is contained in Claassen's Fig. 4 which is not reproduced here. The sums of the fundamental frequencies, including degeneracies, for a particular molecule are plotted against the number of fluorine atoms. Do this

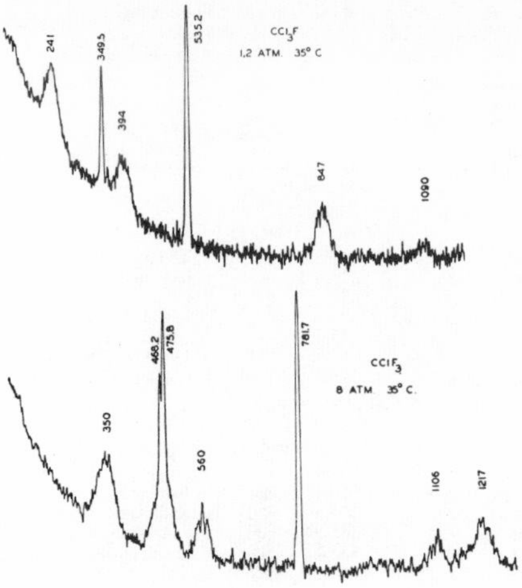

FIG. 2. Raman spectra of gaseous CCl_3F and $CClF_3$.

yourself and see if you obtain a linear plot. This result then tends to give us a bit more confidence in the assignment of the 446 cm^{-1} band for CCl_2F_2 if the other proposed assignment, that at 473 cm^{-1}, were to cause a noticeable inflection in the curve. Would it?

114. "Vibrational Spectra of Molten Halides of Mercury. I. Mercuric Chloride, Mercuric Bromide, and Mercury Chlorobromide" G.J. Janz and D.W. James 38, 902 (1963).

The observed frequencies, intensities, and polarization data for these three molecules in the molten state are:

$HgCl_2$ 313 cm^{-1} (10,p), 376 (0,dp)

$HgBr_2$ 195 cm^{-1} (10,p), 271 (0,dp)

HgBrCl 111 cm^{-1}(vw), 203(vw), 236(m), 319(vw), 335(m)

where the relative intensity is listed for the bands of the first two molecules, together with p for polarized(meaning the depolarization ratio $\rho<6/7$; see sections 3-6 and 7-8 of the book by Wilson, Decius, and Cross referenced on p. 209 herein as well as IRRS, section III-1) and d for depolarized(condition on ρ?). The line intensities for the mixed halide salt were too weak for satisfactory polarization measurements.

As these two references indicate, only totally symmetric vibrations(those that retain the symmetry of the static molecule throughout the course of the vibration) will give polarized Raman lines. Is this fact compatible with the assignments made for ν_1 for HgX_2 in Table I for these measurements on the molten state? It then follows that fundamentals that are not totally symmetric will be depolarized. Is this then compatible with the assignment of the 376 and 271 cm^{-1} weaker("zero intensity") lines as the asymmetric stretch ν_3? If these two molecules belong to point group $D_{\infty h}$, as indeed is assumed in this work, will they have a center of symmetry? What then about the principle of mutual exclusion? Now it should be pointed out that the gaseous state HgX_2 values of ν_2 and ν_3 taken from ref. 2 were determined from the infrared and optical data. To what do

TABLE I. Vibrational assignment and force constants for $HgCl_2$ and $HgBr_2$.

Salt	Frequencies (cm^{-1})			Force constants ($\times 10^{-5}$) dyne/cm^2		
			Gaseous state[a]			
	ν_1	ν_2	ν_3	k_1	k_{12}	k_δ/l^2
$HgCl_2$	360	70	413	2.67	0.037	0.039
$HgBr_2$	225	41	293	2.32	0.067	0.023
			Molten state[b]			
$HgCl_2$	313	(100)	376	2.11	−0.07	(0.08)
$HgBr_2$	195	(90)	271	1.86	−0.07	(0.08)
			Solid state[c]			
$HgCl_2$	314	116	375	2.12	−0.06	0.10
$HgBr_2$	184	...	...	...	...	...

[a] See reference 2.
[b] This work; the values for ν_2, given in brackets, were estimated by comparison with the value observed for HgBrCl in the molten state where this frequency is allowed.
[c] See references 6, 9, 10 for assignment; force constants, this work.

the authors attribute the occurrence of these Raman-forbidden bands in the spectra of the molten HgX_2 compounds? Is there any supporting evidence from the solid-state spectra? Notice in the Table footnote that the ν_2 frequencies were not observed but estimated.

Now let us consider the question of linearity vs nonlinearity. It is admittedly an artificial issue as the authors do not even mention the possibility that the molecules HgX_2 are nonlinear; what exactly is their statement? We will nonetheless pursue the question much as we did in exercise no. 102 for XeF_2. For bent HgX_2 which normal modes of the fundamentals are totally symmetric? The two observed bands for each molecule could serve as the symmetric and antisymmetric stretches if the frequencies and p-dp results are compatible. Are they? Explain. What about the intensities? If the molecules were bent, would the bending mode ν_2 be totally symmetric? polarized? Does its nonappearance or at least very low intensity in the spectra rule strongly in favor of linearity?

Discuss the authors' assignment of the five Raman lines for HgBrCl. Do they all seem reasonable?

From Table I are the stretching force constants more dependent on the state of the material than on the halogen, Cl or Br? Is this true for the interaction and bending force constants as well?

115. "Raman Spectra of Molten Silver Nitrate" G.E. Walrafen and D.E. Irish <u>40</u>, 911 (1964).

The planar NO_3^- ion belongs to the D_{3h} point group. By the standard reduction method, referenced on p. 209 herein, show that there are one fundamental each of a_1' and a_2'' species and two of e' species for a total of six including degeneracies.

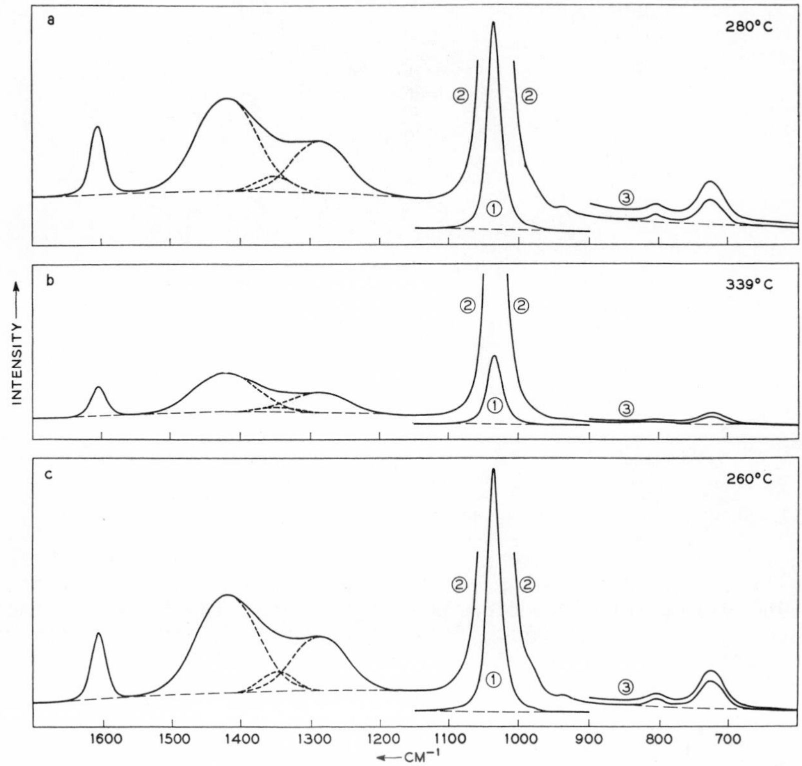

FIG. 1. Photoelectric Raman traces of molten silver nitrate. (1) single beam, slitwidth 10 cm^{-1}, amplification 1000. (2) double beam, slitwidth 10 cm^{-1}, amplification 3000. (3) double beam, slitwidth 16 cm^{-1}, amplification 3000. Time sequence *a, b, c*.

Fig. 1 represents photoelectric Raman traces obtained in the time sequence a,b,c of about 1.5 h duration. Let us see if we can find the nine bands that the authors do; for comparison we will take the fundamental frequencies 1050, 831, 1390, and 720 cm^{-1}, for ν_1-ν_4 respectively, from IRRS p. 178(ref. on p. 302). First, there are two unresolved bands near 727 cm^{-1}, indicated by the marked asymmetry of the absorption; can this be interpreted as a small splitting of the $\nu_4 e'$ band? This would mean that in the melt the nitrate ion has a symmetry lower than planar symmetric, but physically the ion would be "almost planar" corresponding to what probable point group? The next band at 805 cm^{-1} is weak. This is associated with the Raman-forbidden (to first order, cf IRRS p. 242) $\nu_2 a''$. Is its weak intensity compatible with the proposed structure of the ion in the melt? The intense, slightly asymmetric band at 1037 cm^{-1} is said to contain a weak component at $\sim$1015 cm^{-1}(not resulting from other mercury exciting lines which produce the weak bands below 1037; can you pick out one of these in Fig. 1?). Can the totally symmetric vibration $\nu_1 a_1'$ be split? Two possible explanations have been advanced for the weak band at 1015 and/or the slight asymmetry of the intense 1037 band; two forms of nitrate(ref. 7) and intermolecular association(ref. 6,8, and 9). The second explanation would also require the 1350 cm^{-1} band to be dealt with; the authors suggest that Fermi resonance(IRRS p. 215ff) involving the overtone of $\nu_4 e''$ may be its source. The bands at 1290 and 1420 cm^{-1} may be due to the splitting of the $\nu_3 e'$ fundamental. Finally the overtone $2\nu_2$ is allowed in the Raman spectra(show that its species is a_1') and may be responsible for the 1605 cm^{-1} band. Note here that the second harmonic, or first overtone, would then be much more intense than the fundamental. Review the suggested reason for this from a combination of things presented in this paragraph.

Does the absorption intensity increase or decrease with increasing temperature?

After puzzling over these spectra, we may feel some empathy with Banquo, of Shakespeare's "Macbeth", who upon being confronted by the three witches said to them "You should be women but your beards forbid me to interpret you as such...".

116. "Raman Spectrum of the Liquid $PbCl_2$-KCl System" K. Balasubrahmanyam and L. Nanis <u>40</u>, 2657 (1964).

We wish to discuss just one or two points in the Raman spectrum of pure solid lead chloride. The three frequency shifts are observed at 148, 302, and 352 cm^{-1}. The first of these is very intense and the third is weak and diffuse. The Raman line at 302 cm^{-1} could not be definitely identified with Hg λ5461 excitation since it falls very near the broad and intense mercury line at λ5550. Verify this statement. Is it necessary to treat the 302 cm^{-1} line as being Stokes or anti-Stokes? What is the difference? Which is the more common? The authors say that the presence of the Raman line at 302 cm^{-1} was established by scanning the spectrum on the anti-Stokes side of Hg λ5790; the Stokes side of this Hg line contains several additional mercury lines.

Find the energy levels between which each one of these Hg transitions occur if you can. An excellent source is the Grotrian diagram for mercury on p. 202 of "Atomic Spectra and Atomic Structure" by G. Herzberg (Dover Publications New York 1944).

117. "Raman Spectral Studies of Water Structure" G.E. Walrafen <u>40</u>, 3249 (1964).

The photoelectric Raman spectra of water and heavy water are shown in Fig. 1. Are the four bands assigned above 1500 cm^{-1} for H_2O clearly recognizable from the spectrum? Are the bands obviously the three fundamentals plus one overtone of the H_2O molecule? Are the symmetry species correct? These are the intramolecular bands. Answer the same questions for the D_2O bands above 1100 cm^{-1}. Notice the combination band of one of the fundamentals with one of the intermolecular librations or restricted rotations. Which of the fundamentals is present only marginally?

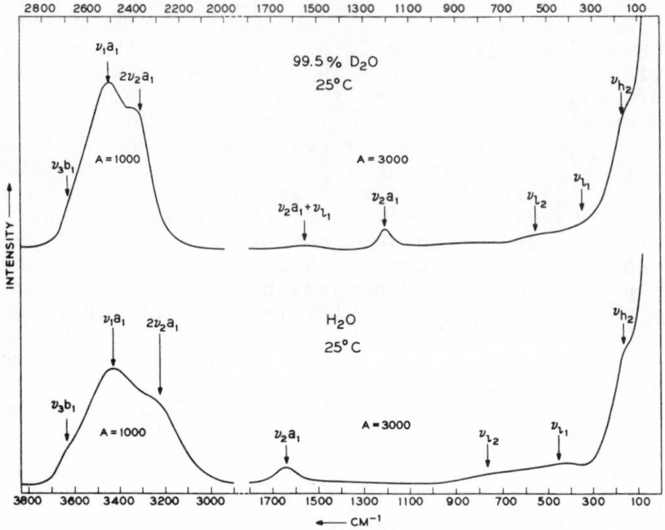

FIG. 1. Photoelectric Raman spectra of water and heavy water. Identical sample geometries were employed in both cases. A = amplification.

The intermolecular frequencies are summarized in Table I.

TABLE I. Raman frequencies, polarizations, and infrared frequencies of intermolecular bands of water and heavy water.

H_2O

DESIGNATION	RAMAN (N.)	RAMAN (B.)	RAMAN*	POLARIZATION	INFRARED
ν_{h_1}	60 CM⁻¹	60	60	—	—
ν_{h_2}	175	172	175	wp/dp**	167 (C.)
ν_{l_1}	475	510	450	wp/dp	500 (C.)
ν_{l_2}	758	780	~780	wp/dp	710 (F+G) 705*

D_2O

DESIGNATION	RAMAN (N.)	RAMAN (M.)	RAMAN*	POLARIZATION	INFRARED
ν_{h_1}	60 CM⁻¹	—	60	—	—
ν_{h_2}	176	170	175	wp/dp	167 (C.)
ν_{l_1}	300	350	~375	wp/dp	357 (C.)
ν_{l_2}	450	~500	~550	wp/dp	530 (F+G) 525*

*—this work.
**—wp/dp—highly depolarized but not necessarily maximum polarization, wp—weakly polarized, dp—depolarized. B.—Ref. (8), C.—Ref. (9), F. & G.—Ref. (10), M.—Ref. (11), N.—Ref. (12).

There are two librational modes assigned for each liquid, H_2O and D_2O. In the article it is stated that there are three restricted rotations or librations of the individual molecules. Do you see how this arises? The h_1 and h_2 frequencies refer to bands produced by hydrogen-bond stretching and bending motion respectively. The former bands are not shown in Fig. 1.

Fig. 4 illustrates a possible five-molecule water structure "entity"; though the author recognizes that it is only an approximation it is nonetheless useful for interpreting the spec-

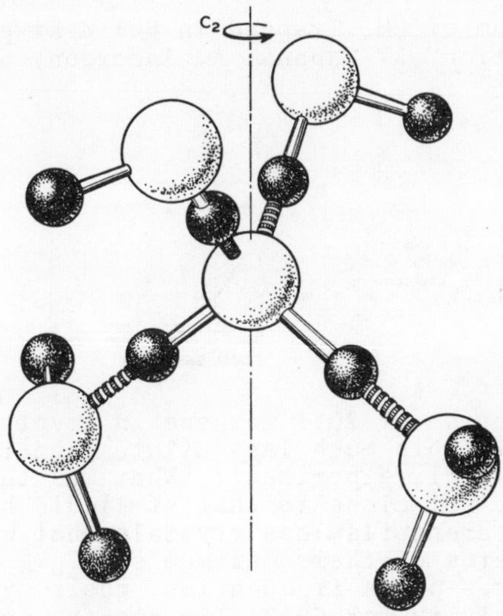

FIG. 4. Five-molecule hydrogen-bonded structure of C_2 symmetry. Small spheres, H atoms; large spheres, O atoms. Disks refer to hydrogen bonds.

trum and in any case must represent the main features of the structure of liquid water. Why should this be so? Verify that the entity as drawn belongs to point group C_2. What changes would have to be made in the individual geometries of the four "ligand" molecules so that the entity would have its symmetry raised to that of C_{2v}?

The intramolecular vibrational assignments made on the bases of both of these symmetries are listed in Table II. Are

TABLE II. Vibrational assignments for the five-molecule hydrogen-bonded structure of water according to C_{2v} and C_2 symmetries. p-polarized, dp-depolarized.

$(H_2O)_5$

SYMMETRY	C_{2v}				C_2	
SPECIES	A_1	A_2	B_1	B_2	A	B
NONSYMMETRIC VALENCE			175 $\nu_7 b_1$	175 $\nu_9 b_2$		175 $\nu_8 b + \nu_9 b$
SYMMETRIC VALENCE	175 $\nu_1 a_1 + \nu_2 a_1$				175 $\nu_2 a + \nu_3 a$	
NONSYMMETRIC DEFORMATION		450 $\nu_5 a_2$	~780 $\nu_6 b_1$	450-780 $\nu_8 b_2$	450 $\nu_1 a$	~780 $\nu_6 b$ 450-780 $\nu_7 b$
SYMMETRIC DEFORMATION	60 $\nu_3 a_1 + \nu_4 a_1$				60 $\nu_4 a + \nu_5 a$	
POLARIZATION	p	dp	dp	dp	p	dp

the numbers of frequencies correct for the five-particle "molecule"? Are the polarization assignments correct? Are they not inconsistent with those of Table I? Make up a list of assignments similar to that of Table II for $(D_2O)_5$. Rationalize those frequencies that are similar in both species and also the appearance of the factor $\sim 1/\sqrt{2}$ where it occurs.

Fig. 6, not reproduced, shows the temperature dependence of the three intermolecular bands of Fig. 1 in terms of their intensities. Predict this behavior, explain your reasoning, and check yourself from the article. Do it in this order.

118. "Raman Spectrum of CH_4 Trapped in Solid Krypton. Evidence for Molecular Rotation" A. Cabana, A. Anderson, and R. Savoie 42, 1122 (1965).

FIG. 1. The Raman spectrum of a solution of 5% methane in krypton recorded at 13°K.

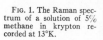

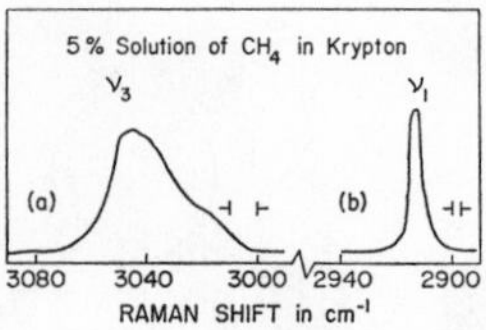

The Raman spectrum of 20:1 methane in krypton at 13°K is shown in Fig. 1. Is this much less dilute than the usual low-temperature inert matrix experiment? What was the reason? Why was the temperature not close to that of liquid helium? How big were the transparent flawless crystals that were grown from the solution, in terms of their maximum size?

What are the gas-phase frequencies, their symmetry species, and their Raman activity for CH_4? The authors point out that ν_2 has roughly a tenth of the intensity of ν_3.

What is the general Raman selection rule for ΔJ? From their ref. 4, and IRRS, for the totally symmetric lines in the Raman spectrum of tetrahedral molecules only transitions with $\Delta J=0$ can occur. Thus the O,P,R and S branches would be completely suppressed for ν_1. Does this appear to be the case in Fig. 1? The ν_2 fundamental was not detected. The authors state that the ν_3 region of the spectrum reproduced in Fig. 1 has been smoothed. The Q branch of this fundamental was expected to lie near 3017 cm^{-1}; why? They suggest that at this low temperature only P(1) would be strong enough to be observed; would this be on the low-frequency side of the Q branch? They further suggest that R(0), R(1), S(0) and S(1) should be observable on the high-frequency side. Why the limitation to these five rotational lines?

Upon what two points then do the authors base their support of the concept of free rotation for the methane molecule in the krypton matrix?

119. "Vibrational Spectra of ReF_7" H.H. Claassen and H. Selig 43, 103 (1965).

This molecule is one of the two known binary heptacoordinate compounds; do you happen to know the identity of the other one?

The Raman data of the present work coupled with earlier infrared results suggest that the point group of this compound is D_{5h}. Sketch the molecule showing the correct bond angles.

As the volatility of the compound was low, no attempt was made to observe the vapor spectrum. Because of the yellow color of the liquid, Hg λ5461 excitation was used. What exactly does the yellow color have to do with the wavelength of the exciting line? Why was it necessary to jacket the sample tube with a larger tube through which water at 55°C was circulated?

Fig. 2 shows the Raman spectrum of the liquid. Would we

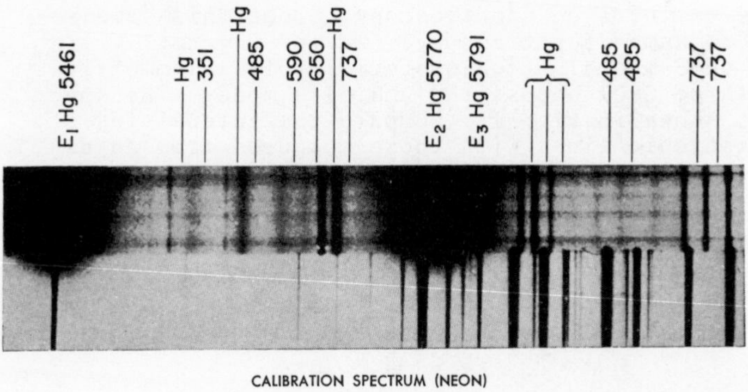

FIG. 2. Raman spectrum of liquid ReF_7.

move the identifying leaders and their labels to the left about 1/16" for a better match? Notice that the Hg wavelengths and frequency shifts in cm^{-1} are both shown without labels; this doesn't cause us any confusion though. What about the lines labeled "Hg"? What is different about the 737 shifts, compared to the other Raman lines? The 590 line should have the same appearance for the same reason if the assignments of Table I are correct. The authors point out this and other puzzles, but state that the pentagonal biprism(D_{5h}?) model fits the data better than any of the models considered in ref. 1(infrared work).

TABLE I. Infrared and Raman spectra of ReF_7.

Raman (liquid)	Infrared (vapor)	Assignment
		e_1'
	299 s	a_2''
351 m		e_2'
	353 s	e_1'
485 s		e_1''
590 m		a_1'
650 w		e_2'
	703 vs	a_2'' and e_1'
737 vs		a_1'
	850 w	485+353= 838
	945 w	590+353= 943
	1082 w	737+353=1090
	1180 w	485+703=1188
	1286 m	590+703=1293
	1350 w	650+703=1353
	1438 m	737+703=1440

120. "Performance of a New Photoelectric Detection Method for Optical Spectroscopy. II. Application to Laser Raman Spectroscopy" J.E. Griffiths and Y.-H. Pao 46, 1679 (1967).

In the paper immediately preceding this one the authors discuss theoretical and experimental characteristics of a new photoelectric system particularly suitable for low-level light conditions. Why is Raman spectroscopy an obvious application choice? Other than long photographic exposure times, what other method had been used to compensate for low-level light signals in the Raman effect? What did this due to the attainable resolution?

What has been a principal disadvantage of the application

of laser sources to infrared spectroscopy? Does this same restriction apply to Raman spectroscopy? Why or why not?

The $\nu_3(a_1)$ fundamental of chloroform involves symmetric bending of the three C-Cl bonds; the umbrella mode. The spectrum obtained is shown in Fig. 1. Compare the intensities of the individual isotopic lines with those computed from an assumed random distribution of ^{35}Cl and ^{37}Cl. Show that these last quantities are their values $HC^{35}Cl_3$ (42.9%), $HC^{35}Cl_2{}^{37}Cl$ (42.0%), $HC^{35}Cl^{37}Cl_2$ (13.6%), and $HC^{37}Cl_3$ (1.5%). Are we surprised that the last-named species is not detected distinctly?

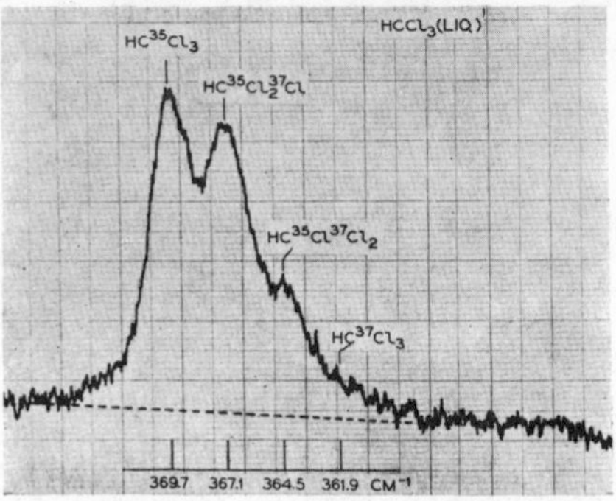

FIG. 1. The $\nu_3(a_1)$ fundamental of $HCCl_3$ [$P=19$ mW (6328 Å); slitwidth = 1 cm^{-1}×18 mm; scan rate = 1.4 cm^{-1}/min; time constant = 2 sec].

Show that for the breathing mode $\nu_1(a_1)$ the statistical distribution of intensities for the tetrachlorides MCl_4, with M a group IVA element, is $M^{35}Cl_4$ (31.6%), $M^{35}Cl_3{}^{37}Cl$ (42.2%), $M^{35}Cl_2{}^{37}Cl_2$ (21.1%), $M^{35}Cl^{37}Cl_3$ (4.7%), and $M^{37}Cl_4$ (0.4%). The last one can be done in your head to one significant figure; $(1/4)^4 \sim 0.4\%$. The isotopic fine structure for carbon tetrachloride had been observed in 1931(ref. 7); the others are reported in Table I with the spectra themselves shown in Fig. 2. Which of the four compounds has a fine structure appreciably different from that of the other three? Is that of the other three approximately in accord with your(and their) calculated intensity pattern? The anomalous pattern of $SiCl_4$ reveals a real clue about the structure of this liquid; a clue that was certainly hidden heretofore.

TABLE I. Raman displacements (cm^{-1}) of isotope lines in $\nu_1(a_1)$ of MCl_4 molecules.

Species	Si[a]	Ge	Sn
$M^{35}Cl_4$	425.8	399.2	371.7
$M^{35}Cl_3{}^{37}Cl$	423.4	396.6	368.8
$M^{35}Cl_2{}^{37}Cl_2$	420.8	393.8	366.0
$M^{35}Cl^{37}Cl_3$	...	391.0	363.1
$M^{37}Cl_4$	...	...	...

First compare the pattern of $SiCl_4$ in Fig. 2 with that of the three-chlorine species chloroform in Fig. 1. The marked

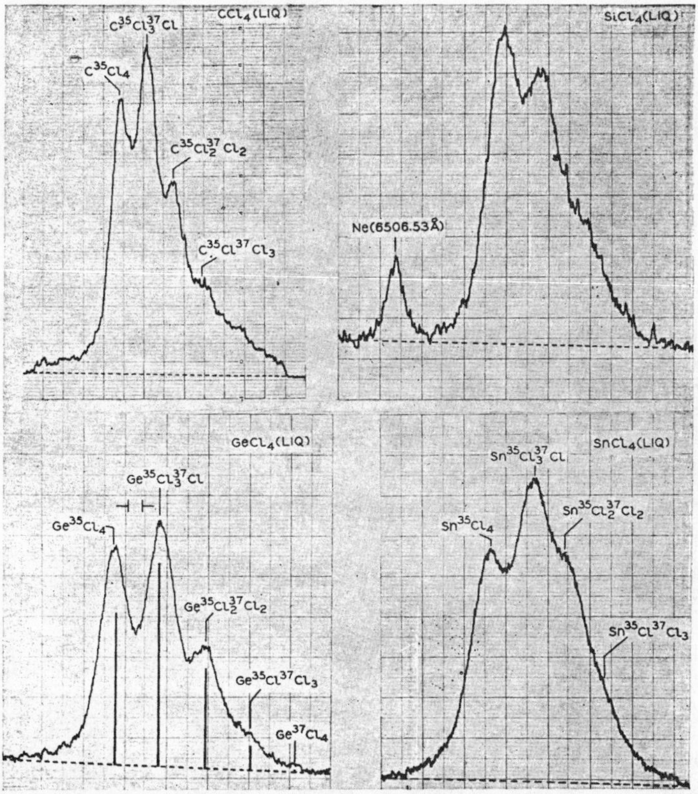

FIG. 2. The $\nu_1(a_1)$ fundamentals of: (a) CCl_4 [$P=16$ mW (6328 Å); slitwidth $=1$ cm$^{-1}\times 18$ mm; scan rate $=1.2$ cm^{-1}/min; time constant $=2$ sec]. (b) $SiCl_4$ [$P=20$ mW (6328 Å); slitwidth $=1.0$ cm$^{-1}\times 9$ mm; scan rate $=0.6$ cm^{-1}/min; time constant $=2$ sec]. (c) $GeCl_4$ [$P=20$ mW (6328 Å); slitwidth $=1.0$ cm$^{-1}\times 9$ mm; scan rate $=0.5$ cm^{-1}/min; time constant $=2$ sec]. (d) $SnCl_4$ [$P=20$ mW (6328 Å); slitwidth $=1.0$ cm$^{-1}\times 9$ mm; scan rate $=0.6$ cm^{-1}/min; time constant $=2$ sec].

similarity suggests that a three-chlorine species is responsible for the absorption of $SiCl_4$ near 424 cm^{-1}. The authors suggest that we envisage a dimerlike structure in which the two silicon atoms are bonded through two chlorine bridges forming a species with six chlorine atoms of one general type(normal Si-Cl covalent bonds) and two chlorine atoms of another(bridging chlorine atoms). Would silicon need to assume a quasipentavalent state in order to this? In the language of hybridization, what orbitals would it call upon? The authors point out that for this proposed model to satisfy the observed spectrum it is only necessary to require that the two $SiCl_3$ units in the associated complex be vibrationally decoupled.

Can you suggest a spectral test of the above interpretation? Their results are shown in Fig. 3. Show how the spectra given here render convincing evidence for their suggested liquid structure for liquid $SiCl_4$. How did they establish that the anomalous fine structure was not due to impurities, including the most likely one? What especially convincing experiment did they perform with practical grade $SiCl_4$?

For the spectrum of the breathing mode $\nu_1(\Sigma_g^+)$ of CS_2 in Fig. 4, what is the demonstrable resolution for the two weaker lines? Have you seen any resolution approaching this in any of the exercises in this chapter on Raman spectra? In any other Raman spectra?

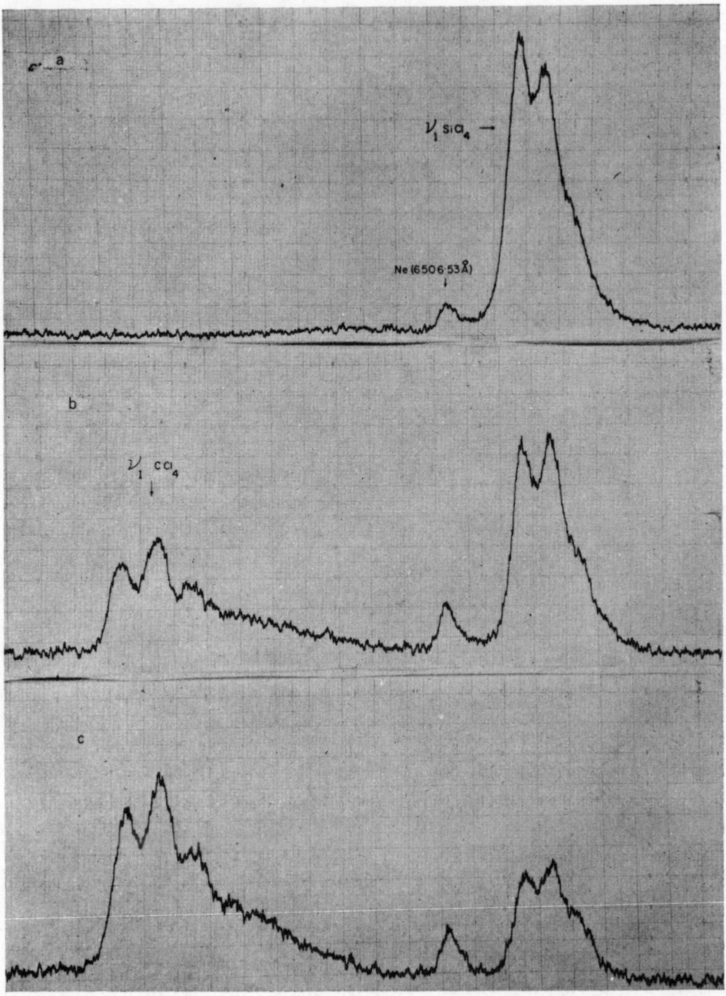

FIG. 3. The $\nu_1(a_1)$ fundamental of $SiCl_4$ in CCl_4 solutions: (a) 100 mole % $SiCl_4$; (b) 50 mole % $SiCl_4$, and (c) 20 mole % $SiCl_4$. Power=18–20 mW (6328 Å); slitwidth=1.0 cm^{-1}×9 mm; scan rate=0 6 cm^{-1}/min; time constant=2 sec. The plot is intensity vs $\Delta\nu$(cm^{-1}).

Finally we have in Fig. 6 a spectrum of the symmetric C-C stretch, or ring-breathing mode, $\nu_2(a_{1g})$ for benzene. Our last activity in this exercise will be to compute the statistical relative intensities of the two isotopes shown, $^{12}C_6H_6$ and $^{12}C_5{}^{13}CH_6$, and then approximate the areas under the two curves. Compare both of your answers with their value of 6% abundance; do you suspect that this was calculated or measured? The discussion at the first of exercise no. 93 may prove helpful with the statistics.

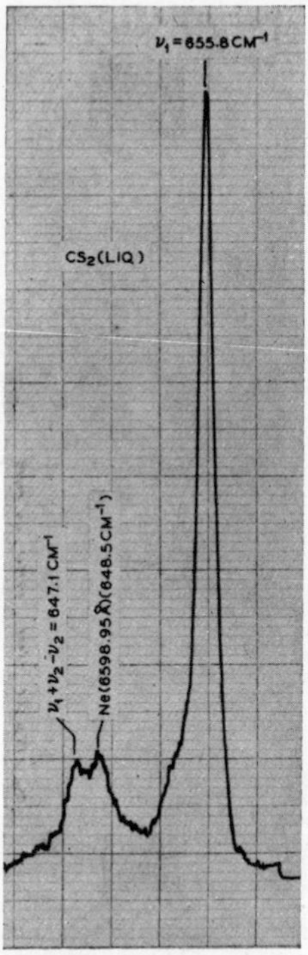

FIG. 4. The $\nu_1(\Sigma_g^+)$ fundamental of CS_2 [P=19 mW (6328 Å); slitwidth=0.35 cm^{-1}×18 mm; scan rate=3.3 cm^{-1}/min; time constant=2 sec].

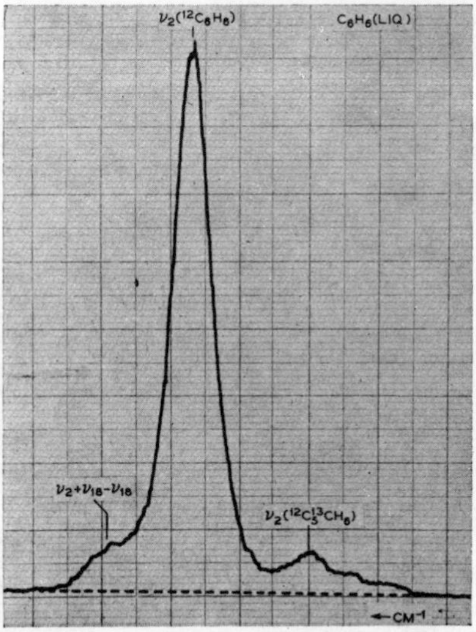

FIG. 6. The $\nu_2(a_{1g})$ fundamental of C_6H_6 [P=20 mW (6328 Å); slitwidth=1.5 cm^{-1}×18 mm; scan rate=1.5 cm^{-1}/min; time constant=2 sec].

121. "Resonance Raman Effect and Resonance Fluorescence in Halogen Gases" W. Holzer, W.F. Murphy, and H.J. Bernstein 52, 399 (1970).

Raman and fluorescence spectra of the colored halogen gases were excited with an argon-ion laser whose output is at 5145, 5017, 4965, 4880, and 4765 Å. You should be able to identify the ionic transitions responsible for these lines from Moore's work referenced on p. 70 herein. These diatomics were chosen because the excitation energies are close to or within absorption bands and because their spectroscopic constants are accurately determined. Excitation diagrams are shown in Fig. 9.

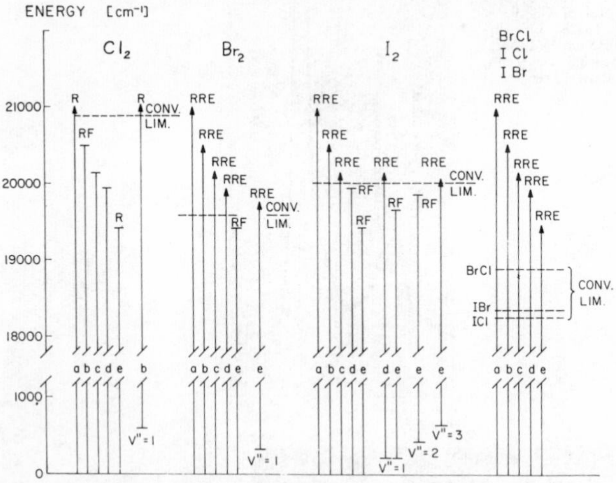

FIG. 9. Diagram of the excitation of the halogen molecules by argon-ion laser radiation at 4765 Å (a), 4880 Å (b), 4965 Å (c), 5017 Å (d), and 5145 Å (e). The convengence limit and the observed spectra (RF, RRE, R) are indicated.

RF and RRE spectra are those due to resonance fluorescence and the resonance Raman effect; characteristics of each are presented in Table I. R in Fig. 9 refers to the usual Raman ef-

TABLE I. Differences in the observation of resonance fluorescence (RF) and resonance Raman effect (RRE).

	RF	RRE
Band envelope	Very sharp doublet lines at low pressure	Broad regular Q branch and rotational wings
Overtone pattern	Irregular overtone sequence of doublets, some lines might be completely missing	Continuous broadening of the Q branch and continuous decrease in peak intensity with higher order
Depolarization ratio	Depolarized	Bands expected to be polarized in Raman effect are polarized
Behavior with increasing gas pressure	Intensity decreases, (quenching) doublet of sharp lines changes into multiplet structure	Intensity increases, shape of the band does not change considerably
Behavior with foreign gases added	As above	Intensity and band shape do not change considerably

fect. Discuss the differences in the behavior of RF and RRE for each of the five points in Table I; for example two ideas that you may bring in are collisional quenching and the Franck-Condon principle.

Fig. 1,2 and 3 illustrate examples of RRE, RF and R spectra

for, as it turns out, three different halogens. For the first

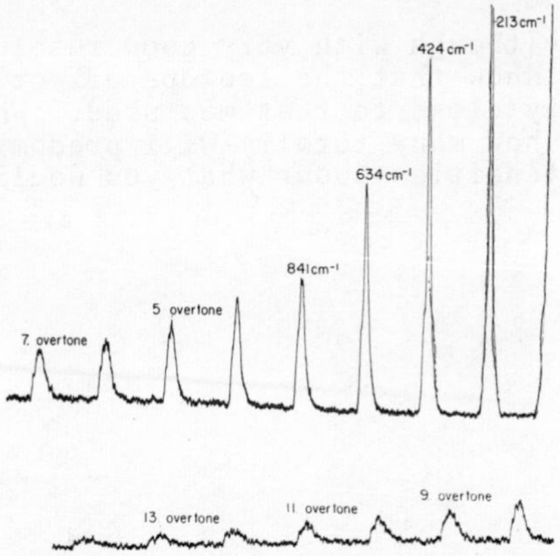

FIG. 1. The resonance Raman spectrum of I_2 gas. I_2, 15 torr; slits, 7.5 cm^{-1}; time constant, 2 sec; 11.5 Å/min; 4880-Å excitation.

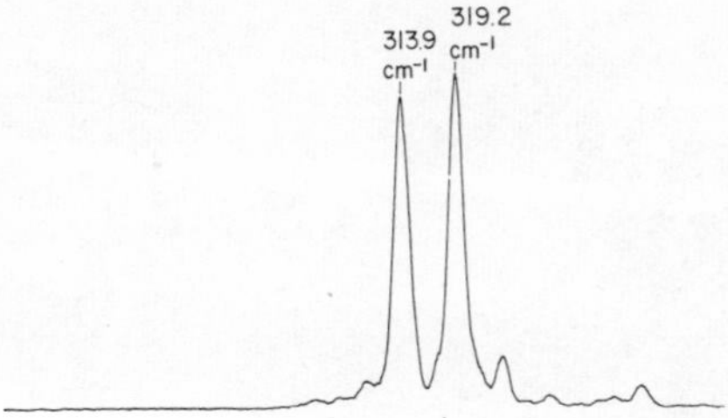

FIG. 2. The resonance fluorescence doublet ($v''=1$) of Br_2 gas. Br_2, 4 torr; slits, 1.25 cm^{-1}; time constant, 1 sec; 2.0 Å/min; 5145-Å excitation.

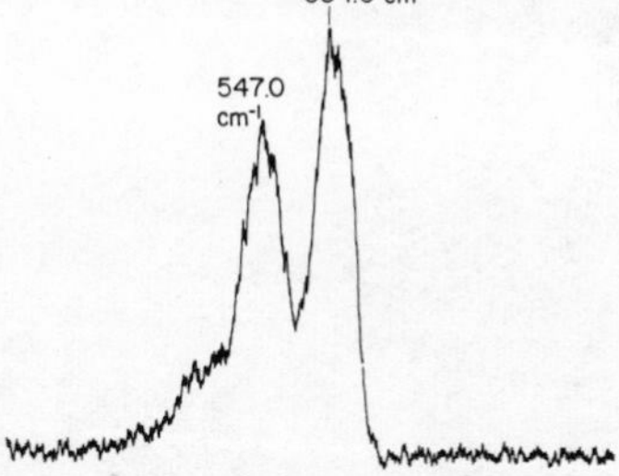

FIG. 3. The Raman fundamental band of Cl_2 gas at higher resolution, showing the isotope structure. Cl_2, 1 atm; slits, 1.25 cm^{-1}; time constant, 2 sec; 1.25 Å/min; 4880-Å excitation.

two verify as many of the features listed in Table I as you can. For I_2 determine ω_e and $\omega_e x_e$ for comparison with optical results. Is the numbering of the overtones correct? How many are observed? How many "harmonics"?

In Fig. 2, to what are the two lines of the doublet due? isotope effect? P and R branches? instrumental slit effect?

combination of these things? Notice the slit width given in the caption. It may help to draw a synthetic spectrum or stick plot.

In the ordinary(though with very good resolution) Raman spectrum of Fig. 3, show that the isotope effect does indeed give a splitting very close to that measured. Which two isotopes of Cl_2(out of how many total?) will predominate? Are the experimental intensities about what you would predict?

CHAPTER X

OPTICAL SPECTROSCOPY

122. "Wavelength Shifts in the near Ultraviolet Spectra of Fluorinated Benzenes" H. Sponer <u>22</u>, 234 (1954).

It is appropriate that our first exercise in this chapter on visible and ultraviolet spectroscopy be based upon an article by one of the pioneers in the field who also authored an early text(1935). She is of course the codeveloper of the Birge-Sponer extrapolation method for dissociation limits. Notice that in this work she has chosen molecules of reasonable size; let us try to be somewhat versatile in this chapter as well.

Table I presents positions of the 0,0 bands(meaning exactly

TABLE I. Positions of 0,0 bands (vapor) and intensities in the near ultraviolet spectra of some fluorinated and chlorinated benzenes.

Molecule	0,0 band cm^{-1}	Intensity $f \times 10^3$
Benzene C_6H_6	38 089 calc	1.6
Fluorobenzene C_6H_5F	37 819[a]	8.9
m-difluorobenzene $C_6H_4F_2$	37 909[b]	9.6
p-difluorobenzene $C_6H_4F_2$	36 843[6]	22.4
1,2,4-trifluorobenzene $C_6H_3F_3$	37 123[7]	19.2
1,3,5-trifluorobenzene $C_6H_3F_3$	38 527[3] calc	~2.0
1,2,4,5-tetrafluorobenzene $C_6H_2F_4$	36 605/12[c]	
Benzotrifluoride $C_6H_5CF_3$	37 819[d]	7.1
1,4-bis-trifluoromethylbenzene $C_6H_4(CF_3)_2$	37 460[5]	12.0
1,3,5-tris-trifluoromethylbenzene $C_6H_4(CF_3)_3$	~38 100[5] calc	2.1
m-fluorobenzotrifluoride $C_6H_4FCF_3$	37 355[e]	(14.7)
p-fluorobenzotrifluoride $C_6H_4FCF_3$	37 866[e]	(2.4)
2,5-difluorobenzotrifluoride $C_6H_3F_2CF_3$	36 800 sol[f]	27.0
Perfluorotoluene $C_6F_5CF_3$	37 700 sol[f]	21.0
Chlorobenzene C_6H_5Cl	37 052[g]	3.0
o-dichlorobenzene $C_6H_4Cl_2$	36 265[h]	4.0
m-dichlorobenzene $C_6H_4Cl_2$	36 186[h]	4.1
p-dichlorobenzene $C_6H_4Cl_2$	35 743[h]	~6.8
1,2,4-trichlorobenzene $C_6H_3Cl_3$	35 108[i]	~6.0
1,3,5-trichlorobenzene $C_6H_3Cl_3$	35 498[4] calc	2.1

[a] S. H. Wollman, J. Chem. Phys. **14**, 123 (1946).
[b] V. Ramakrishna Rao and H. Sponer, Phys. Rev. **87**, 213(A) (1952).
[c] Results of K. N. Rao in our laboratory.
[d] H. Sponer and D. S. Lowe, J. Opt. Soc. Am. **39**, 840 (1949).
[e] W. T. Cave and H. W. Thompson, Disc. Faraday Soc. **9**, 35 (1950).
[f] H. B. Klevens and Lois J. Zimring, J. chim. phys. **49**, 377 (1952).
[g] H. Sponer and S. H. Wollman, J. Chem. Phys. **9**, 816 (1941).
[h] H. Sponer, Revs. Modern Phys. **14**, 224 (1942).
[i] Hedwig Kohn and H. Sponer, J. Opt. Soc. Am. **39**, 75 (1949).

what?) and intensities in the near uv spectra of benzene and some fluorinated and chlorinated derivatives. Is it not unexpected that a spectroscopist would try to obtain spectra in the vapor phase and thus eliminate condensed phase interactions?

See from the Table that with but two exceptions substitution decreases the energy of the transition. These shifts are suggested to be the combined result of two effects. The halogen atom because of its electron affinity withdraws σ electronic charge from the ring, called the inductive effect. Will this effect be greater for chlorine or fluorine and why? The second effect, that of migration, is that of charge from the nonbonding 2pπ electrons of the substituent migrating into the ring. See the daggered(†) footnote in the article. One of these effects will increase the potential field for the π electrons in the ring and the other will decrease it; which effect does which? Should we think of the added negative charge of the migration effect as having the effect of binding the π electrons less tightly, and thus causing a red shift in the 0,0 band? Give the corresponding argument for the blue shift.

Note that for ring chlorination the red shift is a strictly monotonic function of the number of substituents, with secondary variations in the transitional frequency because of position of substitution for a given number of chlorine atoms. For the trifluorinated compounds we see that in one of the cases the migration effect dominates and the induction effect in the other. Can you rationalize this result in terms of different symmetries? Which of the two effects dominates for less than or more than trisubstitution with fluorine?

In all of this discussion have we taken for granted that the same electronic transition is involved? Familiarity with the Hückel treatment for benzene tells us immediately with which transition we are concerned; it is the $\tilde{A}^1B_{2u}\leftarrow\tilde{X}^1A_{1g}$. Is there another excited state lying lower than the one of this transition? What is its multiplicity?

123. "Vacuum-Ultraviolet Absorption Spectra of the Cyclic Ethers: Trimethylene Oxide, Tetrahydrofuran, and Tetrahydropyran" G.J. Hernandez 38, 2233 (1963).

Earlier work(ref. 14-17) had suggested that the trimethylene oxide molecule had C_{2v} symmetry. Sketch it on this basis.

Fig. 1 is the absorption spectrogram for this molecule.

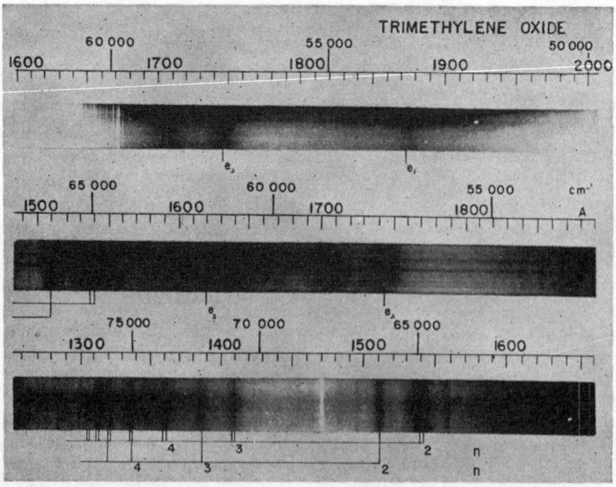

FIG. 1. Absorption spectrum of trimethylene oxide. Upper: hydrogen continuum background; middle: xenon continuum background; lower: krypton continuum background.

Is it a reproduction of the negative or a positive print? Fig. 2 is made from a densitometer tracing. Can you match up the

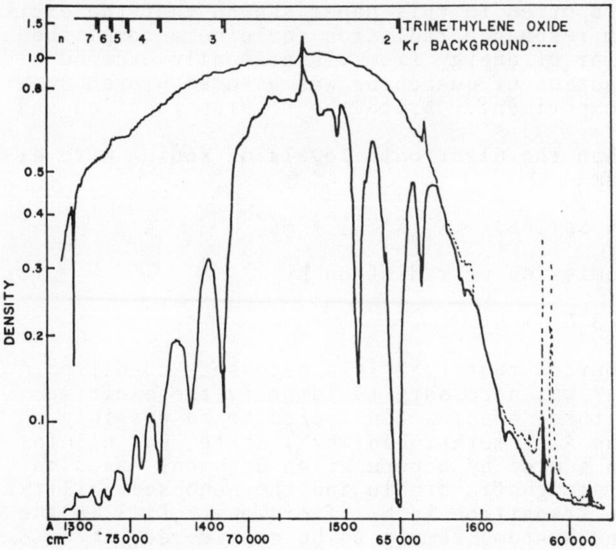

FIG. 2. Absorption spectrum of trimethylene oxide in the region 60 000 to 80 000 cm^{-1}.

absorptions in the common regions of the two spectra? From which one would it be easier to measure line position? The n

TABLE I. Rydberg series (cm^{-1}).

				Trimethylene oxide							
n_I	$\nu_{obs.}$	$\nu_{calc.}$	I	n_{II}	$\nu_{obs.}$	$\nu_{calc.}$	I	n_{III}	$\nu_{obs.}$	$\nu_{calc.}$	I
2	64 910	65 371	10	2	65 020	65 521	9	2	66 300	66 724	8
3	70 990	70 947	8	3	71 070	71 097	6	3	72 250	72 300	6
4	73 500	73 502	7	4	73 630	73 652	4	4	74 880	74 855	6
5	74 880	74 881	6	5	74 960	75 031	3	5	76 270	76 234	4
6	75 700	75 709	6	6	75 790	75 850	1				
7	76 250	76 244	4								
8	76 620	76 611	5								
9	76 860	76 872	3								
∞	77 980	77 980		∞	78 130	78 130		∞	79 330	79 330	
	9.66_7 eV				9.68 eV				9.83 eV		

values are those of two Rydberg series whose members are shown in Table I, abbreviated to include only trimethylene oxide. The four lines assigned to n_{III}, a third possible Rydberg series in Table I, are not all labeled in Fig. 2. Between both Fig. 1 and 2 see how many of the listed lines in Table I you can measure the position of and identify. Is it disconcerting that the three Rydberg series all have different limits? What is the likely explanation, assuming each is a series with running term $-R/(n+0.95)^2$? What is R? Try the formula for each series. Which value is indicated to be <u>the</u> ionization potential of TMO?

124. "Excitation of Electronic Levels of Sodium by Vibrationally Excited Nitrogen" W.L. Starr <u>43</u>, 73 (1965).

The experiment reported in this paper showed that the excitation of sodium atom resonance radiation could be accomplished by collisional transfer of energy from vibrationally excited $N_2{}^{\dagger}$. The reverse reaction of quenching was assumed proven by the success of this experiment. Write the reverse reaction and discuss the proof.

In this paper then the electronic levels of sodium were excited by the mechanism

$$N_2(X^1\Sigma_g^+,\ v''>7) + Na(3^2S) \rightarrow N_2(X^1\Sigma_g^+) + Na^*$$

and detected by the emission of radiation by

$$Na^* \rightarrow Na(3^2P \text{ or } 3^2S) + h\nu.$$

Show from standard sources that $v''>7$ is a necessary condition.

Experimentally it was necessary to identify the exciting species. One of the three species considered to be possibly responsible, N_2 in the $A^3\Sigma_u^+$ metastable(why?) state, was eliminated as an excitation source by a population argument based on the observability of a magnetic dipole and the nonobservability of an electric dipole transition in N_2 afterglows. Fill in the details. Only $N_2{}^{\dagger}$ and N atoms remain to be considered; to determine which one is responsible it was necessary to remove one from the flowing system. The author's ref. 4 and 10 showed that if N_2O were added to the afterglow of a nitrogen flow, the N_2O will deactivate $N_2{}^{\dagger}$ but will not react with atomic nitrogen. The result of this procedure is shown in Fig. 1. Is the identity of the exciting species established?

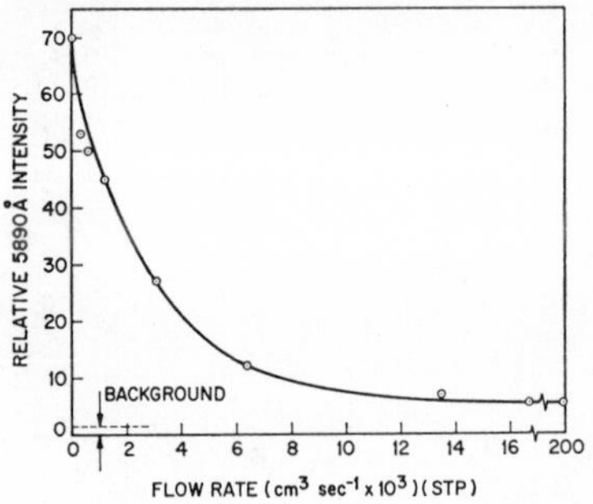

FIG. 1. Relative intensity of the sodium 5890-Å *D* line vs the flow rate of N_2O. The data were obtained from the N_2O deactivation of the N_2 afterglow.

Does there appear to be anything unique about the nitrogen-sodium system? Could it be concluded that the transfer of vibrational energy of a diatomic molecule to electronic energy of an atom is a general phenomenon? Is a close match between the vibrational and electronic levels required? Could there be, or were there, a number of sodium electronic levels excited in the process?

125. "Photoluminescent Properties of Some Europium-Activated Gadolinium and Yttrium Compounds" A. Bril, W.L. Wanamaker, and J. Broos 43, 311 (1965).

Try to translate the authors' affiliation; does it correlate with some of the applications mentioned below? Is there any connection between the dopant and the geographical area where the work was carried out?

Notice the date of ref. 1, in which the luminescence of Gd_2O_3-Eu was first described.

The excitation for the spectra in emission of Fig. 1 was supplied at 254-nm. What is the reason for the name phosphors

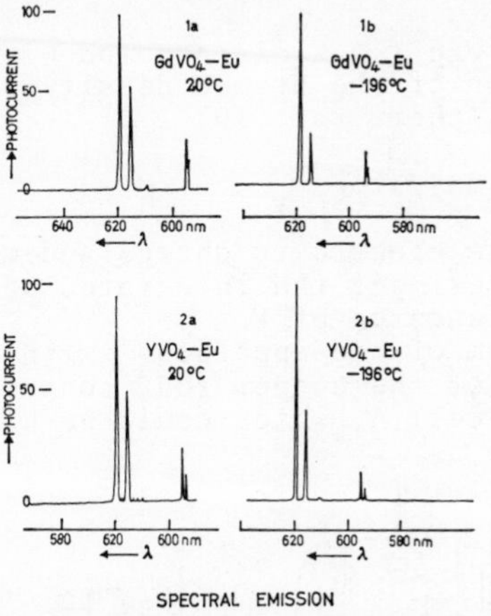

FIG. 1. Spectral-energy distribution curves of the emission of $GdVO_4$–Eu (Curve 1a at 20°C and 1b at −196°C) and YVO_4–Eu (Curve 2a at 20°C, and 2b at −196°C).

applied to these activated compounds? Look at the emission wavelengths in Fig. 1. Would you say each of the four spectra were produced by the same species? Do we see the use of the term activated now, if not before? What will be the color of the emission?

The quantum efficiencies(for what process?) and the Eu con-

FIG. 2. Quantum efficiency as a function of the Eu concentration (expressed as gram atoms Eu per mole of the host lattice). (a) Gd_2O_3–Eu, (b) Y_2O_3–Eu, (c) YVO_4–Eu.

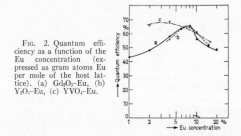

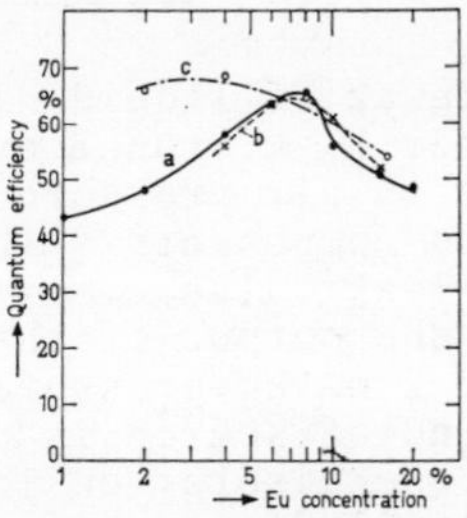

centration range are shown in Fig. 2. These quantum efficiencies are quite high. These phosphors also turn out to have a relatively high brightness at elevated temperatures. These properties mentioned here suggest what applications for these materials? Two or three are mentioned in the article.

Does ref. 5 seem to be a bit unusual in a research journal? Is the relation between technology and science particularly strong and well-defined in this case? Can it be high praise for a scientist's methods to be described as Edisonian? Can you elaborate a bit upon this description?

126. "Measurement of the Density of Saturated Cesium Vapor by an Optical Method" M. Rozwadowski and E. Lipworth 43, 2347 (1965).

What was the reason for entering into this set of vapor-density measurements?

Beer's law $I=I_0 \exp(-k_\nu x)$ is assumed with the absorption coefficient a function of the atomic density ρ and the oscillator strength f(e.g. their ref. 10)

$$\int k_\nu d\nu = \pi e^2 \rho f/(mc).$$

Here e,m and c are the electronic charge and mass and the speed of light. In the experiment the integrated absorption is determined giving ρ as a function of f.

The block diagram of the apparatus is shown in Fig. 1. What was the purpose of the copper rod, connected by a metal ring surrounding the cell A, which could be brought through the

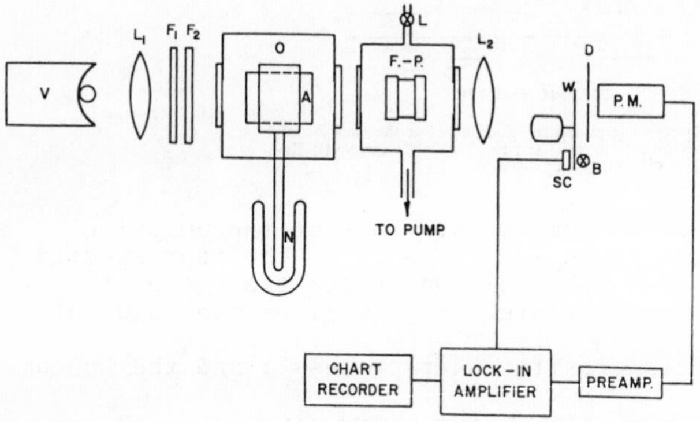

FIG. 1. Apparatus schematic.

oven wall O to a dewar of liquid nitrogen N? How was the Fabry-Perot interferometer F.-P. tuned over the absorption line? What was the difference in air pressure corresponding to the two principal hyperfine components? This pressure was controlled with the pump and leak valve L.

Fig. 2 shows the output of the cesium lamp(V in Fig. 1) with neon added to 2 mm Hg pressure. What is the purpose of the neon? What kind of source is desired in spectroscopy? One that will blanket the absorption lines of interest? Part (b) gives the energy level diagram of states associated with the D_1 resonance line(s) of atomic cesium. What do you think is the upper term in the D_2 resonance line? $6^2P_{3/2}$? Given that the transitional frequency of the cesium atomic clock is 9120 MHz, can you pick out the pair of states involved? Can you give two reasons why cesium was chosen for this standard? What about its isotopic pattern? Melting point for ease in handling? This same transition occurs in atomic hydrogen and is responsible for the 21-cm galactic radiation; there the transition is between the states F=0 and 1. Show that the four hyperfine states are associated with the correct F values. Restate the Russell-Saunders coupling scheme for the case of J and I rather than L and S. Have you already determined the cesium nuclear spin? The theoretical line intensities are from their ref. 14. Are they achieved in the lamp output? Why not? Apply the sum rule, from ref. 14 p. 238, to these intensities. The arithmetic is that $7=\frac{1}{4}(21+7)$ and $9=\frac{1}{4}(15+21)$.

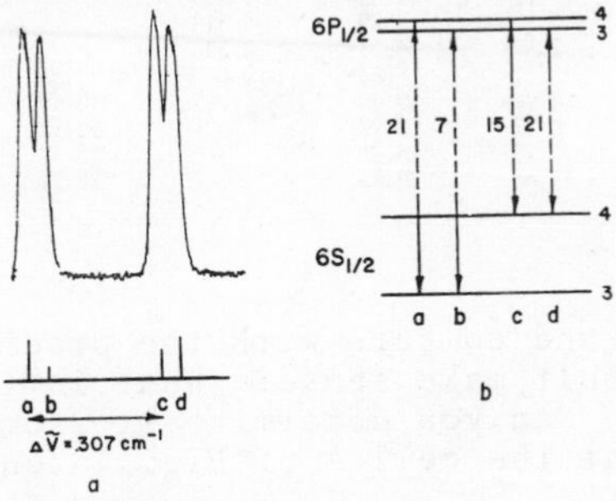

FIG. 2. (a) Profile of $6^2P_{\frac{1}{2}}-6S_{\frac{1}{2}}D_1$ resonance line of cesium using Varian lamp source at 2-mm Hg neon pressure. The lower part of the figure shows the theoretical relative intensities of the hyperfine components. (b) Energy-level diagram of states leading to D_1 resonance line of cesium. The numbers on the vertical are the theoretical relative line intensities.

In Fig. 3 I_0 is the lamp output for 10 mm Hg of neon; from its pressure-broadening we can roughly extrapolate past 2 mm of neon to zero pressure and see how the theoretical intensities could be achieved. In passing, what pressures were required to satisfactorily broaden the rotational lines of NO in exercise no. 104? It is also a case of source vs sample. Curve I is

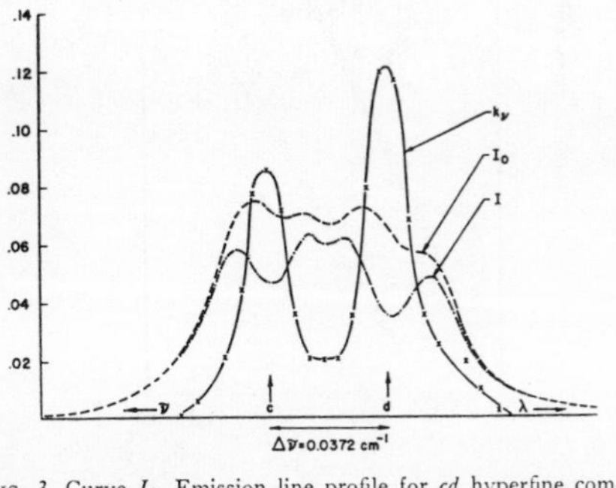

FIG. 3. Curve I_0. Emission line profile for *cd* hyperfine components with cesium lamp pressure of 10 mm neon. Curve *I*. Transmitted intensity through cesium-filled absorption bulb, length 5 cm, at 20.5°C. Curve k_ν. Absorption coefficient for *cd* hyperfine components.

the transmitted intensity; from Beer's law check the value of k_ν at each hyperfine peak and right in between. What are its units? Approximate the integral for the temperature shown and compare your result with that in Table I. Do each hyperfine

Table I. Results.

t [°C]	$\int k_\nu d_\nu/10^8$ [sec^{-1}]	$\rho/10^{10}$ [atoms/cm^3] exptl.	$\rho/10^{10}$ [atom/cm^3] Calc. from Taylor–Langmuir formulas $T<302$°K	$\rho/10^{10}$ [atom/cm^3] Calc. from Taylor–Langmuir formulas $T>302$°K
14.5	0.734	1.55	1.539	
16.8	0.952	2.01	1.906	
20.8[a]	1.39[a]	2.92[a]	2.988	
23.6[a]	2.08[a]	4.12[a]	4.000	
24.6[a]	2.15[a]	4.54[a]	4.428	
26.8	2.50	5.29	5.518	
28.2[a]	2.89[a]	6.11[a]	6.320	
30.4[a]	3.71[a]	7.83[a]		7.55
33.6	4.96[a]	10.49[a]		10.49
35.2	6.12	12.92		12.12
38.3	7.27	15.36		15.57
45.3	13.66	29.85		28.24

[a] Average values.

component separately and compare with the predicted ratio from Fig. 2. Does the result make sense? What is the vapor pressure of cesium at 20.5°C? Can you determine it from Table I? We failed to mention that the cell A of Fig. 1 contained only cesium.

The f value for the c and d lines was determined to be 0.179 from the result(ref. 15 and 16) $f_{D_1}+f_{D_2}=0.98$ and $f_{D_1}/f_{D_2}=2.03$. Make this calculation yourself.

Now compute values of the density ρ at a couple of the temperatures of Table I and compare. The final results are also

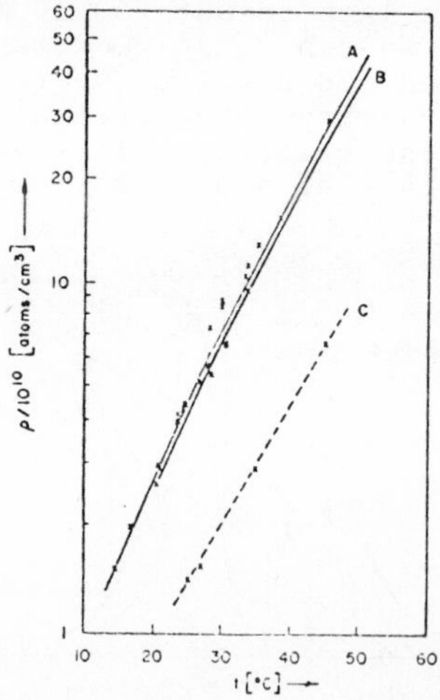

FIG. 5. The points show the experimentally obtained values for the cesium density ρ in units of 10^{10} atom/cm^3. Curve A is drawn using the Langmuir-Taylor data of Ref. 18; Curve B from data given in Ref. 19. Curve C, shown for comparison, is the result of a density determination of cesium above a mixture of rubidium and cesium.

plotted in Fig. 5. Assuming the present results are correct to within ∿5% as the authors do, what do you think of the hot-wire method of ref. 18(1937)? Can you find out what method was employed in the work quoted in ref. 19? Suggest an explanation for the position of curve C.

127. "Electronic Spectra of Polyatomic Molecules-A Brief Survey" D.A. Ramsay 43, S18 (1965). Supplement to 15 November issue.

The author points out that grating instruments have effectively infinite resolving power in that the spectra obtainable are not instrument-limited but molecule-limited. What does this mean in terms of the Doppler width of the rotational lines?

An absorption spectrum of azulene is shown in Fig. 1. This was the largest molecule to date for which a reasonably detailed rotational fine structure had been observed. Do you see the individual rotational lines? Notice that band is singular in the caption! What is the difference in wavenumbers between the two labeled wavelengths? Is azulene an asymmetric top with three unequal moments of inertia and therefore a very complicated rotational energy level pattern to be expected? Does it appear that there are regions of quite regular spacing of the rotational lines? Suggest an explanation. Let us make the crudest sort of measurement of moment of inertia, hoping for order of magnitude results. Take perhaps half an inch or a centimeter of spectrogram and count the lines; just once to save your eyes. Assume that the line spacing is ∿B(or A or C), the rotational constant. From this compute the moment of inertia and compare with a very approximate one obtained from a sketch of azulene. See if the author and I.G. Ross have published results for this molecule since the time of this article.

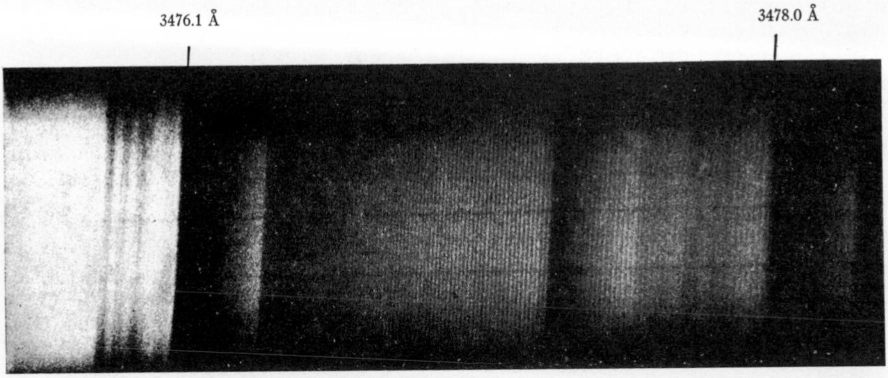

FIG. 1. 3476-A band of azulene photographed in absorption using the 17th order of a 24-ft Ebert spectrograph with a resolution of ∼600 000 and a dispersion of 0.12 Å per mm. The pressure was the saturated vapor pressure at room temperature and the path length was 24 m. This spectrogram was taken in collaboration with I. G. Ross of Sydney University, Australia, during his recent visit to Ottawa.

What is the maximum structural analysis that may be derived from a rotational analysis, without isotopic substitution? How does the latter technique give new information? Can you think of any other method of determining structure that will provide information about the geometrical structures of excited states of molecules and ground states of radicals? What about matrix-isolation spectroscopy? Can we classify that as the same method?

The Doppler broadening problem has been attacked recently with the technique of Lamb-dip spectroscopy; P.R. Bunker in Paper A1, given at the Symposium on Molecular Structure and Spectroscopy at Ohio State University, September 8, 1970. A resolution of 7 MHz has been obtained at 6328 Å. What are some of the resolutions that we have seen in the infrared, laser Raman, or the cesium resonance(with hyperfine) spectra in these exercises? How many MHz are there per cm^{-1}? Like in/ft to the spectroscopist?

128. "Electronic Structure of Cage Amines: Absorption Spectra of Triethylenediamine and Quinuclidine" A.M. Halpern, J.L. Roebber, and K. Weiss <u>49</u>, 1348 (1968).

The cagelike structures and high symmetry of triethylenediamine(1,4-diazabicyclo(2,2,2)octane or DABCO) and quinuclidine (1-azabicyclo(2,2,2)octane or ABCO) are shown in Fig. 1. Are

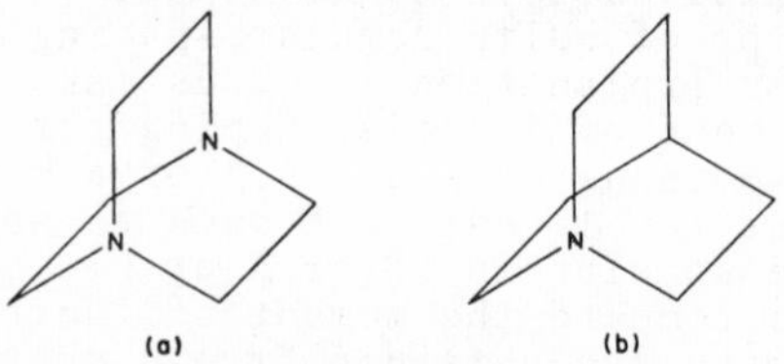

FIG. 1. Structures of (a) DABCO, $C_6H_{12}N_2$ and (b) ABCO, $C_7H_{13}N$.

these compounds isoelectronic? Should their electronic spectra be quite similar? Is it from Fig. 2? Would these compounds be colored in the vapor phase? Is the ordinate k exactly the same quantity as k_ν of exercise no. 126? For a vapor would the k in the present case be a unique function of concentration for the fixed temperature of 300°K? Would it be if you knew that the sample cell had an attached sidearm containing crystals of the amine? Why are the wavenumber scales displaced from one another in each case? If you would see scales such as these labeled kk could you interpret the unit? The compound labeled TMED is tetramethylethylenediamine. The sample cells were 7.12 cm long.

The authors state that spectral comparison of DABCO and ABCO provides a further means of examining the interaction between two nitrogen chromophores situated in a fixed molecular framework. To which point group would each molecule belong?

In Fig. 2, do you see three distinct bands for ABCO, with well-resolved vibrational structure evident for each? The oscillator strength of the first strong(43-53 kk) and weak long-wavelength(39-43 kk) bands are said to differ by more than an order of magnitude. Do you agree? What one piece of information would you need to compute the oscillator strength f for each electronic transition by the method of exercise no. 126?

The first strong absorption band of ABCO vapor is shown in Fig. 3. With assignments made of the two intense vibrational progressions, do you see that the 0-0 band is assigned at 43.75 kk? In passing note that the authors call the <u>band system</u> mak-

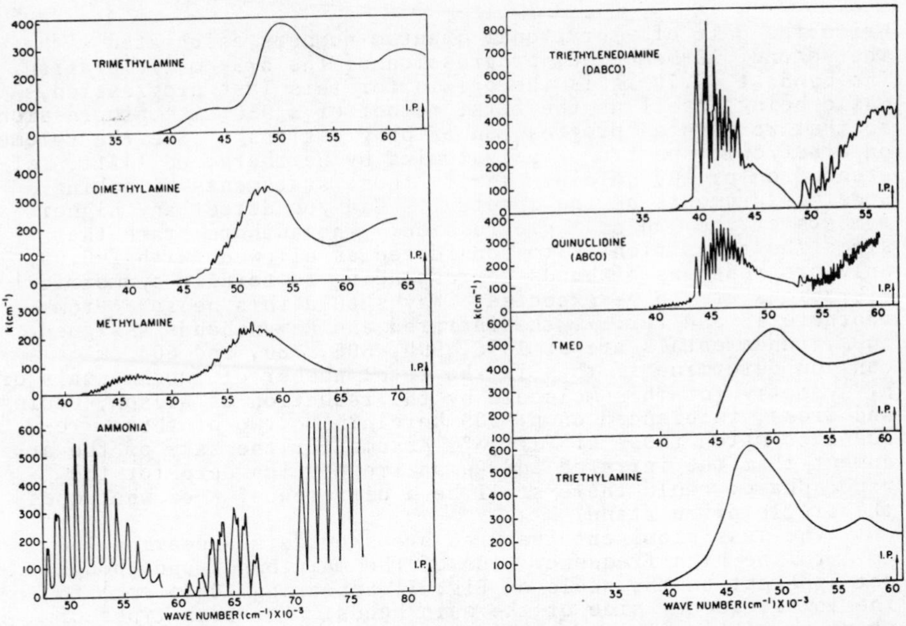

FIG. 2. The vapor-phase electronic absorption spectra of ammonia, methylamine, dimethylamine, trimethylamine, triethylamine, TMED, ABCO, and DABCO at 300°K.

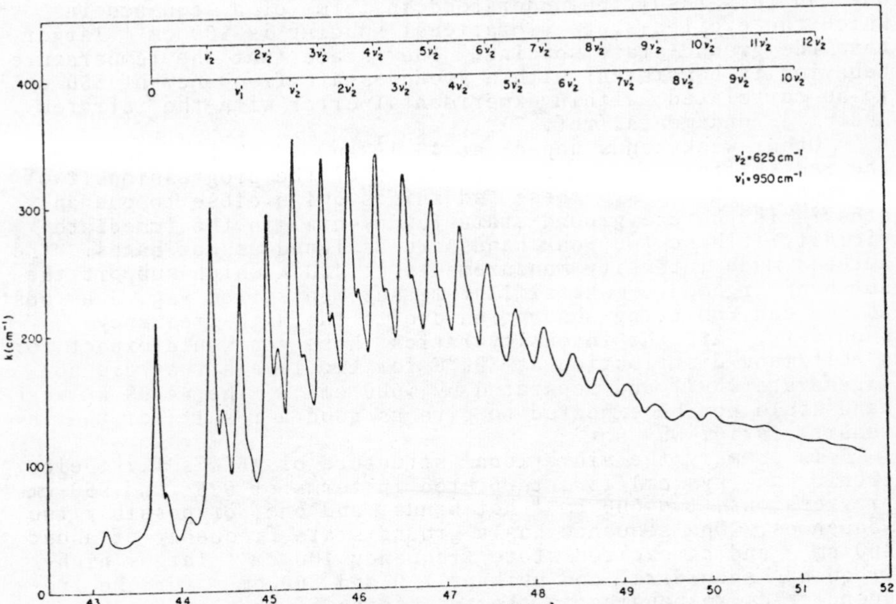

FIG. 3. The first strong electronic absorption band of ABCO vapor at 300°K.

ing up an electronic transition simply a band, which should hardly cause confusion. Is the 12-term progression in ν_2' completely analogous to what one would expect in a diatomic spectra?

Write the pair of vibrational quantum numbers associated with the second, 10-membered, progression of the 625-cm^{-1} vibration. The band at 44.70 kk is the origin for this last progression, while being itself as the first member of a 950-cm^{-1} progression so that we have a "progression of progressions". See the volume on electronic spectra of polyatomics by Herzberg, pp 142ff, referenced on p. 109 herein. Verify these statements by making some measurements on the spectrum. Can you detect any higher members of the 950-cm^{-1} progression? The authors state that since the transition is to considered as allowed, with f=0.06, only progressions of bands corresponding to totally symmetrical vibrations are to be expected. Why should this be so? From footnote 15 and ref. 7, the infrared and Raman bands assigned to a_1 fundamentals are at 1055, 990, 805, 780, and 604 cm^{-1}. Can you determine if this is the total number of fundamentals of a_1 symmetry for this molecule by the reduction of Wilson, Decius and Cross, referenced on p. 209 herein? Do two of these frequencies match those of Fig. 3? Assume for the sake of the argument that the infrared and Raman frequencies were for the vapor phase; could there still be a difference? For what does the single prime stand?

The less prominent features are shoulders appearing 100 cm^{-1} to the high-frequency side of the main bands and(though not apparent on the scale of Fig. 3) appearing 40-50 cm^{-1} to the low-frequency side of the main bands. For both types of shoulders the ratios of intensities to that of the main bands increases with temperature; for the high-frequency-side bands why can this result be understood in terms of a sequence in which the excited-state vibrational spacing is 100 cm^{-1} larger than the ground-state spacing? They state that the temperature behavior is consistent with a ground-state frequency of 350 cm^{-1} to be correlated, within experimental error with the infrared 408 cm^{-1} fundamental(ref. 7).

Other weak bands appear at 43.15 and 44.09 kk. What are the red shifts from the first members of the progressions(two) mentioned above? Are these red shifts quite close to one another? Is there a ground-state fundamental in the immediate vicinity? These two weak bands are assigned as hot bands. The authors made intensity measurements at 400°K which support the hot-band argument; make similar measurements from Fig. 3 as best as you can, omitting absorption due to the high-frequency shoulders. Are the intensity ratios those you would expect for a Boltzmann distribution at 300°K for two levels that are nondegenerate(why?) and separated by $\sim$600 cm^{-1}? The 44.09 kk weak band would not be expected to give as good a result for the intensity ratio; why not?

In summary the vibrational structure of ABCO's first electronic band(system) is interpreted in terms of 625- and 950-cm^{-1} progressions, two 600 cm^{-1} hot bands, and one, or possibly two, sequences. One sequence has a ground-state frequency of about 400 cm^{-1} and an excited-state frequency 100 cm^{-1} larger(high-frequency shoulders) and perhaps 250 and 200 cm^{-1} for the frequencies corresponding to the two states if the low-frequency shoulders form the second sequence.

The structure of the second strong band of ABCO is considerably more complicated, with a set of four long progressions in the 625-cm^{-1} vibration; the weak long-wavelength band also shows four progressions in this frequency. It turns out, as can be partially verified from Fig. 3, that the three electronic transitions of ABCO do not have their maximum intensities in the 0-0 bands, indicating a sizeable change in the positions of the nuclei in the upper electronic states relative to their positions in the ground state(ref. 20). In the case of ABCO, the presence of so many long progressions in the 625-cm^{-1} vibration suggested that this change in geometry is most nearly represented by this vibrational mode. What feature of the spectrum in Fig. 3 permits almost certain identification of the excited state 625-cm^{-1} vibration with the 604-cm^{-1} mode of the ground state? This last vibration has been assigned(ref. 7?) to a cage-squashing mode in which the nitrogen atom and the CH group opposite along the CH-N axis, with the rest of the cage distorting in a manner which accommodates this motion. Approximate this normal mode with a sketch. What would you estimate a C-N-C angle to be in the ground electronic state? Would all three of these be equal by symmetry? The authors consider that in the excited electronic state(s) of ABCO the nitrogen may be approaching planar. Is the comment about the excited vibrational state of ammonia, in footnote 21, relevant and why?

DABCO's spectra are analyzed in a similar fashion, though the conclusion is not nearly so clear cut. What is the basic reason for this(p. 1354)? Some argument is presented leading to the conclusion that these electronic transitions are those of the $n \rightarrow p$ Rydberg type. What exactly does $n \rightarrow p$ mean? What are Rydberg transitions(exercise no. 123)?

129. "Dipole Moments of the Excited States of Azulene" R.M. Hochstrasser and L.J. Noe 50, 1684 (1969).

In a dilute(10^{-5}-10^{-6} mole azulene/mole naphthalene here) mixed crystal where the guest-guest interactions are vanishingly small, the resultant splitting of spectral lines $\Delta\varepsilon = E_\xi G \Delta\mu$; it is directly proportional to E_ξ, the local electric field, the geometric factor G, and the change in dipole moment $\Delta\mu$ between the ground and excited electronic states. The authors state that certain problems having to do with field-induced moments may be eliminated by dispersing the dipolar molecule in a nonpolar lattice. Does naphthalene meet this requirement? The ground-state dipole moment for azulene had been determined(ref. 10) to be 0.79 D. The sign of $\Delta\mu$ could not be determined in this work; how could this be done in principle and why could not the observation be made in this experiment(p. 1688)? It seemed reasonable to the authors to assume that the dipole moment is decreased in both of the electronic states studied, relative to that of the ground state, with support from ref. 7 and 8. Therefore the sign of $\Delta\mu$ is taken as negative.

The spectra of the first[0,0 band at λ6823.76, 14 650.0 cm^{-1}] and the second[0,0 band at λ3564.30, 28 048.0 cm^{-1}] $^1\pi\pi^*$ excited states of azulene in the host crystal are given in Fig. 1. Notice the splitting of the 0,0 bands in the high-field spectra. Can you correlate the very faint blue color of the substitutional crystal with one of the absorptions? Does this spectrum prove that the faint blue color is due to azulene rather than the host naphthalene? What about the name of the former? Notice the ordinate; we see the transition from film to densitometer reading.

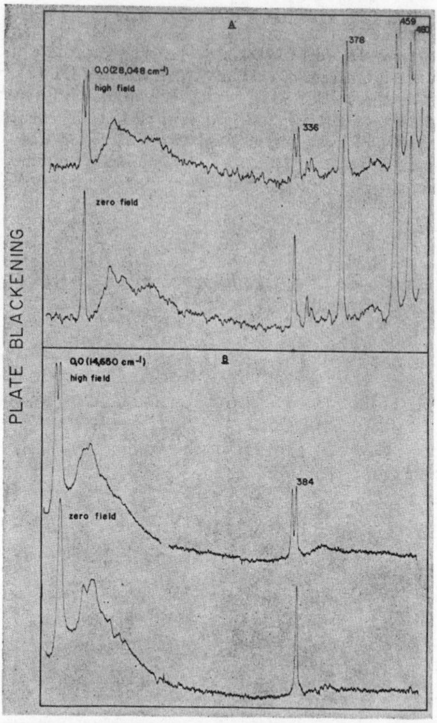

FIG. 1. Absorption spectra of the first and second $^1\pi\pi^*$ states of azulene in a substitutional solid with naphthalene at 4.2°K. Part A of this figure corresponds to the second $^1\pi\pi^*$ state and Part B, to the first $^1\pi\pi^*$ state. The origin (0, 0) and the first several absorption bands (3) are seen at high electric field directed along the **c**′ axis and zero field as noted on the spectra.

Fig. 2 shows the splitting of one of the 0,0 bands at several values of the applied field directed along one of the axes. Notice the last sentence of the caption.

That the splitting is linear in the applied electric field is shown in Fig. 3. Comparing linewidths of the two transitions in Fig. 1, one sees that that of the first is somewhat greater. This fact prevented the obtaing of reliable splitting patterns for that state with the applied field E along the a or b axes. Determine the slope, splitting energy per unit applied voltage, for each of the four cases.

To apply the equation given at the first of the article, we need to know the relation between the applied electric field E and the local field E_ξ along each of the crystal axes, as well as the three geometric factors. We have used the subscript ξ in a sense that is trivially different from that of the authors.

The numerical results are:

$$E_{\xi a} = 1.516\ E \qquad G_a = 0.876$$
$$E_{\xi b} = 1.655\ E \qquad G_b = 0.421$$
$$E_{\xi c'} = 1.850\ E \qquad G_{c'} = 1.758.$$

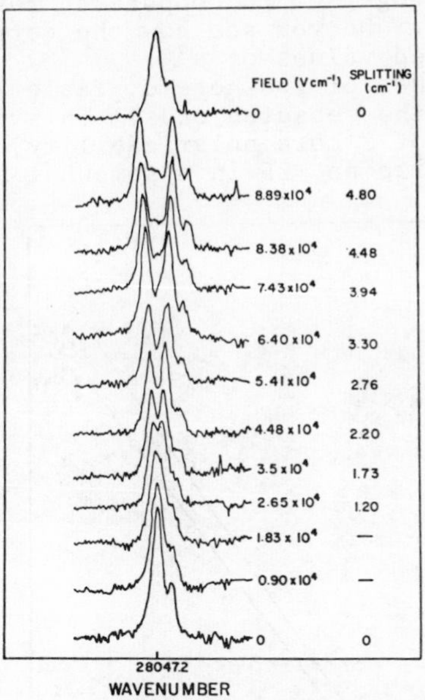

FIG. 2. Absorption spectra at high resolution of the second $^1\pi\pi^*$ state of azulene as a function of field strength. The electric field is directed along the **c**′ axis and the field strength and splitting are noted on the spectra. The 0,0 band is a doublet even in zero field; the splitting is 2.5 cm^{-1}, but the reasons for this splitting are not clearly understood.

The reason that the third axis is labeled c' rather than c, the crystal axis closest to the direction of the longest in-plane axis of naphthalene molecules, is contained in footnote b of Table I. That the electric field along c'(∿ along c) gives the largest effect for both excited electronic states can be checked in Fig. 3.

TABLE I. Dipole moments for the excited states of azulene.[a]

State	Field direction	Present work: μ (debye unit) $\Delta\mu$	μ^*	$\mu^{*\,b}$	Previous work μ (debye unit) μ^*
$^1\pi\pi^*(^1B_1)$	c′	−1.10±0.05	−0.30	−0.42	+0.02 to −1.02[c,d]
(14 650.6 cm^{-1})					−1.36[d,e](calc)
$^1\pi\pi^*(^1A_1)$	a	−1.09±0.05	−0.29	−0.31	+0.23 to −0.53[c,d]
(28 048.0 cm^{-1})	b	−1.10±0.05	−0.30	−0.31	−0.07[f]
	c′	−0.99±0.05	−0.19	−0.31	−2.07[d,e](calc)

[a] The excited-state dipole moments in this table are all negative, based on the assumption that the dipole switches direction in going from the ground to the excited state (see Footnote e).

[b] These are the preferred values of the excited-state dipole moments. Since the **a** and **b** electric field measurements give the most reliable result for $\Delta\mu$, the **c**′ magnitude of μ^* being low, the value of the second-state moment is calculated to be an average of the **a** and **b** field direction results. We assume the same reasoning for the first state, that the **a** and **b** field direction measurements would give the most accurate result even though these measurements cannot be reliably interpreted (see discussion under Stark Effects) and correct the first state moment to correspond to the ratio $\frac{1}{2}[\Delta\mu_a+\Delta^a{}_b]:\Delta\mu_c{}'$ given by the second state. These considerations arise because we understand that the azulene molecules are slightly misoriented with respect to the host molecules as discussed in the text.

[c] W. W. Robertson, A. D. King, Jr., and O. E. Weigang, Jr., J. Chem. Phys. **35**, 464 (1961). Results are based on a frequency-shift-solvent effect.

[d] The values reported have been adjusted to a ground-state dipole of +0.80 D.

[e] R. Pariser, J. Chem. Phys. **25**, 1112 (1956). Results are based on quantum-mechanical calculation for π electrons.

[f] H. Sauter and A. C. Albrecht, Chem. Phys. Letters **2**, 8 (1968). Results are based on an electric field broadening technique.

Put the numbers together to compute $\Delta\mu$ for each of the four cases in Table I. Do you see how the correction was made to obtain the preferred values of μ^*?

The computed values of footnote e, Table I(ref. 8) are quite different than the measured ones. This is discussed in the article in terms of σ core polarizability, which topic was touched upon in exercise no. 25 in a somewhat different context.

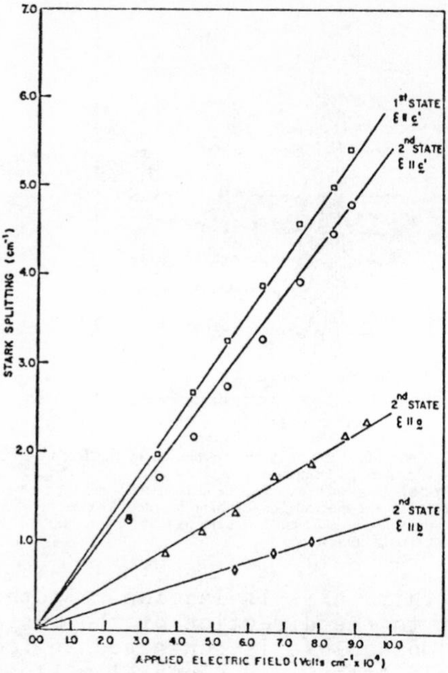

FIG. 3. Stark splitting versus applied electric field curves for the first and second $^1\pi\pi^*$ states of azulene. The various field direction(s) and states are noted on the curves. The curves drawn through the raw data points have been adjusted with a single constant least-squares procedure.

130. "Absorption Spectrum of the Argon Molecule in the Vacuum-uv Region" Y. Tanaka and K. Yoshino 53, 2012 (1970).

Nine discrete band systems were identified in the absorption spectrum of diatomic argon. Five of these gave the most reliable values for the vibration constants and dissociation energy for the ground state of the argon molecule, $X^1\Sigma_g^+(0_g^+)$. What does each of the quantities mean in this term symbol? The constants and values of D_0° and D_e are given in Table IX. Fig. 12 shows the potential energy curve for the ground state calculated as a Morse function(exercise no. 73 and Herzberg's volume on diatomics, referenced on p. 93 herein and referred to in this exercise as SDM, p. 101). In part (b) the curves marked AM and CJAM are those determined from high-energy atomic beam experiments in ref. 8 and 11. The value of r_e, 3.81 Å, was taken from theoretical work of ref. 3. What is required to determine this quantity experimentally? Notice that the bond

TABLE IX. Ground state vibration constants and dissociation energy (cm^{-1}).

System	I	II	III[a]	VI	VII	Average	III[b]
ω_e	30.88	30.40	31.05	30.88	30.18	30.68±0.37	67.0
$\omega_e x_e$	2.57	2.43	2.72	2.62	2.45	2.56±0.12	4.03
D_0^0	77.3	79.0	73.8	77.0	77.6	76.9±1.9	$D_0=423$
D_e	92.1	93.6	88.7	91.8	92.1	91.6±1.9	

[a] The band identification in system III is not as accurate as in others.

[b] Constants of the upper state of system III.

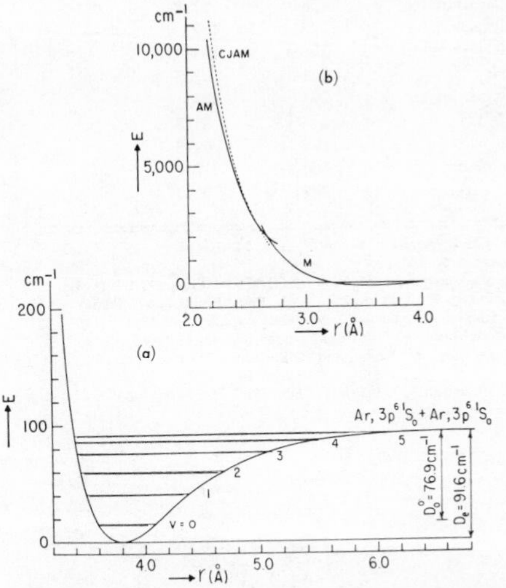

FIG. 12. The potential energy curve for the ground state. (a): calculated as a Morse function with presently obtained vibration constants. The value of the equilibrium distance, $r_e=3.81$ Å, is borrowed from previous work (see Ref. 3). (b): "M" is a Morse curve equivalent to (a), the potential well depth somewhat exaggerated. The curve is displaced horizontally until it matches "AM" at $r=2.63$ Å (see text). The equilibrium distance of the curve at the new position is $r_e=3.60$ Å. For "AM" see Ref. 8. "CJAM" agrees much more closely with that of the Morse curve than "AM" does, at arou nd $r=2.63$ Å.

energy is just a fraction of a kcal/mole. See if you can match the value of the Morse curve at r=3.4 and 5.4 Å. Are the dissociation products the correct ones?

Table X gives well depths from several sources. See if the method used in exercise no. 47 to "prove" that He_2 does not have a bound vibrational state, computing the zero-point energy from the harmonic frequency of a Lennard-Jones 6-12 potential, will yield one or more bound vibrational states for Ar_2. Use either the LJ(6-12) or the spectroscopic well depth.

Fig. 13 gives energy level diagrams for states of the atom and the diatomic. Because of the very low dissociation energy the two manifolds have virtually the same origin on the spectroscopic scale. Is the height of the vibrational state density just about to scale for the ground molecular state? Do you see that the nine excited vibrational state densities are all greater than that of the ground electronic states? What does this say about the dissociation energies of these upper electronic states wrt that of the ground state? Discuss in terms of molecular orbital theory. For the atomic states, what is the difference between the primed and the unprimed terms? See p. 2013.

Table X. Depth of potential well for the ground state of Ar_2.[a]

Potential type	Well depth (cal/mole)	Well depth (cm^{-1})	Reference
Multiparameter	286	100	b
Exp-6	302	106	c
Kihara	293	103	c
Kihara	284	99.3	d
Kihara	274	95.8	e
Multiparameter	278	97.2	f
Multiparameter	292	102	g
Multiparameter	296	104	h
Morse	264	92.3	i
Multiparameter	294	103	j
LJ(6–12)	234	81.8	c
SW	185	64.8	c
Sutherland	606	212	c
This work	262	91.7	

[a] This table is essentially a copy of Table I in Ref. 36.
[b] Reference 5.
[c] A. E. Sherwood and J. M. Prausnitz, J. Chem. Phys. **41**, 429 (1964).
[d] J. A. Baker, W. Fock, and F. Smith, Phys. Fluids **7**, 897 (1964).
[e] J. C. Rossi and F. Danon, Discussions Faraday Soc. **40**, 97 (1965).
[f] M. L. McGlashan, Discussions Faraday Soc. **40**, 97 (1965).
[g] J. H. Dymond, M. Rigby, and E. B. Smith, J. Chem. Phys. **42**, 2801 (1965).
[h] R. J. Munn and F. J. Smith, J. Chem. Phys. **43**, 3998 (1965).
[i] Reference 37.
[j] Reference 6.

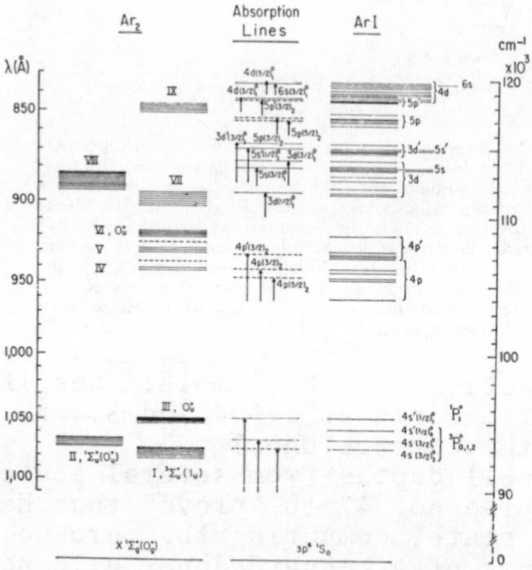

Fig. 13. Energy level diagram of atomic and molecular argon. See Ref. 26 for the term designations for these atomic levels. The horizontal dashed lines in ArI, except for one at the bottom, indicate levels derived from $3p^5np$ configurations with $J=2$. For all these, transitions from the ground level are observed. Similar levels derived from $3p^5nf$ configurations with $J=2$ are not shown but transitions to these from the ground level are also observed. Each of the two dashed horizontal lines in Ar_2 indicates the location of a cluster of several very diffuse bands. Vibrational levels and dissociation energy are here ignored in the ground state of Ar_2. All are shown at the same level as the ground level of ArI.

As you might expect, the molecular band systems did not appear in the spectra until pressures of the order of a few torr(relation to mm Hg?) were reached. Fig. 3 shows the spectrum of systems I and II. Why do the v' numberings begin simply

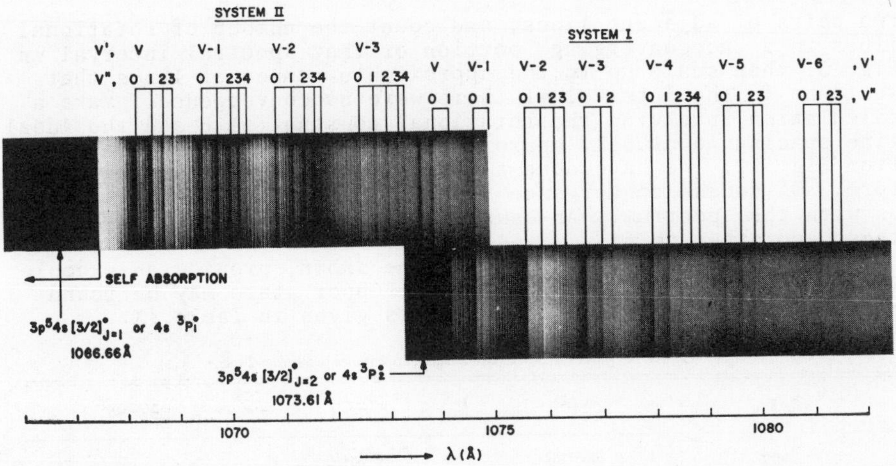

FIG. 3. Absorption spectrum of Ar_2, systems I and II. All exposures were taken with the argon background, in the second order, and a 60-min exposure time. The cell containing the argon was cooled by liquid nitrogen. The argon pressures were 10 and 20 torr for the top and bottom exposures, respectively. The absorption at the resonance line is caused partly by the cell argon (low temperature), and partly by the source argon (high temperature).

with v? Verify as many of the band positions of system II as you can, given Table II to work with. Show how the $\Delta G_{v+\frac{1}{2}}$ are then plotted in Fig. 5(SDM p. 438) and the dissociation energy of the ground state determined. Are we using the method credited in part to the author of the article of exercise no. 122? Determine this dissociation energy to compare with that of Table IX. Have you found D_0° or D_e and why? Can you also determine from Fig. 5 the dissociation energy of the upper state of system II? Why or why not?

Do you see the rotational lines in Fig. 3? From the question asked at the bottom of p. 244, or rather from its answer, we know that determination of the internuclear distances in the upper and lower states of a band system would follow a rotational analysis. Let us make the drastic assumption that these distances are the same in the lower and upper states of system II. To show how primitive this assumption is, choose ten different diatomics from Table 39 of SDM showing as much variation as you can between internuclear distances in the ground and some excited state. Now from Table II pick a spacing between one or

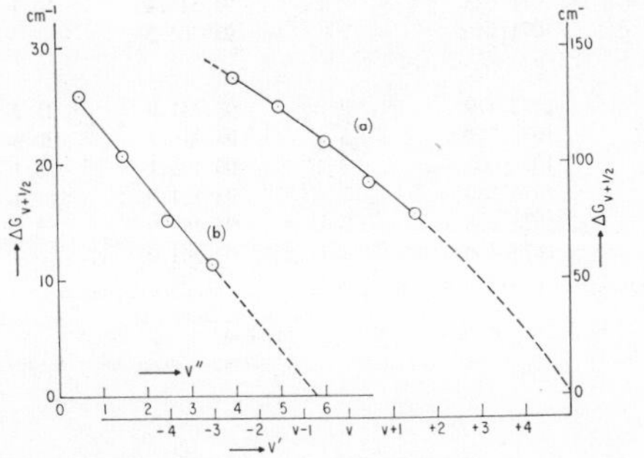

FIG. 5. $\Delta G_{v+1/2}/v$ curves of the states involved in system II. Curve (a): the upper state, $0_u^+(^1\Sigma_u^+)$; curve (b): the ground state, $0_g^+(^1\Sigma_g^+)$.

two pairs of adjacent lines, and count the number of rotational lines in a "nonconverging" portion of that spectral interval in Fig. 3; then scale up to the approximate number of lines that would be in the interval if there were no convergence. Make a calculation involving the rotational constant(s) B and the ideal line spacing to crudely approximate r_e and compare with that from Fig. 12. We have done a similar stunt once or twice before. Of course convergence isn't all "bad"; how does it help us with the spectrum? As usual SDM provides ample discussion of these and other points.

Fig. 6 and 7 and Table III, none shown, provide an example where the dissociation energy of the upper state may be found by this extrapolation. The result is given in Table IX.

TABLE II. Band system II (bandheads), appearance pressure ~2 torr.[a]

v'	v''	λ (Å)	I[b]	ν (cm⁻¹)	$\Delta G_{v''+1/2}$	$\Delta G_{v'+1/2}$
$v+x$[c]	?	1065.137[d]	3[e]	93 884.6		
	?	1065.227[d]	3[e]	93 876.7		
	?	1065.323[d]	3[e]	93 868.2		
	?	1065.409[d]	3[e]	93 860.6		
	?	1065.457[d]	3[e]	93 856.4		
$v+2$[c]	0?	1066.306	3[e]	93 781.7	25.0	77.8
	1?	1066.597	6[e]	93 756.1		77.8
						⟨77.8⟩Av
$v+1$[c]	0	1067.192	10[e]	93 703.9	25.6	91.1
	1	1067.483	3[e]	93 678.3		92.0
						⟨91.6⟩Av
v	0	1068.230	8	93 612.8	26.5	110.1
	1	1068.533	6	93 586.3	19.6	108.3
	2	1068.756	4	93 566.7	15.5	109.1
	3	1068.933	3	93 551.2	11.1	107.8
	4	1069.060	1	93 540.1		108.9
						⟨108.8⟩Av
$v-1$	0	1069.488	6	93 502.7	24.7	121.2
	1	1069.770	5	93 478.0	20.4	122.6
	2	1070.004	4	93 457.6	14.2	125.8
	3	1070.166	4	93 443.4	12.2	127.0
	4	1070.306	2	93 431.2		125.5
						⟨124.4⟩Av
$v-2$	0	1070.876	4	93 381.5	27.5	139.6
	1	1071.191	4	93 354.0	22.2	137.3
	2	1071.446	5	93 331.8	15.4	135.7
	3	1071.623	3	93 316.4	10.7	135.4
	4	1071.746	2	93 305.7		135.8
						⟨136.8⟩Av
$v-3$	0	1072.479	2	93 241.9	25.2	
	1	1072.769	3	93 216.7	20.6	
	2	1073.006	3	93 196.1	15.1	
	3	1073.180	3	93 181.0	11.1	
	4	1073.308	2	93 169.9	8.0?	
	5	1073.400	1	93 161.9?		

[a] As in Footnote b in Table I.
[b] Same as Footnote c in Table I.
[c] These band groups are readily obscured by the broadened argon resonance line absorption.
[d] Diffuse.
[e] Values not so reliable as for the rest of the bands.

CHAPTER XI

ROTATIONAL SPECTROSCOPY

131. "Microwave Spectra and Structures of Methyl Mercury Chloride and Bromide" W. Gordy and J. Sheridan 22, 92 (1954).

The exercises of this chapter and the next two are all from articles based upon research in the radiofrequency, microwave, or millimeter wave portion of the spectrum. Does the latter region join with the far infrared region at some indefinite value? Does nature divide the spectrum into regions or do we? Generally based upon what? Where did the detection methods used in these next three chapters have their historical roots? Their importance to problems of molecular structure was mentioned in SDM pp 59-60. See if in the present chapter we don't generally restrict ourselves to transitions for which $\Delta J=+1$.

Fig. 1 and 2 show these transitions, for the given J values in the captions, for the compounds of the article title. For

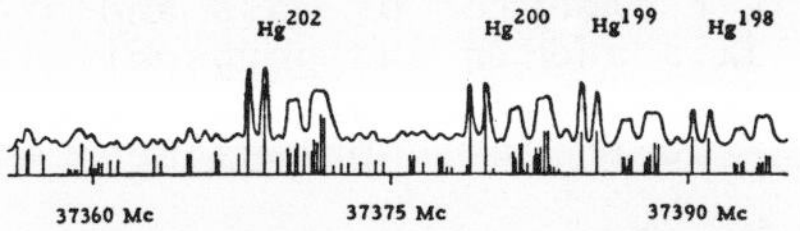

FIG. 1. Calculated and observed pattern of the 8→9 rotational transition of methyl mercury chloride. Theoretical patterns are not plotted for Hg^{201} and Hg^{204}.

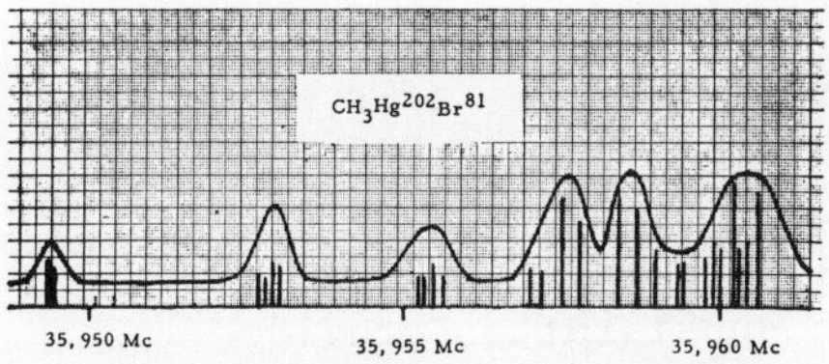

FIG. 2. Calculated and observed pattern of the 15→16 rotational transition of $CH_3Hg^{200}Br^{81}$ for $K \leq 6$.

the chloride spectrum notice how the mercury isotopes have complicated, or enriched, the pattern. Table I, not included, tells us that this spectrum is that of ^{35}Cl with the ^{37}Cl spectrum about 1300 Mc(or MHz) away; toward higher or lower frequencies? Fig. 2 is limited to a single isotope each of Hg and Br, but is still complicated by the presence of the quadrupole hyperfine structure due to the nuclei; which of the four nuclei has a quadrupole moment? which do not and why?(neglect ^{13}C).

Check the positions of the main features of the observed pattern in Fig. 2 by means of the information in Table II and

TABLE II. Observed spectral constants of CH_3HgBr.

Br	Hg	ν_0(Mc/sec) $J=15\rightarrow16$	B_0(Mc/sec) (assuming $D_J=0$)	D_{JK}(kc/sec)
79	198	36 571.55	1142.86	8.2
	199	36 547.33	1142.10	8.2
	200	36 523.48	1141.36	8.2
	202	36 476.28	1139.88	8.2
81	198	36 008.79	1125.28	8.0
	199	35 984.46	1124.51	8.0
	200	35 960.40	1123.76	8.0
	202	35 912.50	1122.27	8.0

a formula you should derive from the expression(IRRS p. 26),

$$F(J,K) = BJ(J+1) + (A-B)K^2 - D_J J^2(J+1)^2 - D_{JK}J(J+1)K^2 - D_K K^4$$

for the rotational energy levels as a function of the quantum numbers J and K(what is its meaning?), for the frequency of the transition between two energy levels in absorption(15→16) given the selection rules that $\Delta J=\pm1$ and $\Delta K=0$. The agreement won't be exact because the ν_0 values have been corrected for the observed nuclear quadrupole perturbations.

The presence of so many isotopic molecules allows more structural information to be obtained from the spectra than would be the case otherwise; why? Table IV contains bond distances and the <HCH obtained(still required some assumptions?) in this way. Reverse the process(easier?) to determine B_0 for one of the CH_3HgBr from the data given. Set up a coordinate system with origin at the center of mass and axes along the three-fold symmetry axis and in and perpendicular to a symmetry plane. Was the <CHgBr assumed or was it definitely determined from the experiment? Do the assumptions seem eminently reasonable?

132. "The Microwave Spectra, Structure, and Dipole Moment of Stable Pentaborane" H.J. Hrostowski and R.J. Myers <u>22</u> 262 (1954).

Fig. 2 shows the structure of B_5H_9 from the microwave data. With the aid of Table IV and the choice of the B_1-H_1 and B_2-H_2 distances as 1.22 Å, make section views of the pentaborane molecule filling in all bond distances and angles; is there enough information in Table IV to do this?

Again isotopic substitution provided information for determination of the structural parameters from the moments of inertia. What isotopes of boron would occur naturally? Would you expect a statistical distribution of these isotopes in the five (two nonequivalent) boron positions? Would the statistically computed line intensities and changes in the approximate rotational constants be an excellent guide in making line assignments? In addition the prepared B_5D_9 gave the authors another tool with which to determine the structure.

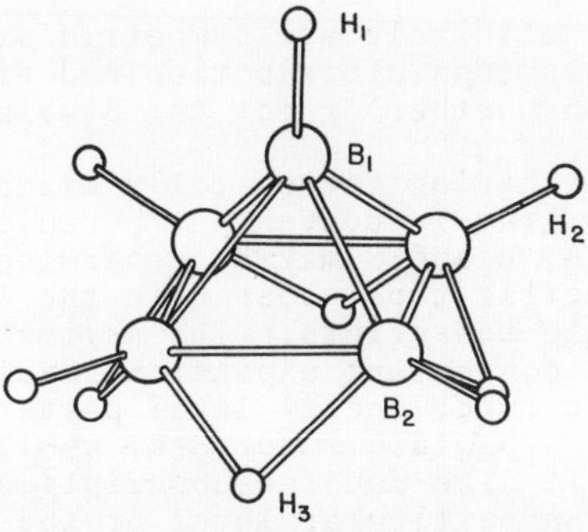

FIG. 2. A C_{4v} structure for stable pentaborane consistent with the microwave data.

TABLE IV. Structural parameters for stable pentaborane.

		Microwave		Elec. diff.[b]	X-ray[c]
B_1-B_2[a]		1.687 ±0.005		1.700 ±0.017	1.66 ±0.02
B_2-B_2		1.800 ±0.003		1.805 ±0.014	1.77 ±0.02
B_1-H_1 and B_2-H_2	(1.18)	(1.22)	(1.26)	1.234 ±0.066	1.21 and 1.20 ±0.05 ±0.07
B_2-H_3	1.39	1.35 ±0.02	1.31	1.359 ±0.077	1.35 ±0.04
$\angle B_1-B_2-H_2$	134°20′	136°10′ ±30′	138°0′	120° ±20°	115° ±5°
$\angle B_1B_2B_2-B_2B_2H_3$ (external dihedral)	192°	196° ±2°	200°	187° ±10°	190° ±5°

[a] The hydrogen and boron designations are those of Fig. 2.
[b] Reference 3.
[c] Reference 4.

TABLE II. Some observed lines for B_5D_9 and B_5H_9.

Observed frequency[a] mc/sec	Calculated frequency[b] mc/sec	Assignment Transition	Molecule	
20 845.4	(20 845.4)	1–2	$B_5{}^{11}D_9$	C_{4v}
21 051.8	(21 051.8)	1–2	$B_4{}^{11}B^{10}D_9$	C_{4v}
21 117.2	(21 117.2)	$1_{11}-2_{21}$	$B_4{}^{11}B^{10}D_9$	
21 032.1	(21 032.1)	$1_{10}-2_{20}$	C_s	
20 938.8	(20 938.8)	$1_{01}-2_{11}$		
21 329±1	21 329	$1_{11}-2_{21}$	$B_3{}^{11}B_2{}^{10}D_9$	
21 241±1	21 242	$1_{10}-2_{20}$	C_s	
21 146±1	21 147	$1_{01}-2_{11}$	B^{10} at apex	
	21 215	$1_{11}-2_{21}$	$B_3{}^{11}B_2{}^{10}D_9$	
21 211±1	21 212.5	$1_{10}-2_{20}$	C_s	
	21 210	$1_{01}-2_{11}$	B^{11} at apex	
42 017.3±0.1	(42 017.3)	2–3	$B_5{}^{11}H_9$	C_{4v}
42 5398	(42 539.8)	2–3	$B_4{}^{11}B^{10}H_9$	C_{4v}
42 776.1	(42 776.1)	$2_{21}-3_{31}$		
42 563.1	42 563.1	$2_{20}-3_{30}$	$B_4{}^{11}B^{10}H_9$	
42 525.3±0.4	42 525.6	$2_{12}-3_{22}$	C_s	
42 488.1	42 488.1	$2_{02}-3_{12}$		
42 286.5	(42 286.5)	$2_{11}-3_{21}$		

[a] Unless specified, ±0.2 mc/sec.
[b] Calculated using the parameters in Table III.

In Table II, why do the lines in the second set have a transitional frequency about twice that of the members of the first set? Is this correct in the sense of the rotational level spacing of symmetric tops? Slightly asymmetric tops? The double subscripts on the J values are characteristic of asymmetric top or rotor levels as will be discussed shortly. Show that all of the molecules listed definitely are symmetric or asymmetric tops, from the boron isotope distribution and Fig. 2, corresponding perfectly with whether or not the J values are double-subscripted.

It turns out that having isotopically mixed asymmetric top molecules around is extremely convenient because then all of the rotational constants A, B and C may be determined; this is not the case for the symmetric tops as seen in the last exercise.

Now for the double subscripts. The asymmetry parameter $\kappa \equiv (2B-A-C)/(A-C)$ is a convenient expansion parameter for the complicated asymmetric rotor energy level pattern. As usual $A \geq B \geq C$. Show that for a prolate rotor(B=C) $\kappa=-1$, whereas for an oblate rotor(A=B) $\kappa=+1$. The double subscripts on the J value are the values that the particular level of the asymmetric top take for the quantum number $K(\geq 0)$ in the prolate(first) and oblate(second) limits. An illustration and authoritative discussion is given in Chapter 4, esp. p. 86 of "Microwave Spectroscopy" by C.H. Townes and A.L. Schawlow (McGraw-Hill Book Co. Inc. New York 1955). Examine our Table II and I(not included) while referring to their Fig. 4-1, if possible.

133. "Microwave Spectrum, Spectroscopic Constants, and Electric Dipole Moment of Li^6F^{19}" L. Wharton, W. Klemperer, L.P. Gold, R. Strauch, J.J. Gallagher, and V.E. Derr 38, 1203 (1963).

Why was this particular isotope of Li selected rather than the more abundant Li^7? Transitions between the J=1 and J=0 states, for each of the four lowest vibrational states, were measured with the electric beam resonance method. From the results obtained and shown in Tables II and IV compute the quantity μ_v/er_v for v=0-3. This quantity is the actual dipole moment divided by the dipole moment for the completely ionic molecule at the same internuclear distance, and is called the %, or fractional, ionic character. What is the behavior of this quantity with increasing v? Is it as you would predict or not and why? Discuss. For the diatomic with the greatest difference in electronegativities or greatest ionic character, is dissociation into ions or neutral atoms the preferred process energetically? What two quantities must be compared?

What was the method used to determine ω_e and $\omega_e x_e$? The infrared vibration-rotation spectrum of LiF from ref. 2 could then be correctly interpreted with the knowledge of B and its variation with vibrational quantum number. The centrifugal distortion constant $D_e=4B_e^3/\omega_e^2$(derivation in SDM, p. 104) could then be determined from the values of B_e and ω_e. Compute D_e in this way to compare with the value given in Table II.

Finally from both Tables II and IV compute the quantity $1/\theta \equiv [r_e \cdot (\partial\mu/\partial r)/\mu_e]$; $1+(1/\theta)$ occurs as a linear factor in the Stark energy. What is the approximate size of $1/\theta$ in relation to unity? The authors state that the nonlinear variation of dipole moment with vibrational quantum number has been determined significantly beyond experimental error.

We thus see that this method, called MBER spectroscopy, is able to provide information about individual vibrational levels. What property of the molecule must be nonzero for this method, and for microwave spectroscopy in general, to yield information at all? This property is not absolutely required to be nonvanishing, as witness the detection in recent years of the quadrupole transition in H_2(literature reference?). Do you know of any other observed transitions for nonpolar molecules? How about for atoms? What about Cs in exercise no. 126?

TABLE II. Spectroscopic constants Li^6F^{19}.

$\omega_e = 964.07\ cm^{-1}$	$D_e(\sim -Y_{02}) = 443.14$ kc ± 0.15 kc
$\omega_e x_e = 8.895\ cm^{-1}$	$\beta_e(\sim -Y_{12}) = -5.60$ kc
$Y_{01}(\sim B_e) = 45.2308111$	Gc$\pm$5 kc
$Y_{11}(\sim -\alpha_e) = -0.7223089$	Gc$\pm$5 kc
$Y_{21}(\sim \gamma_e) = 0.0058270$	Gc$\pm$5 kc
$B_0 = 44.8711134$	Gc$\pm$5 kc
$B_1 = 44.1604585$	Gc$\pm$5 kc
$B_2 = 43.4614577$	Gc$\pm$5 kc
$B_3 = 42.77416$	Gc$\pm$300 kc

TABLE IV. Dipole moments of Li^6F^{19} $J=1$ state in different vibrational states.

	Accuracy ($D \times 10^5$)	Precision ($D \times 10^5$)
$\mu_e =$ 6.28446 D	$\pm$100	...
$\mu_0 =$ 6.32764 D	$\pm$100	$\pm$10
$\mu_1 =$ 6.41511 D	$\pm$100	$\pm$10
$\mu_2 =$ 6.50341 D	$\pm$100	$\pm$10
$\mu_3 =$ 6.59326 D	$\pm$100	$\pm$12
$\mu^I = +0.08612$ D	$\pm$3	...
$\mu^{II} = +0.00060$ D	$\pm$2	...

$$\mu_v = \mu_e + \mu^I(v+\tfrac{1}{2}) + \mu^{II}(v+\tfrac{1}{2})^2$$

134. "Microwave Spectrum of Aluminum Monofluoride" D.R. Lide,Jr. 38, 2027 (1963).

In this exercise we will partially come to grips with the problem of determining the quadrupole coupling constant eqQ from the microwave spectrum, though still not completely in terms of the theory.

Table I gives us the observed spectral transitions and the

TABLE I. Observed spectrum and constants of AlF.

Trans. ($J=0\rightarrow 1$)	$\nu(v=0)$	$\nu(v=1)$
$F=5/2\rightarrow 3/2$	32 981.8$\pm$0.1 Mc	...
$F=5/2\rightarrow 7/2$	32 978.5$\pm$0.1	32 681.6$\pm$0.5
$F=5/2\rightarrow 5/2$	32 970.6$\pm$0.1	32 673.9$\pm$0.5

$B_0 = 16\,488.30 \pm 0.05$ Mc $B_e = 16\,562.5 \pm 0.3$
$B_1 = 16\,339.90 \pm 0.25$ $\alpha_e = 148.4 \pm 0.3$

$(eqQ)_{Al^{27}} = -37.6 \pm 1.0$ Mc
$\mu = 1.4 \pm 0.1$ D
$r_e = 1.65437$ Å

derived constants. It would seem to be quite routine and automatic to ask you to verify the value of r_e from B_e, or B_v from B_e and α_e, but you've <u>done</u> those things before so let us do something different, even if it appears to be just formula-plugging.

From p. 151 of the volume by T&S referenced on p. 252 herein we find that the quadrupole energy W_Q is given by minus the product of eqQ and the Casimir function f=f(I,J,F); the latter function is tabulated in Appendix I of T&S for the three quantum numbers.

Now looking at Table I you see that there is no mention of a quadrupole coupling constant for F(would it be ^{19}F and why?). Why should this be? Is the nuclear spin of ^{27}Al obviously 5/2 and why? From Appendix I we find that the Casimir function f is 0 when J=0, regardless of the value of F=I(then). Does this tell us that the quadrupole hyperfine splitting must be present in the upper states of these three transitions? For I=5/2 and J=1 the function f takes the values 0.0500, -0.1600, and 0.1400 for F=7/2, 5/2, and 3/2 respectively. Note that this order of ascending F is not that of ascending energies of transition as in Table I. Now apply the basic spectroscopic formula that transitional energy is equal to the difference in level energies for the three cases and solve for eqQ; do you have two checks including the J=0→1 transitions in the excited vibrational state? Does vibrational excitation seem to affect the eqQ value? Do you match Lide's value in the Table? Can you give a classical description of the case J=0 in terms of eqQ?

Check his statement that the value determined in this work was almost identical to the eqQ in the $^2P_{3/2}$ state of atomic Al (ref. 4). Together with the relatively small dipole moment this fact leads him to indicate to us that the bonding in AlF is primarily covalent. Do you agree? Is the fraction ionic character less than 20%?

135. "Millimeter Wave Spectroscopy of Unstable Molecular Species. I. Carbon Monosulfide" R. Kewley, K.V.L.N. Sastry, M. Winnewisser and W. Gordy <u>39</u>, 2856 (1963).

The frequency range of the spectrometer described in the article is 60 000 to 300 000 Mc; what is the wavelength range in mm and the wavenumber range?

Fig. 2 is a photograph of the central part of the instrument. Can you identify the items mentioned in the caption, and others when possible, from the block diagram of Fig. 1? How was the pressure kept constant at 5×10^{-3} mm Hg in the free space cell during a run? What is the function of the small dewar flask in the foreground of Fig. 2? How was the CS produced?

Why do the authors not classify CS as a free radical? What is its ground state term? The same as that of CO? Why then does it require for its detection experimental methods similar to those employed for free radicals?

Let us now examine Fig. 3 and the authors' accompanying discussion, and see if we can appreciate the extreme accuracy inherent in instruments used in these frequency ranges. Actually we will of course discuss precision. How is the accuracy checked in this, and similar, experiments? By an unusual type of "radio" station? From its first call letter you would expect

FIG. 2. Photograph of central part of the millimeter wave spectrometer showing the free space cell, discharge tubes, *G*- to *S*-band horns, and part of the vacuum system.

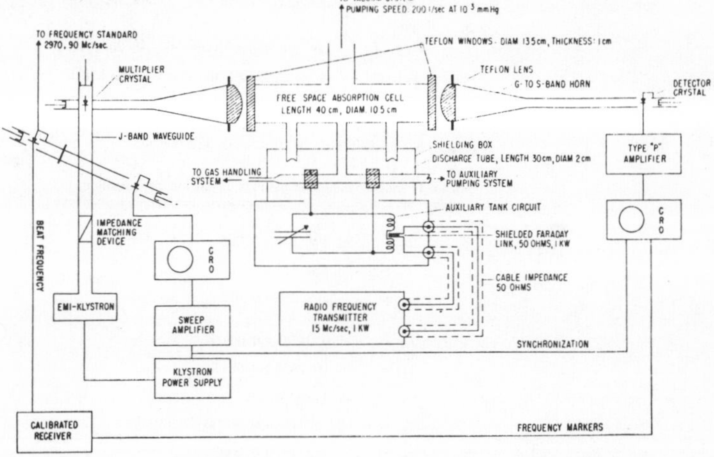

FIG. 1. Block diagram of the millimeter wave spectrometer and the radiofrequency discharge setup.

it to be east of the Mississippi; is it in fact at the present time?

In Fig. 3(a) the strong absorption peak represents the $J=1\rightarrow2$ transition of $C^{12}S^{32}$ at the frequency shown and also listed in Table I. Does this transitional frequency correspond exactly to the third harmonic of the klystron (Fig. 1) fundamental frequency at 32 660.336 Mc/sec? The weaker line in (a) shows the $J=3\rightarrow4$ transition of the same isotopic molecule, at the

(a) $J=3\to4$ transition at 195 954.16 Mc/sec using the 6th harmonic of klystron fundamental frequency.

$J=1\to2$ transition at 97 981.007 Mc/sec using the 3rd harmonic of klystron fundamental frequency.

Spacing between the two lines is due to centrifugal distortion terms.

(b) Absorption peak for the $J=1\to2$ transition recorded with wide band preamplifier.

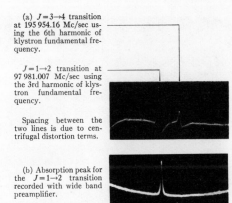

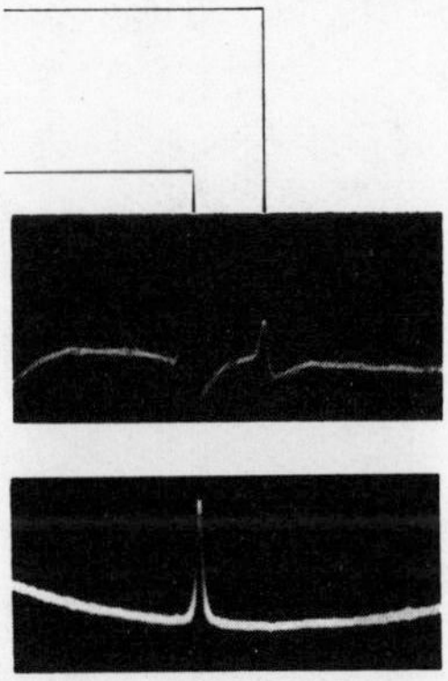

FIG. 3. Part (a) shows the centrifugal behavior of CS^{32} and (b) the true line shape of the $J=1\to2$ rotational transition.

TABLE I. Frequencies of CS lines from molecules in the ground vibrational state.

	Transition	Observed frequency (Mc/sec)	Calculated frequency (Mc/sec)
$C^{12}S^{32}$			
	$J=0\to1$	48 991.000±0.006[a]	48 991.013
	$J=1\to2$	97 981.007±0.008	97 981.009
	$J=2\to3$	146 969.039±0.030	146 969.021
	$J=3\to4$	195 954.162±0.050	195 954.174
	$J=4\to5$	244 935.737±0.050	244 935.734
$C^{12}S^{34}$			
	$J=0\to1$	48 206.948±0.006[a]	48 206.948
	$J=1\to2$	96 412.953±0.015	96 412.952
	$J=2\to3$	144 617.117±0.008	144 617.117

[a] Previous measurements made by Mockler and Bird (Ref. 6).

TABLE II. Spectroscopic constants of carbon monosulfide.

CS^{32}			
	B_o =	24 495.592±0.006	Mc/sec
	B_e =	24 584.367±0.006	Mc/sec
	α_e =	177.550±0.012	Mc/sec
	$(D_e+\frac{1}{2}\beta_e)\simeq D_e$ =	0.04285±0.00166	Mc/sec
	H_e =	0.0653±0.0085	kc/sec
	ω_e =	1 242.4±20	cm^{-1}
	I_e =	20.563108±0.000005	amu Å^2
	r_e =	1.53492±0.0007	Å
CS^{34}			
	B_0 =	24 103.554±0.006	Mc/sec
	$(D_e+\frac{1}{2}\beta_e)\simeq D_e$ =	0.03979±0.0017	Mc/sec
	H_e =	0.0625±0.009	kc/sec

frequency shown and listed in Table I also; this was done with the sixth harmonic of the same klystron sweep. Show that the spacing between the two lines, at the klystron frequency, is 1.312 Mc/sec. If there was no centrifugal distortion term, would the lines coincide?

From Table I are the frequencies of each isotopic molecule what you would expect for a rigid rotor to three significant figures? From the expression for the energy levels in rotation of a symmetric top given on p. 250 herein, set K=0 for the CS diatomic(why? is this true for all diatomics?) and add a term for higher centrifugal effects $+H_vJ^3(J+1)^3$. Subscripts v may be added to B and D, though as shown in Table II the differences between D_e and D_0 and also H_e and H_0 are negligible. Show that the centrifugal distortion constants are consistent with the transitions observed in Table I. Could B_e and D_e together serve to determine the value of ω_e shown? What is the value from the ultraviolet data(ref. 11, SDM)? cf exercise no. 133.

Two last questions; can you reconcile their statement that the measured half-width(meaning exactly what?) of the CS^{32} line (may we assume about the same for all?) is 144 kc/sec with the spacing shown in Fig. 3(a)? Also discuss the problems of isotopic abundance for both sulfur and carbon and the apparent lack of interference from nuclear quadrupole hyperfine effects.

136. "Microwave Spectrum of 1,3-Cyclohexadiene" S.S. Butcher 42, 1830 (1965).

Chemical studies(ref. 1) have suggested that the 1,3-cyclohexadiene ring in 9,10-dihydrophenanthrene is nonplanar. The author states that the character of the microwave spectra of this and similar(article immediately following this one is the microwave spectrum of 1,3,5-cycloheptatriene) compounds can be very sensitive to the symmetry of the equilibrium conformation of the ring. In favorable cases the conformation of the ring may be determined from a study of the normal isotopic species if bond lengths and angles are assumed to be transferable from similar molecules in which they are known. These were chosen from values summarized by ref. 7.

Referring to Fig. 1, the assumptions were: $d(C_3'-C_3)=1.47$, $d(C_3-C_2)=1.34$, $d(C_2-C_1)=1.50$, $d(C_1'-C_1)=1.50$, $d(C\text{-}H)_{aliphatic}=1.10$, and $d(C\text{-}H)_{olefinic}=1.086$ Å; $<C_3'C_3H_3=116°$, $<C_3C_2H_2=122°$, $<H_1C_1H_1=109°28'$, and $<C_3'C_3C_2=<C_3C_2C_1$; and that each of the ethylene groups is individually planar. Write these values in on the appropriate places in the plan view of Fig. 1; will one more parameter completely determine the conformation of the molecule at equilibrium? The one chosen as a measure of the nonplanarity of the ring was the torsional angle θ, defined as the angle one ethylene group is rotated relative to the other about the $C_3'-C_3$ bond. How was the final value of θ determined to be 17.5°±2° from the experimental result of the analysis of this asymmetric rotor's spectra that $(I_a+I_b-I_c)/2=6.1881$ amu·Å²? The answer of course is that given the other structural parameters θ was varied until this function of the moments of inertia matched the experimental value; the results are shown in Table V where the a, b and c principal inertial axes are designated by y, x and z as you should quickly verify.

Prove that the function of the moments of inertia given above is equal to $\sum m_ic_i^2$ in the general case. Compute this quantity for the case of a planar carbon skeleton with four out-

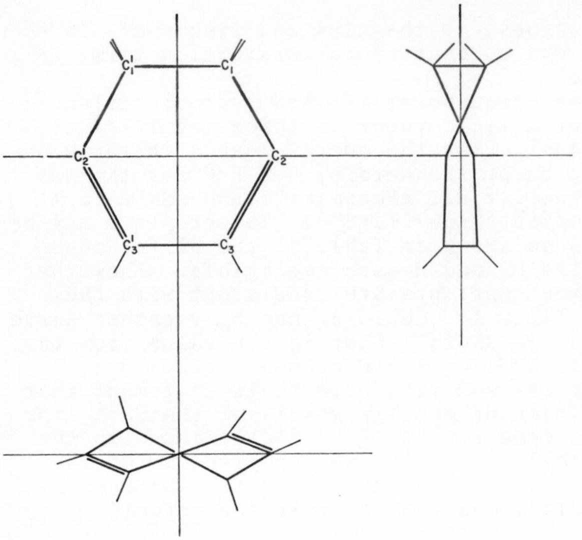

FIG. 1. Projections of the 1,3-cyclohexadiene ring.

TABLE V. Moments of inertia
(in atomic mass units × square angstroms).

	Observed	Calculated ($\theta = 17.5°$)
I_x	99.8607	99.02
I_y	99.6313	99.44
I_z	187.1159	186.10
$(I_x+I_y-I_z)/2$	6.1881	6.18

of-plane hydrogen atoms. Compare with his value of 3.25 amu·Å^2. In either event this quantity is considerably less than the experimental result. He states that the final value of θ is relatively independent of all of the structural assumptions except those of $d(C_1\text{-}H_1)$ and $<H_1C_1H_1$. With uncertainties in these of 0.015 Å and 5°, the resultant uncertainty in θ turns out to be only 1°.

He states that the nonplanarity of the ring is likely to be the result of two contributing factors, the angle strain in the C-C-C angles which would exist if the ring were planar, and the interaction of the nonbonded protons in the methylene groups which would be eclipsed in the planar conformation. The derived values for the nonplanar ring are 120°10' for $<C_3'C_3C_2 = <C_3C_2C_1$, and 110°30' for $<C_1'C_1C_2$. Are these values close to what you might expect for "strain-free" angles and why? For the nonplanar ring the derived $C_1'\text{-}C_1$ torsional angle is 45°. Draw in the missing C-H methylenic bonds in the lower left projection of Fig. 1. If the two methylenic carbon valences were tetrahedral, show that the nonbonded proton interactions would be minimized for a torsional angle of 60° (problem of staggered ethane?).

The present spectroscopic evidence was not sensitive to slight deviations from planarity of the ethylene groups.

From the moments of inertia of Table V compute the asymmetry parameter κ; is the rotor nearly oblate? For benzene show that $I_a+I_b=I_c$. Is this general for planar molecules?

Refer to exercise no. 98.

One of the easiest questions in this entire reader: what factor would tend to keep the carbon skeleton planar? An additional piece of evidence for the C_2 symmetry shown in Fig. 1 is the dipole moment component along the c axis μ_C; experimentally it was found to be $<2x10^{-4}$ D, small enough to be considered zero. Is this condition necessary or sufficient for the molecule to have C_2 symmetry?

137. "Microwave Spectrum and Structure of Nitric Acid" A.P. Cox and J.M. Riveros 42, 3106 (1965).

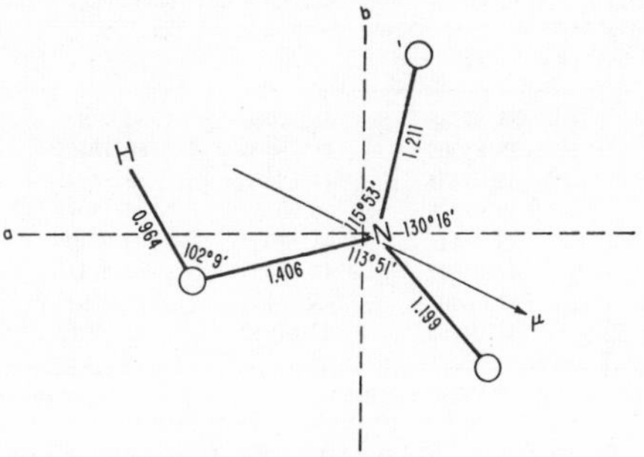

FIG. 1. Structure of nitric acid.

The results of the structural determination, using the isotopic molecules listed in Tables II and III, are shown in Fig. 1 from which you should verify the authors' statement that there is a 2° tilt of the NO_2 group away from the hydrogen atom. What is the distance of separation of the hydrogen and cis-oxygen atoms? Compare your answer with the authors' of 2.14 Å. They suggest that because of the smallness of this distance the tilt is more likely to be due to a repulsive rather than an attractive mechanism. Give an argument to support this conclusion. They also state that the direction of the tilt makes the possibility of internal hydrogen bonding, suggested by ref. 2 as an explanation of the high barrier to internal rotation, seem very unlikely. Is the argument similar? Is the tilt in the proper place in Fig. 1? What identity among the angles can you check immediately? The bond distance of 1.406 Å, when compared with the expected value of 1.44 Å(ref. 21) for a single N-O bond, would indicate considerable double bond character in the former. Discuss in terms of resonance structures. Formaldoxime is quoted as a similar case(p. 3112), in terms of three structural parameters. Fuel for hybridization arguments is provided by the planar angle of over 130°. Table VII(omitted) compares NO_2 groups.

The rotational constants of Table II look to be rather similar for the different isotopic molecules, but the asymmetry parameter κ varies quite a bit. Confirm this behavior by computing κ from A, B and C for its lowest and highest values.

Our use of Table III will be a bit different than heretofore. First try the conversion factor in the caption to bring one or two pieces of data over from Table II. Next demonstrate

TABLE II. Rotational constants of isotopic species (Mc/sec).

	A	B	C	κ
$H^{18}ONO_2$	12 970.84	11 273.96	6023.17	0.511525
cis-$HON^{18}OO$	12 318.60	11 847.31	6030.18	0.850093
trans-$HON^{18}OO$	12 714.55	11 493.29	6028.32	0.634693
$D^{15}NO_3$	12 971.72	11 308.51	6033.97	0.520533
cis-$HO^{15}N^{18}OO$	12 317.55	11 847.23	6030.60	0.850382
HNO_3[a]	13 011.15	12 099.93	6260.60	0.730031
$H^{15}NO_3$[a]	13 012.35	12 096.97	6260.09	0.728864
DNO_3[a,b]	12 970.71	11 312.82	6034.86	0.521920

[a] Values from Ref. 1.

[b] On the basis of measurements made in the present work, we feel that $A=12\,970.67$, $B=11\,312.69$, $C=6034.90$ Mc/sec give a better over-all fit to the DNO_3 data. These lead to a value of $\Delta=0.1058$ amu·Å² which is in better agreement with the more accurate $D^{15}NO_3$ value. Since it can be shown that the moments of inertia derived from these constants give almost exactly the same hydrogen coordinates, the values of Ref. 1 have been retained in the structural calculations.

TABLE III. Moments of inertia of nitric acid and isotopic species. Conversion factor 5.05531×10^5 (Mc/sec)amu·Å².[a]

	I_A	I_B	I_C	$\Delta=I_C-I_B-I_A$
HNO_3[b]	38.85368	41.77966	80.74801	0.1147
$H^{15}NO_3$[b]	38.85008	41.78990	80.75465	0.1147
DNO_3[b]	38.97481	44.68656	83.76847	0.1071
$D^{15}NO_3$	38.97178	44.70359	83.78083	0.1054
$H^{18}ONO_2$	38.97442	44.84059	83.93105	0.1160
cis$HON^{18}OO$	41.03802	42.67053[c]	83.82473	0.1162
trans$HON^{18}OO$	39.76004	43.98488	83.85935	0.1144
cis$HO^{15}N^{18}OO$	41.04152	42.67082	83.82765	0.1153

[a] Values of amu based on ^{16}O scale. [b] Values from Ref. 1. [c] See also footnote to Table II.

the origin of this factor. Finally for each isotopic molecule, beginning with $H^{15}NO_3$, rationalize the changes in the three principal moments of inertia wrt the first set, those of $^1H^{14}N^{16}O_3$. Recognize that while the bonds distances and angles are assumed to be exactly invariant with isotopic substitution, the position of the center of mass and the orientation of the a-b axes(but not the c axis?) may both shift a bit relative to the nuclear framework.

By means of Table V, let us examine the properties of principal axes in more detail. In order to separate off completely

TABLE V. Coordinates of atoms in principal axis system of HNO_3.[a]

	I		II		III	
	a	b	a	b	a	b
H	1.6771	0.4937	1.6803	0.4997	1.6779	0.4828
O(hyd)	1.2289	−0.3566	1.2313	−0.3570	1.2276	−0.3623
N	−0.1241	0.0216	−0.1247	0.0211	−0.1564	0.0308
O(*cis*)	−0.3409	1.2110	−0.3415	1.2131	−0.3194	1.2171
O(*trans*)	−0.8851	−0.9044	−0.8865	−0.9062	−0.8770	−0.9122
	$\Sigma_i m_i a_i b_i=-0.0102$		$\Sigma_i m_i a_i b_i=0.0347$		$\Sigma_i m_i a_i b_i=0.2132$	

[a] (I) Coordinates obtained by using I_A and I_B in Kraitchman's equations. (II) Coordinates obtained from Kraitchman's equations using I_A and I_B corrected for the shortening of the average bond lengths in the isotopic species. (III) Coordinates obtained by using I_B and I_C in Kraitchman's equations. Reasons are given in the text for considering these an unpreferred set.

the translational energy of the molecule, we place the origin for the coordinate system, in terms of which the rotational energy is discussed, at the center of mass. Then a necessary and sufficient condition for a set of orthogonal axes to be principal axes is that the products of inertia, $I_{xy}=-\sum m_i x_i y_i = I_{yx}$, etc., vanish. You may read about this in e.g. the book by H. Goldstein, "Classical Mechanics" (Addison-Wesley Pub. Co. Inc. Reading, Mass. 1959) Chapter 5. Verify from Table V that for the preferred set I the product of inertia in the last row has the value shown. What about the two other products of inertia? Ideally all three(or six) would be zero; the small residual value shown is effectively that. What advantage is there then in expressing the total rotational energy in terms of the principal axis system, first classically and then quantum mechanically? See IRRS, WDC, or T&S.

The Stark effect for several transitions was studied in order to determine the three components of the dipole moment. An upper limit of 1.1×10^{-3} D was given for μ_c; to what type of nonplanarity does this result place an upper limit of 0°2'? (p. 3108). The other two components are shown in Table IV. Verify their statement that the total dipole moment vector is inclined at 23°57' to the a axis. Why should the decrease in

TABLE IV. Stark coefficients and dipole moment of HNO_3.

Transition		$\Delta\nu/E^2$[(Mc/sec)/(V/cm)^2]$\times10^6$ Observed	Calculated
$0_{00}\rightarrow1_{01}$	$M=0$	30.33[a]	30.80
$0_{00}\rightarrow1_{11}$	$M=0$	17.03[a]	17.02
$3_{03}\rightarrow3_{22}$	$M=3$	4.15±0.08	4.132
$3_{13}\rightarrow3_{22}$	$M=3$	4.10±0.08	4.076
$2_{12}\rightarrow2_{11}$	$M=2$	8.51±0.17	8.506
$1_{11}\rightarrow2_{12}$	$M=1$	47.5 ±0.9	47.30
$\mu_a=1.986$ D		$\mu_b=0.882$ D	$\mu=2.17\pm0.02$ D

[a] Values from Ref. 1.

intensity for μ_b transitions, both with deuteration and trans-^{18}O substitution, confine the direction of the total dipole moment to the second and fourth quadrants of Fig. 1? Why would the authors suggest that the negative end of μ is directed away from the H atom? Do you see any unique way to decompose μ into vector sums of lone-pair and bond moments? Can it in fact be done for even a triatomic, bent or linear?

138. "Electronic Structure of SrO" M. Kaufman, L. Wharton, and W. Klemperer 43, 943 (1965).

For spinless nuclei, the resonance frequencies at low electric fields are those associated with the energy levels predicted(ref. 14) for a $^1\Sigma$ molecule in an electric field of strength E(their equation [1]):

$$W = B_V J(J+1) + \frac{\mu_V^2 E^2}{B_V}\left(\frac{J(J+1)-3M_J{}^2}{2J(J+1)(2J+3)(2J-1)}\right) + \frac{\mu_V^4 E^4}{B_V{}^3} f(J,M_J).$$

A typical low-field resonance was shown in their Fig. 3. Given the value of μ from Table III, verify that the transition between the Stark levels occurs at the frequency shown. $f(J,M_J)$ is said to be 10^{-2} or smaller; does this information allow you to determine the relative contributions of the second and third terms to the transitional frequencies? The linewidth($\Delta\nu_{\frac{1}{2}}$, at half peak-height) is shown; verify it from the scale. Assume for the moment that the ground state is $^3\Pi$. Could you estimate the splitting of this line to be expected in the earth's magnetic field, and could it be detected with the resolution indicated in Fig. 3?

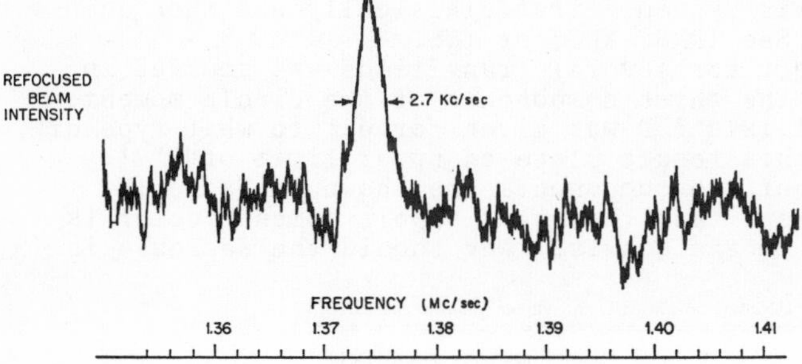

FIG. 3. $^{88}Sr^{16}O$ rf spectrum at $E_{Stark} \approx 500$ V/cm. $J=3, |M_J| = 0 \rightarrow 1$ transition.

TABLE III. Some properties of ^{88}SrO and ^{138}BaO.

	^{88}SrO	^{138}BaO
B_e (cm^{-1})	0.33798	0.31259
r_e (Å)	1.9199	1.9397
μ_e (D)	8.913[a]	7.933
μ_e/er_e	0.97	0.85
$\mu_1-\mu_0$ (D)	−0.026	+0.042
ω_e (cm^{-1})	653.3	669.8
$\omega_e x_e$ (cm^{-1})	4.0	2.05
k_e (dyn/cm)	3.41×10^5	3.56×10^5
$D_0°$ (eV)	4.3	5.7

[a] Assuming linearity of μ with v.

From Table III, might you need to revise your identification of the world's most ionic diatomic, used in exercise no. 133?

Why was SrO more difficult to study than BaO? Altogether four experiments were performed, of which the rf spectroscopy was only one.

Interesting descriptions of bonding in the alkaline earth oxides are given.

139. "Microwave Spectrum and Structure of Germyl Silane" A.P. Cox and R. Varmi 46, 2007 (1967).

GeH_3SiH_3 showed the microwave spectrum of a symmetric top. The spectrum was rather weak, which was consistent with an expected low dipole moment. Estimate this quantity yourself from the values of methyl silane, 0.73 D, and methyl germane, 0.635 D to compare with their approximation, given in the article. Molecules of the three principal isotopes of Ge, together with the dominant Si isotope, gave lines in the spectrum that could be assigned. The most abundant Ge isotope(which one?) yielded a value of B_0 for the molecule of 3655.119±0.005 MHz. From the centrifugal distortion constant D_J of 1.8±0.1 kc/sec, determine an approximate value for the fundamental stretching frequency of the Ge-Si bond. What assumption(s) is involved? No D_{JK} splitting was observed.

The $^{74}Ge/^{72}Ge$ and $^{74}Ge/^{70}Ge$ pairs yield 0.6769 Å for the distance of the Ge atom from the center of mass with better internal consistency than the estimated experimental uncertainty of ±0.0001 Å. Taking parameters from methyl silane and methyl germane, Si-H=1.483 and Ge-H=1.529 Å, $<SiH_2=108°20'$ and $<GeH_2=109°15'$, show that the germanium-silicon bond length is 2.357 Å. The result $\cos\alpha=1-(3/2)\sin^2\beta$ will be helpful, where e.g. $\alpha=<SiH_2$ and $\beta=<HSiGe$. Show that this formula reduces to an identity for $\alpha=\beta=109°28'$. Can you derive it for the general case?

When the differences between this substitution structure and an actual one for the ground vibrational state of some isotopic molecule are taken into account, the error limits on this Ge-Si distance are increased to ±0.004 Å. We have completely neglected this point in the previous exercises of this chapter. The nominal value is very close to the sum of the two covalent radii(ref. 7); the difference of 0.03 Å is taken as possibly indicating some multiple bonding in this central bond. From this, what is the numerical value of the covalent radii sum?

It proved impossible to determine the barrier to internal rotation because of the weakness of both GeH_2DSiH_3 and torsionally excited GeH_3SiH_3 lines.

A prediction is made of this quantity from the information shown in Fig. 1. Discuss the assumptions that are made, assuming that all of the circles correspond to experimental points which they do not. Is the interpolation or extrapolation used here similar to that used for the estimate of μ? Just for an exercise, estimate μ using the scheme implied in Fig. 1.

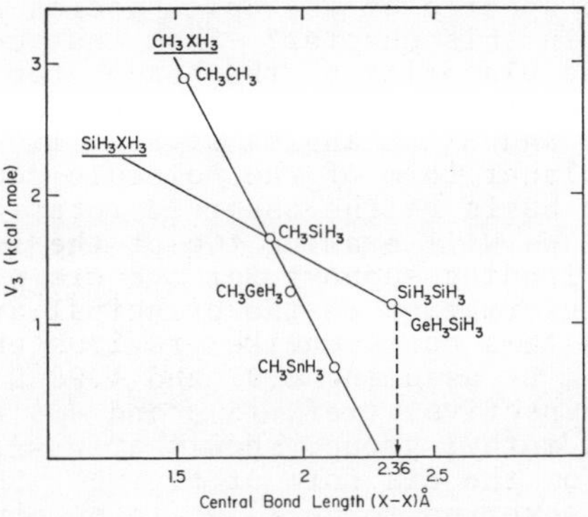

FIG. 1. Prediction of barrier in GeH_3SiH_3.

The author of this reader does not disdain estimates of this sort, having very recently approximated the Sn-Sn stretching force constant of distannane in a quite similar manner using the same reference molecules plus digermane. In passing, what type of spectroscopy could be used to test this prediction? Can you supply the details?

140. "Microwave Spectrum, Rotational Isomerism, and Internal Barrier Functions in Propionyl Fluoride" O.L. Stiefvater and E.B. Wilson, Jr. 50, 5385 (1969).

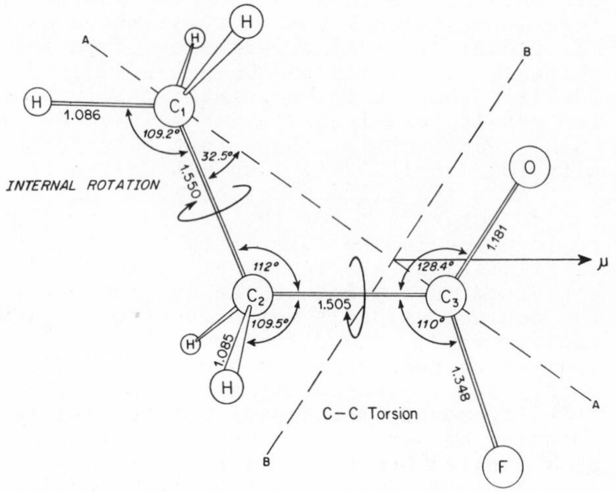

FIG. 1. Proposed structure of *cis*-propionyl fluoride. From observed rotational constants and related molecules.

The microwave spectrum proved the existence of two stable rotameric(discuss this adjective) conformations. Fig. 1 shows the proposed structure of the more stable of the two(by how much? see Fig. 6), cis-propionyl fluoride. Why was it necessary to borrow structural parameters from related compounds, and not make a study of the spectra of isotopic species as we have seen done several times in this chapter? Show that conclusive evidence exists for the planarity of the heavy-atom skeleton of the cis conformation from the moment of inertia result of Table I and the C-H lengths and H-C-H angles of Fig. 1. A trans conformation for the planar form of the molecule could not be ruled out entirely on the basis of the observed rotational constants; why not?(p. 5388). We will examine two of the four pieces of evidence quoted as lending support for the cis planar form. Table II presents information on the principal-axis dipole-moment components. In this case, unlike previous ones, μ_c was assumed to be zero. By assuming 2.97 and 1.95 D for the C=O and C-F bond moments respectively(ref. 15), and 0.4 D for the resulting moment of a methyl group, show that $\mu_a \simeq 2.15$, $\mu_b \simeq 1.92$, and $\mu_{total} \simeq 2.88$ D for the cis form of Fig. 1. Show also that the ratio μ_a/μ_b is observed to be 1.19, to be compared with the calculated cis ratio of 1.12. Now show that the trans form would lead to a calculated value for this quantity of about 10, with $\mu_b \sim 0.3$ D. The other piece of evidence that we will look at is shown in Fig. 3, where the dihedral angle is that between the OCF and CCC planes. The angle θ is 0° for the cis form. Interestingly enough, the script form of theta used in the article is the same as that used in a famous textbook of quantum mechanics coauthored by the present senior author, some thirty-five years ago. Possibly you picked this up while reading the article. In any event, what is the dictionary definition of the word gauche? Do you see how that applies to the conformation

TABLE I. Rotational parameters for the ground state of *cis*-propionyl fluoride.

Rotational constants (MHz)	Moments of inertia (amu·Å²)[a]
$A=10\,042.52_9\pm0.02$	$I_a=50.339_0$
$B=3762.19_4\pm0.02$	$I_b=134.371_3$
$C=2832.69_1\pm0.02$	$I_c=178.463_2$
$\kappa=-0.742157$	$I_a+I_b-I_c=8m_HC_i^2+\Delta$
	$=6.247_1$

[a] Conversion factor: 5.05531×10^5 MHz/amu·Å².

TABLE II. Comparison between calculated and observed Stark coefficients for the ground state of *cis*-propionyl fluoride.

Transition	$\Delta\nu/E^2$[(MHz×10^6) per(V/cm)²] Calc[a]	Obs
$5_{05}\rightarrow6_{06}$		
$M=1$	-0.776	-0.79_2
$M=2$	$+0.542$	$+0.53_5$
$M=3$	$+2.738$	$+2.75_5$
$M=4$	$+5.812$	$+5.85_5$
$6_{24}\rightarrow6_{33}$		
$M=4$	184.95	184.8
$M=5$	290.64	290.9
$M=6$	421.01	418.6
$\mu_a^2=4.94$ D²,	$\mu_a=2.22$ D	
$\mu_b^2=3.48$ D²,	$\mu_b=1.87$ D	
$\mu_{\text{total}}^2=8.42$ D²,	$\mu_{\text{total}}=2.90$ D	

[a] Reference 12.

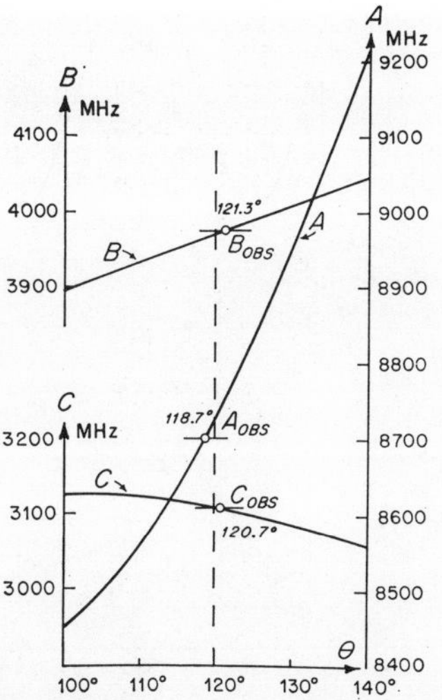

FIG. 3. Determination of the dihedral angle for *gauche*-propionyl fluoride from a comparison between calculated and observed rotational constants.

with a 120° dihedral angle? Why is the identification of the gauche form support for a cis conformation of the planar form? "The identification of transitions arising from the gauche form of propionyl fluoride was not only the major goal but also the most arduous part of this work. It amounted to finding a dozen gauche transitions amongst several thousand cis lines." Fig. 3 shows how they proceeded. Exactly what do the three open circles and horizontal dashes represent?

In reading through some of the articles upon which this chapter is based, you have probably happened across the term satellite lines. What does it mean in the most general terms? Much information is contained in Fig. 6, including the observed spacing of the torsional levels. How could energies of this magnitude be determined in the microwave region of the spectrum? In the infrared, workers use frequencies and very approximate intensity ratios to check on suspected hot-band assignments. Here in the microwave region, known "hot-lines"(not bands? why not?) or satellite lines occurring at known frequencies, together with rotational lines for the ground vibrational states, have their intensity ratios measured very precisely to obtain good estimates of the vibrational spacings. What two assumptions must be made? Are these same two also made in the somewhat different problem encountered in the infrared region?

Now let's take a broad look at Fig. 6. Is the behavior of the potential energy with the dihedral angle θ what we would expect, both as to position of the minima and their relative depth? Is there an obvious zero point energy in the deepest well? Could this be a contributor to the inertial defect Δ of Table I? How? Beginning with the uppermost level of 2110 cal/mole and moving downwards, explain the behavior of the energy levels as a uniform splitting of the degeneracies appears

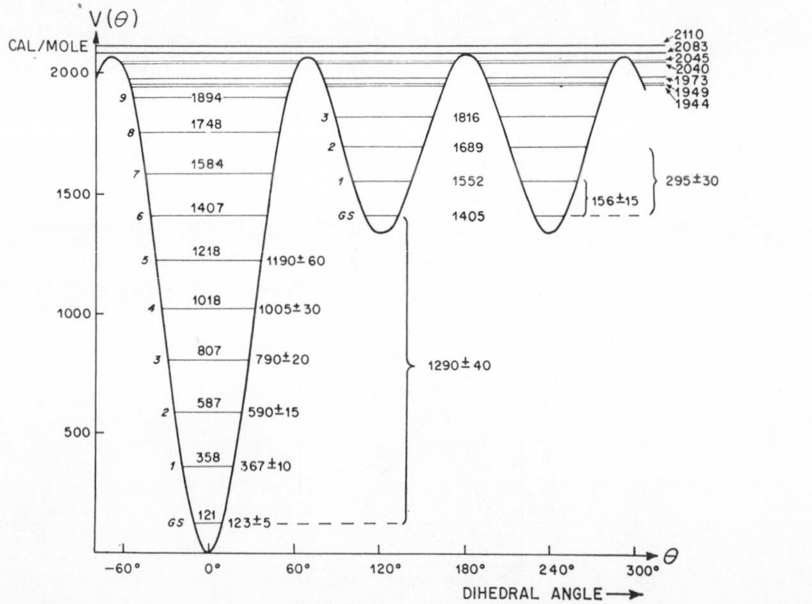

FIG. 6. Potential function $V(\vartheta)$ for rotation around the central carbon–carbon bond, and the calculated and observed torsional levels

followed by a differential splitting. Is the fact that the maxima of the potentials barriers separating the cis from the gauche conformation and the two equivalent gauche conformations from each other turn out to be approximately the same height (how much?) unexpected? If expected, on what grounds?(chance for original research here!)

Show that the function $V(\theta)=\sum \frac{1}{2}V_n(1-\cos n\theta)$, with n taking the values 1 and 4, reproduces the potential curve of Fig. 6. What are the torsional vibration frequencies for cis and gauche propionyl fluoride?

The barrier height to internal rotation of the methyl group about the C_1-C_2 bond was found to be 2400±60 cal/mole, not so different from that in ethane. Will the potential function for this internal rotation be symmetric? How many maxima per 2π rotation? See Fig. 1.

This was part of a series of experiments on related compounds to both establish and predict trends in conformational properties. For example Table VIII, not shown, establishes the fact that the gauche dipole moment is quite similar to that of the cis. This parallels similar results in three other molecules. Can you explain this, and at the same time the direction of μ in Fig. 1, in terms of the two largest(by far) bond moments discussed above?

CHAPTER XII

NUCLEAR MAGNETIC RESONANCE SPECTROSCOPY

141. "The Nuclear Spin Quantum Number of Si^{29} Isotope" R.A. Ogg, Jr. and J.D. Ray 22, 147 (1954).

Vol. No. 69 (1946) of Physical Review reports, in two papers by different sets of authors, the first detection of nuclear resonance effects in bulk matter. From their prediction and detection by physicists, these methods have been used by chemists to attack a seemingly endless array of problems as is well known. You might scan these two articles if you are so inclined; does the timing coincide with the answer to the question asked in exercise no. 131 about the historical events giving rise to the generating and detection systems?

In the article upon which the present exercise is based, the authors state that the nuclear-spin quantum number of Si^{29} was probably ½(ref. 1), and that the NMR experiment should determine this unequivocally. The experiment was the study of the spin-spin interaction splitting of the NMR of an atomic species chemically bound to Si. What is the isotopic distribution for silicon? Is the experiment clearcut as far as this atom is concerned? Why? What practical reasons would lead them to choose silane, SiH_4, as the compound to study? Except for the fine structure due to spin-spin interaction(between which two isotopes?), why would a unique(or single) proton resonance be expected? Why or why not would nuclear quadrupole effects occur? What is the n.b.p. of silane? What method was used to obtain spectra of the liquid? Notice the instrument that was used and its location; is this a respected name in the field today?

In the question of the single proton resonance, because of the tetrahedral symmetry, we know that there is no chemical shift with which we need be concerned. Sketch the expected spectrum of silane, giving the lines the correct intensities. Experimentally a single sharp peak appeared, flanked symmetrically by two weak, equally-spaced side peaks, at approximately twenty milligauss from the center. The weaker side peaks had equal amplitudes. Is all this what you had predicted? Explain each feature of the experimental spectrum. What would be the intensity ratio of one of the weak peaks to the strong central one? Sketch the expected pattern in the region of Si resonance.

Is it possible to evaluate the spin-spin coupling constant J_{12}(which is exactly what?) from the splitting in milligauss? If so, do it. If not, what else is needed and why?

142. "Proton Magnetic Resonance Spectrum of Ammonia and the Interactions Involved in Hydrogen Bond Association" R.A. Ogg, Jr. 22, 560 (1954).

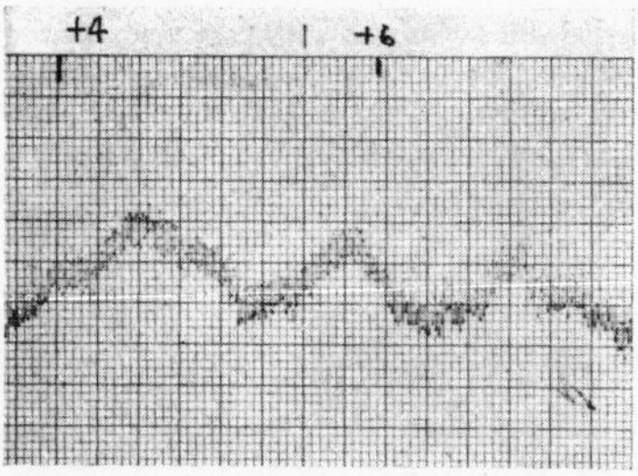

FIG. 1. Proton resonance spectrum of "wet" ammonia vapor. Magnetic field increasing to right. Scale indicates shift in parts per million from water reference. This and all other samples are at room temperature.

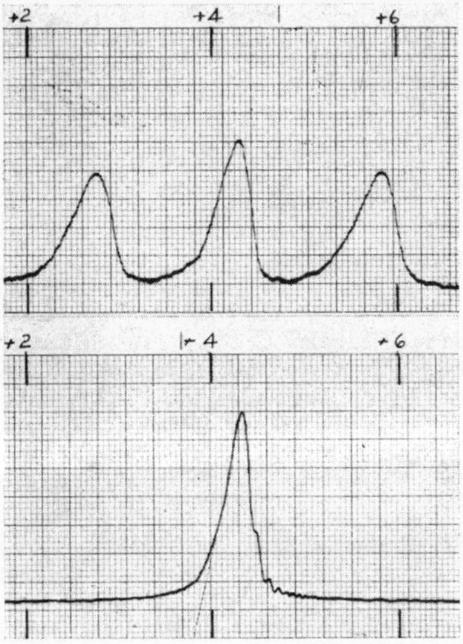

FIG. 2. A. Proton resonance spectrum of "dry" liquid ammonia. Magnetic field scale as in 1. B. Proton resonance spectrum of "wet" liquid ammonia. Magnetic field scale as in 1.

From the captions of Fig. 1 and 2, what is it that causes the disappearance of the proton resonance spin-spin structure in "wet" liquid ammonia? Does the exchange reaction

$$H_2O + NH_3 \rightleftarrows OH^- + NH_4^+$$

for protons seem plausible for the "wet" liquid phase? What about the "wet" vapor phase? ionic reaction...

What is the nuclear spin of N^{14}? Do we see 2I+1 spin-split components in the proton resonance spectra of Fig. 1 and 2(a)? Should these be of equal intensity? The broadening of the two outside peaks, relative to the central one, is due to the quadrupole moment of N^{14} giving rise to a strong relaxation mechanism when the charge around it is sufficiently asymmetric. Is there an asymmetry around the nitrogen nucleus in ammonia? All three peaks are consequently broadened; it turns out that the outer two should be broadened half again as much as the central one because the states of the N^{14} nucleus with $I_z=\pm1$ each have two possible transitions available, $\Delta I_z=-1$ or -2 for the +1 state, and $\Delta I_z=+1$ or +2 for the -1 state, while the $I_z=0$ state may have $\Delta I_z=\pm1$; the ±2 transitions are twice as probable as the ±1 transitions. Show that this leads to the theoretical broadening ratio of 3 to 2 for the outer peaks vs the central one. Is this result approximated in the spectrum of Fig. 2(a)? The above discussion is taken from p. 228 of the excellent reference book "High-resolution Nuclear Magnetic Resonance" by J.A. Pople, W.G. Schneider, and H.J. Bernstein (McGraw-Hill Book Co. Inc. New York 1959), hereinafter referred to as PSB.

From the properties of liquid ammonia, can you suggest a method to be used to remove the last traces of water? They estimate that not more than 1 part in 10^7 is enough to bring about the exchange reaction given above. See if your suggestion corresponded to the method that was used.

A striking illustration that the N^{14} quadrupole broadening is really that rather than reduced resolution is provided by the spectrum of Fig. 3. What is the natural abundance of N^{15}?

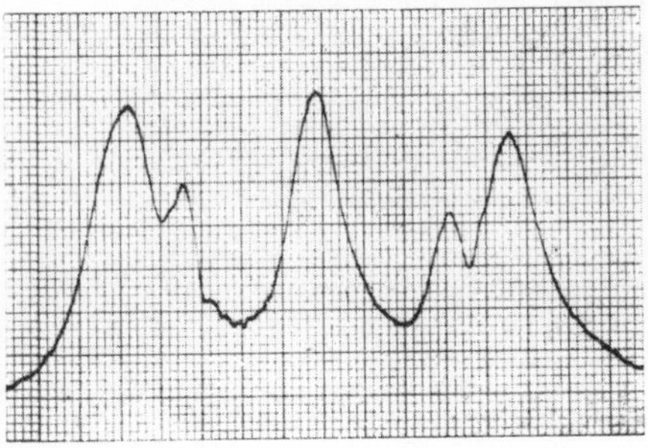

FIG. 3. Proton resonance spectrum of "dry" liquid ammonia enriched in N^{15} isotope.

Its nuclear spin? May it have a quadrupole moment? Why not? What effect would it have on the proton resonance of "dry" ammonia? Do we see the effects of both kinds of nitrogen nuclei in this figure? Would instrumental factors tend to make the weaker N^{15} peaks submerged in the inner shoulders of the outer peaks <u>if</u> these weaker lines had the same "intrinsic" width? Is it obvious that they do not? Estimate the fraction of N^{15} in their enriched sample(15%) from the spectrum.

Fig. 3 has still more to tell us. The author states that the ratio of multiplet separations agrees quantitatively with the ratio of the known gyromagnetic(or magnetogyric, see below) ratios of N^{14} and N^{15} nuclei. From PSB p. 4 and Appendix A (Table of Nuclear Properties) we find the following information, some of which may be familiar to you: The magnetic moment μ and the nuclear spin angular momentum component $Ih/2\pi$ are related through $\mu=\gamma Ih/2\pi$; do you see the reason for the preference of PSB for the phrase magnetogyric ratio used to describe γ? That

γ has the value $ge/(2M_pc)$, with g the nuclear g factor and M_p the mass of a proton, you should verify using an argument very close to that presented in exercise no. 66, p. 136 herein. Show also the relation of γ to the nuclear magneton. Compute the numerical value of the latter; what is its size compared to the Bohr magneton and why? What is the reason for the suffix -ton in magneton? Name some other nouns with this suffix used in this way. What do they all have in common? The magnetic moments for N^{14} and N^{15}, in multiples of the nuclear magneton, are 0.40357 and -0.28304 respectively. Show that the author's statement given at the first of this paragraph is correct. Discuss the effect of the minus sign in μ for N^{15}. What is the common name for the multiplet separation, for a given isotope of N, in Fig. 3? Is this separation field-dependent? Determine the values, proton resonance split by N isotopes, from common sources and compare the ratio with your measured ratio from Fig. 3. The splitting due to N^{14} may be measured absolutely in Fig. 2(a); the numbers on the scale at the top are the chemical shift in ppm wrt liquid water, one unit of displacement corresponding to one part per million(30 cycles). Does this spectrum yield about 47 cps for the spin-spin coupling constant?

Does the vapor resonance occur at higher fields? Fig. 4

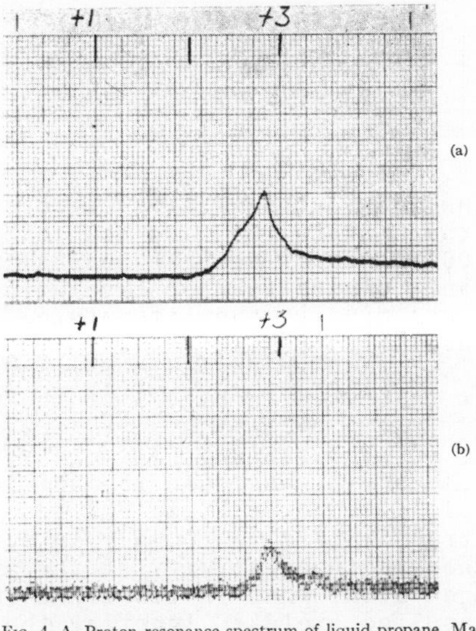

FIG. 4. A. Proton resonance spectrum of liquid propane. Magnetic field scale as in 1. B. Proton resonance spectrum of gaseous propane. Magnetic field scale as in 1.

presents a similar pair of spectra for a compound that does not exhibit hydrogen bonding. Is the proton resonance frequency a function of the phase?

For ammonia are the protons in the liquid less shielded diamagnetically than those of the vapor molecules? Discuss. Does the hydrogen bonding in the liquid polarize the molecules in such a sense as to actually decrease the electron density that would be effective in shielding the protons?

143. "Chemical Shifts of Nitrogen" B.E. Holder and M.P. Klein 23, 1956 (1955).

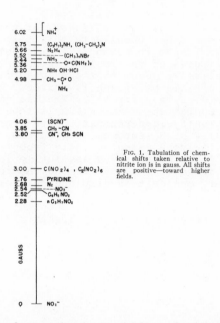

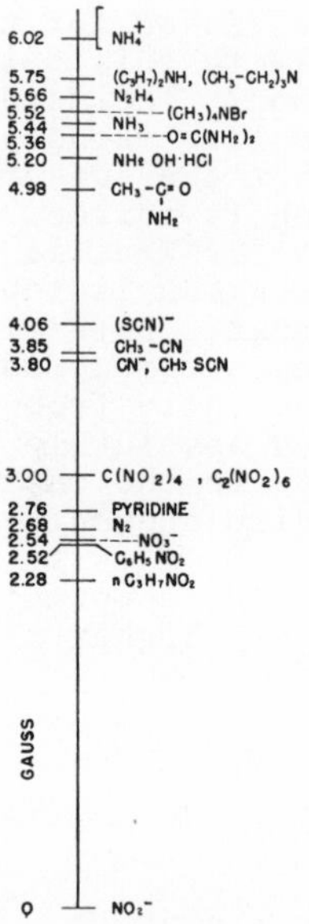

FIG. 1. Tabulation of chemical shifts taken relative to nitrite ion is in gauss. All shifts are positive—toward higher fields.

The units of the chemical shifts of nitrogen in Fig. 1 are gauss, relative to the nitrite ion, and all shifts are normalized to exactly 10 000 gauss. Move the decimal point to obtain the shifts in ppm, and compare four or five with the most recent tabulation that you can find in the literature. The present experiments, as well as their ref. 1, verified that the relative shifts, or chemical shifts, are the same for both nitrogen isotopes. Is this fact indicated in Fig. 3 of the previous exercise and why?

In the simplest interpretation, an increase in electron density around a nucleus will increase the shielding and hence require a higher applied magnetic field for resonance at a fixed frequency. Is the trend in Fig. 1 roughly in accord with this idea? What about the electronegativity of the atoms or groups bonded to the nitrogen atom; what is its trend in moving down the scale to lower applied fields? Ignoring the spin-spin splitting structure, sketch the NMR spectrum of an equimolar solution of NH_4NO_2 and NH_4NO_3 showing the chemically shifted peaks in the correct intensity ratios.

The correct interpretation of the magnetic shielding of nuclei is given in their ref. 3-6. With the shielding divided into a diamagnetic term and a second-order paramagnetic term, the first is dominated by the 1s and 2s electrons for nitrogen. Should this contribution be essentially constant as the nitrogen finds itself in different molecular environments? Why? The contribution of the second term increases as the orbital angular momentum increases. Do we see why the ammonium ion has the greatest positive chemical shift on these grounds? Discuss.

144. "Observation of Chemical Shifts of O^{17} Nuclei in Various Chemical Environments" H.E. Weaver, B.M. Tolbert, and R.C. LaForce <u>23</u>, 1956 (1955).

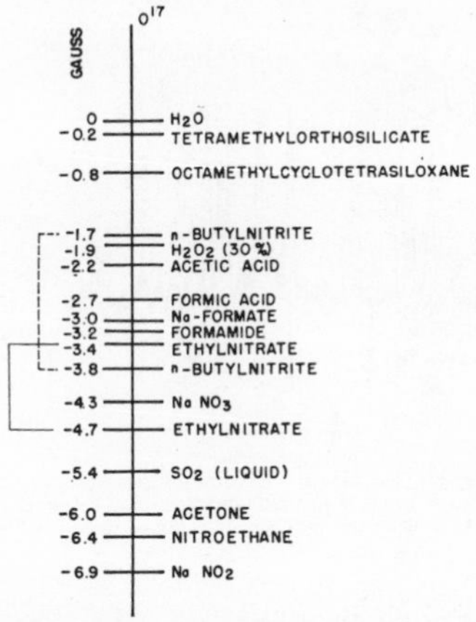

FIG. 1. The chemical shifts are taken relative to water and are based on a resonant field of 10 000 gauss. The brackets to the left of the vertical listing indicate those compounds in which a double resonance was observed. The dashed bracket indicates that the measurement was hampered by very low *S/N* for this compound.

This article immediately follows that of the previous exercise. Notice that the chemical shifts are referred to a compound for which resonance(always for a fixed frequency) occurs at the highest applied field. In recent years probably a larger fraction of chemical shift data for a given nuclei are referred to the same reference molecule; do you happen to know what molecule is now used predominantly as a chemical-shift reference for protons?

Why are there two shifts listed for ethyl nitrate? Given that their intensity ratio is 1:2, with which shift would you associate each of the nonequivalent oxygen nuclei?

Does the trend in Fig. 1 parallel that of the previous exercise for nitrogen? Do we run into extreme difficulty with our first, and simplest, interpretation given there when we see that the nitrite ion has its resonances at the lowest applied fields for <u>both</u> N^{14} and O^{17}? Why? Does the correct interpretation in terms of diamagnetic and second-order paramagnetic contributions hold up in view of this fact? It does if we associate increased electronic orbital angular momentum with increased asymmetries in the electronic structure as we do have here.

Compute the difference in chemical shifts for O^{17} in the nitrate and nitrite ions. Should this difference now be independent of the reference compound? Do the same from Fig. 1 of the previous exercise for N^{14}. Are these two differences the same for the two nuclei, within experimental error? What about the signs? Still the same? Now is this equality an obvious identity or was it a mere coincidence? "...a more detailed theoretical picture is necessary before any significance may be attached to this result."

145. "High-Resolution NMR Spectrum of Chrysene" J.D. Memory and T.B. Cobb 39, 2386 (1963).

The molecular structure and spectrum are given in Fig. 1.

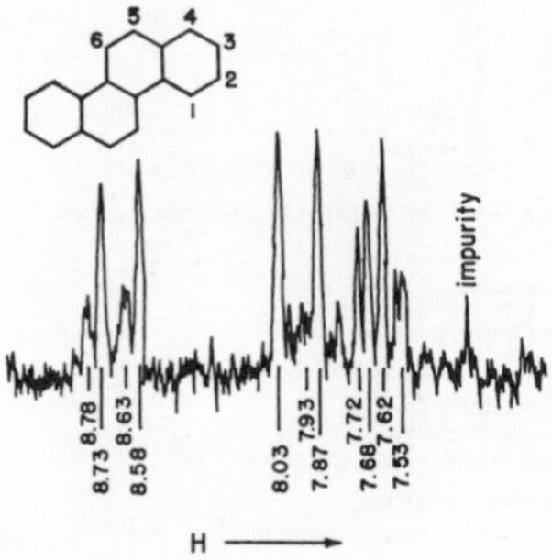

FIG. 1. A 60-Mc/sec NMR spectrum of chrysene in CS_2. Chemical shifts are in parts per million referred to tetramethylsilane. Small line at extreme right was determined to be a solvent impurity.

From lower resolution spectra PSB, p. 248, had reported two lines for chrysene with intensity ratio 2:1 and the weaker line at lower field; they concluded that the weaker line arose from protons 1 and 6, and the stronger one from protons 2-5? All of this is pointed out in the authors' ref. 1. The chemical shifts of the stronger and weaker lines, relative to cyclohexane, were -6.1 and -7.0 respectively. Show that this difference in chemical shifts is exactly what one obtains from the difference between the low- and high-field groups of Fig. 1. Is the convention regarding the sign of the chemical shift opposite to that of the previous two exercises? From the caption we see the standard reference liquid for proton chemical shifts, TMS. From the information that cyclohexane resonates at a field 1.6 ppm lower than does TMS(PSB p. 89) show that the absolute values of the two low-resolution chemical shifts in Fig. 1 and from PSB, mentioned above, are in accord as well as their differences.

In the present work it had been observed(ref. 3) that for polycyclic hydrocarbons inter-ring spin-spin splittings seemed to be absent. The authors thought it likely then that the four lines at 8.73, 8.58, 8.03, and 7.87 ppm arose from protons 5 and 6 with the lower-field lines due to the latter. These two protons are obviously not equivalent and form an AX multiplet. Is there any reason then that the spacing between the 8.58 and 8.03 lines should be the same as that between the other members of the multiplet(really two separated doublets?)? Show from the data in the caption that the spin-spin coupling constant is ≃9 cps. When the authors refer to the low- and high-field groups as bands, do you catch the analogy with vibrational spectroscopy? To what are the individual lines making up the band due in the present case? At any rate, the rest of the low-field band beyond 8.50 ppm was then assigned to proton 1; to what would the complicated splitting pattern be due? What would now account for the remaining lines in the high-field band? Is the intensity ratio observed by PSB preserved? Discuss.

146. "Confirmation of Phase Change in Solid Adamantane by NMR" H.A. Resing 43, 1828 (1965).

Draw the structure of adamantane. Is it symmetrical so that it might be expected to have a reasonably high melting point? At the time this work was performed there was some controversy as to whether a solid-solid phase transformation took place in the neighborhood of 200°K.

Treating the regularity of the temperature behavior of the spin-lattice relaxation time T_1 (PSB p. 23) as indicative of a single phase, the plot of Fig. 1 would certainly indicate that at a temperature just above 200°K two phases do exist in equilibrium with one another.

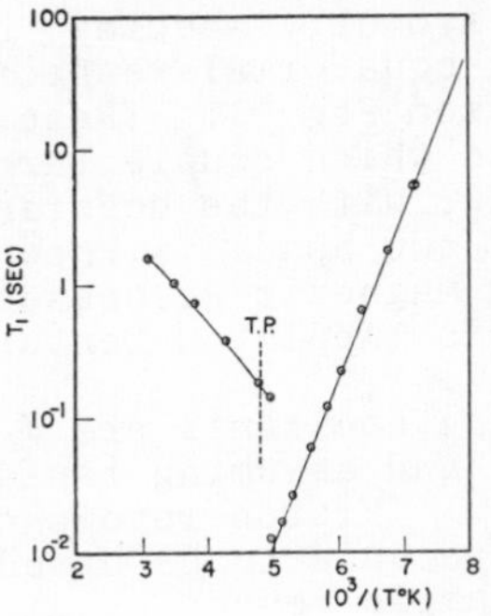

FIG. 1. Spin–lattice relaxation time T_1 vs reciprocal temperature for solid adamantane. The phrase transition temperature is indicated by T. P.

Note that at one point just below the transition temperature each phase existed long enough so that T_1 values could be determined for each. Read how this was done in ref. 6.

Show from Fig. 1 that an energy or enthalpy of activation for the process of relaxation, with rate constant $1/T_1$, is 5.7 kcal/mole in the low-temperature phase. From the author's references see if you can determine why the low-temperature relaxation process is identified as molecular rotation. Are there any other mechanisms by which this molecular crystal can "relax"? Just what is the crystal relaxing towards, at a fixed temperature?

147. "Proton Coupling Constants in Cyclobutanone" L.L. Combs and L.K. Runnels 44, 2209 (1966).

Fig. 1 shows the two possible conformations of cyclobutanone, with numbering of the protons. If the molecule rapidly

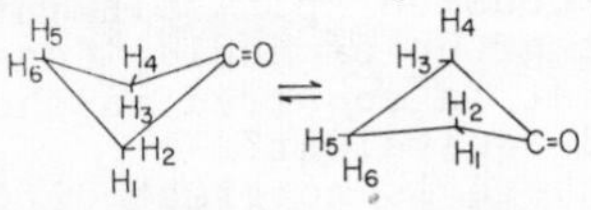

FIG. 1. Ring flips in cyclobutanone.

flips between the two conformations, show that the following pairs of coupling constants are equal on the basis of symmetry alone:

$$J_{52} = J_{54}, \qquad J_{62} = J_{64}, \qquad J_{51} = J_{53}, \qquad J_{61} = J_{63}.$$

By considering the flipping to be sufficiently rapid(exactly what does this imply about the relation between the period of this normal vibrational mode and the mean relaxational and/or radiative lifetime of the excited spin states?) so as to allow only the average values of spin-spin coupling constants to be observed, show that the following additional relations are implied:

$$J_{51} = J_{62} = J_{53} = J_{64} \equiv J_A,$$

$$J_{52} = J_{61} = J_{54} = J_{63} \equiv J_E.$$

The subscripts A and E are not symmetry species, but rather stand for axial-axial and axial-equatorial respectively; the convention should be apparent from Fig. 1. These two coupling constants are members of a class which couple through three bonds, known as vicinal coupling, with the constants sometimes written $^3J_{ik}$. Coupling through two bonds is known as gem coupling, with $J={}^2J_{ik}$; cf "Nuclear Magnetic Resonance Spectroscopy" by R. Lynden-Bell and R.K. Harris (Appleton-Century-Crofts, New York 1969), p. 104.

The three remaining coupling constants are $J_{12}=J_{34}$ and J_{56}. Indicating chemical shifts by δ and assuming rapid flipping, show that $\delta_1=\delta_2=\delta_3=\delta_4$ and $\delta_5=\delta_6$. Proton resonance chemical shifts in ppm relative to TMS are said to be based on the δ scale and are positive if the sample resonates at a higher frequency than does TMS for the same applied field. The other notation for the chemical shift is τ, with the relation between them $\delta+\tau=10.00$. At what value of τ does TMS resonate? cf L-B&H p. 23.

For saturated six-membered rings J_A is in general not equal to J_E(ref. 1, PSB P. 194); the authors give examples in ref. 2-6 of considerable differences reported for $|J_A-J_E|$. To determine if this is also true for cyclobutanone, the authors assume that $J_A=J_E$ and calculate the spectrum using both second-order perturbation theory and an exact method(ref. 7). These calculated results are compared with the experimental spectrum in Fig. 2. The best fit with the experimental spectrum was obtained with $\delta_1=2.98$ ppm and $\delta_5=1.93$ ppm, and with $J=J_A=J_E=7.9$ cps. Show that the δ values are completely compatible with the τ values in Fig. 2. Is it possible to determine the primary resonance frequency from the information given above and in Fig. 2? If so, determine it; if not, what else do you need and why? The authors state that the coupling constants J_{12} and J_{56} are unobservable; does this appear to be the case in all three spectra? Is this surprising in view of the declaration(p. 92-3) of PSB that generally electron-coupled spin-spin interaction between members of an equivalent group of nuclei does not give rise to multiplet splitting? What is required in the present case to make say protons 1 and 2 equivalent?

As seen in Fig. 2 there is no great difference between the exact and the second-order perturbation calculations. The authors state that this is to be expected since the ratio of coupling constant to chemical shift is only 0.12. Check this value; is there a way that it might be changed? If so, how? Will it then make any difference in the last argument, and again

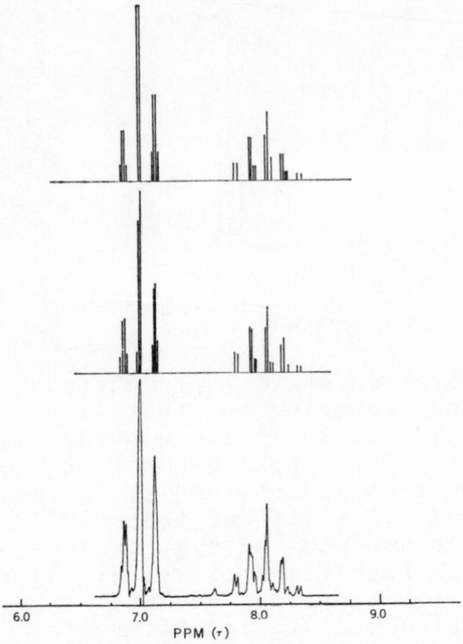

FIG. 2. Proton spectrum of cyclobutanone. Top to bottom are second-order perturbation, exact, and experimental spectra. The weak absorption at about 7.6 τ is attributed to an impurity, also detected by gas chromatography.

if so how? They feel that from the results of Fig. 2 that $|J_A - J_E| < 1$ cps, in disagreement with the results of ref. 2-6, but in accord with previous findings about four-membered rings (ref. 8).

148. "Virtual Coupling as the Cause of Anomalous NMR Spectra of Alkyl Fluorides" D.L. Hooper, N. Sheppard, and C.M. Woodman 45, 398 (1966).

The anomalous spin-coupling effects in the spectra of primary and secondary aliphatic fluorides(ref. 1-3) were here shown to be quite general, having been observed in the NMR spectra of the straight-chain alkyl fluorides $C_nH_{2n+1}F$ with n>3.

The resonance of the α proton or protons occurs as two distinct bands separated by ∿50 cps, due to spin-spin coupling to the ^{19}F nucleus over two bonds; would the coupling constant be written $^2J_{HF}$ in the notation of the previous exercise? Each band then shows further splitting due to coupling to the β protons, but the splitting pattern is different in the two cases. One band(usually the low-field component) is sharp, while the other is much less well-resolved, consisting in extreme cases of only a single broad peak. Fig. 1(a) shows the effect quite well.

The authors show that this additional complexity is a result of "virtual coupling" to the γ protons. The proton spectrum of an alkyl fluoride may be analyzed as the superposition of two subspectra corresponding to the two orientations of the fluorine spin. Do these subspectra correspond to the two bands shown for each of Fig. 1(a) and (b)? The effective chemical

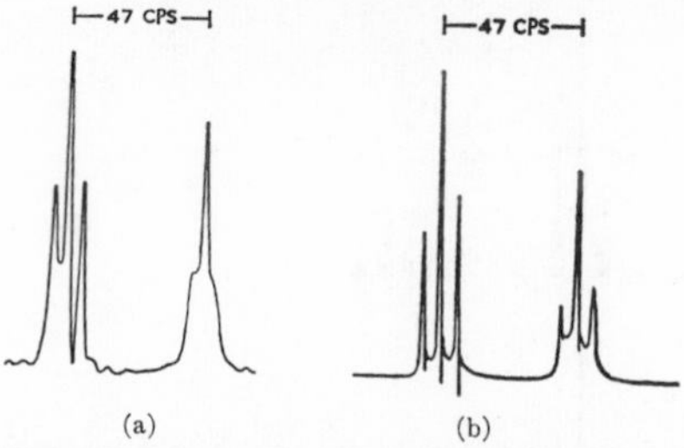

FIG. 1. The α-CH_2 region of the ^1H spectrum of *n*-heptyl fluoride, (a) at 40 Mc/sec, (b) at 100 Mc/sec.

shifts are said to be given by $\nu \pm \frac{1}{2} J_{HF}$. Why should this be so? Is J_{HF} here the vicinal coupling constant $^3J_{HF}$(through three bonds to the β proton)? If so is the ν value that for the β proton chemical shift? The coupling constant $^3J_{HF}$ to the β protons can be as great as 30 cps, and it is therefore possible for the effective chemical shifts of the β and γ protons to be almost equal in one subspectrum(e.g. those on the rhs of Fig. 1(a) and (b), the high-field side), giving rise to virtual coupling of the α to the γ protons and blurring of the first-order splitting pattern, and widely different in the other or low-field subspectrum(band). Through how many bonds would the coupling of α to γ protons take place? What is the meaning of the word virtual as used in this article?

Since the effect is dependent on the relative magnitudes of coupling constants and chemical shifts, it is affected by a change in the frequency of the experiment, and is frequently not observed at 100 Mc/sec.

In the example of n-heptyl fluoride given in Fig. 1, analysis of the complete 100 Mc/sec spectrum(why not the 40 Mc/sec spectrum?) indicated that the true chemical shifts of the β and γ protons(τ values) are approximately 8.35 and 8.60 ppm, and the vicinal coupling constant $^3J_{HF}$ is 24 cps. Show then that the effective chemical shifts of the β protons are 8.47 and 8.23 at 100 Mc/sec, but 8.65 and 8.05 ppm at 40 Mc/sec. Since $^2J_{HF}$ and $^3J_{HF}$ have the same sign(ref. 6), you should show that these results lead to a broadening of the upper band in the α-CH_2 region at 40 Mc/sec with a much less marked effect at 100 Mc/sec.

149. "Magnetic Nonequivalence in the High-Resolution NMR Spectra of Diborane" T.C. Farrar, R.B. Johannesen, and T.D. Coyle 49, 281 (1968).

Earlier work on the NMR spectrum of diborane suffered from the complications of the 7 spin orientations of the less abundant isotope ^{10}B(19.78%, I=3). See their ref. 1-4 and PSB pp. 298-301. Compute the abundance of $^{11}B_2H_6$, $^{11}B^{10}BH_6$, and $^{10}B_2H_6$ to compare with their values of 64, 32 and 4%. What was the method of preparation of isotopically enriched($\geq$98.2% ^{11}B) diborane? The single sharp impurity peak in the proton spectrum of Fig. 1 was thought to be probably due to the presence of about 0.1% ethane. As is often the case, even this impurity has its utility, as its very sharp linewidth of $\sim$0.3 Hz demonstrates quite well that the linewidth in the proton spectrum of

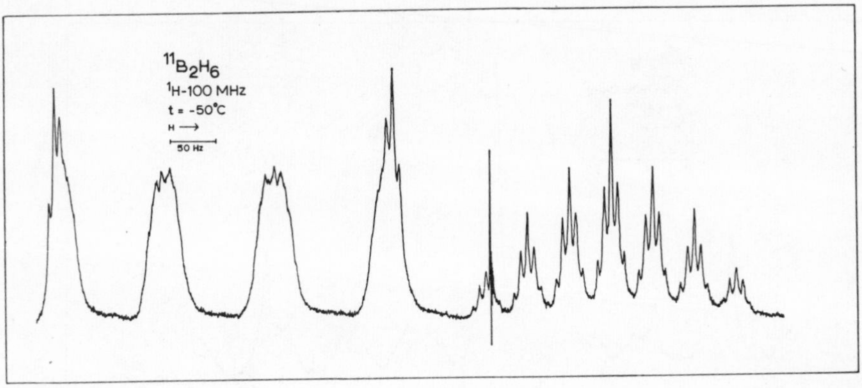

(a)

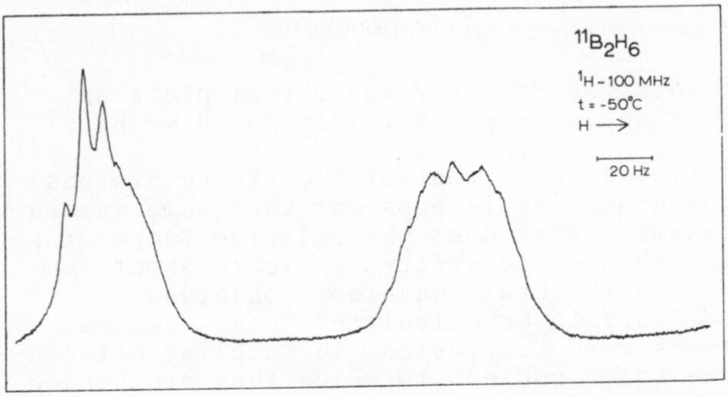

(b)

FIG. 1. (a) Proton magnetic resonance spectrum of $^{11}B_2H_6$ at 100 MHz. (b) Detail of low-field multiplets.

$^{11}B_2H_6$ is not an instrumental effect. Both the proton and ^{11}B spectra are independent of temperature over the -7 to -60°C range studied. Notice how clearly the spectra are labeled.

In (a) the resonance of the bridge protons occurs to the high-field side of the ethane impurity line. The former is a septet of quintets as you should explain from the first-order theory. The nuclear spin of ^{11}B is 3/2. Why do not the other nuclei combine to produce the reverse effect, a quintet of septets?

The terminal proton resonance is a symmetric set of four complex resonances clearly exhibiting considerable fine structure. The outer members of the "quartet" are distinctly different from the inner lines. Show that the first-order analysis would lead us to expect a quartet of triplets(not vice versa?). Why are we justified in assigning the low-field portion of the proton spectrum to the terminal protons? Prove your answer as quantitatively as you can. Observed line shapes for the terminal-proton spectrum are identical at 60 and 100 MHz, except where the lines from bridge and terminal protons overlap at the lower frequency. What does this indicate to us?

Predict the spectrum of ^{11}B. A triplet of triplets? Why? Fig. 2 does show this, but with a more elaborate structure.

An attempt was made to determine the J_{BB} coupling constant by ^{11}B-{^{2}H} double resonance experiments on samples with ~50% of $^{11}B^{10}B^2H_6$. When the ^{2}H nuclei were strongly irradiated(why?) the broad ^{11}B resonance narrowed from a linewidth of about 75 Hz

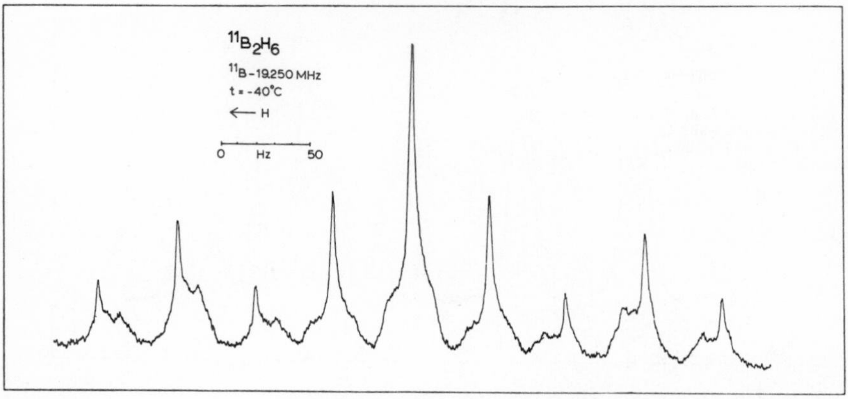

FIG. 2. Boron-11 magnetic resonance spectrum of $^{11}B_2H_6$ at 19.250 MHz.

down to a linewidth of about 18 Hz. Why would this place an upper limit of about 3 Hz on $J_{^{10}B^{11}B}$? Why then would we have $J_{^{11}B^{11}B}$<9 Hz?

From the linewidth(relative to that of the ethane protons) of the bridge proton resonances it is apparent that some quadrupolar broadening is present. What does the relative temperature independence of both the ^{1}H and ^{11}B spectra indicate about the quadrupole contributions to the broad envelopes observed?

For the authors to "explain or calculate" these spectra, the following interactions were considered: i) Coupling between terminal protons and the boron nucleus to which they are bonded, J_{BH_t}; ii) coupling between bridge protons and boron, J_{BH_b}; iii) bridge-proton—terminal-proton coupling, $J_{H_bH_t}$; iv) coupling between the terminal protons and the remote boron atom, J'_{BH_t}; v) coupling between terminal protons on opposite ends of the molecule, $J_{H_tH_t}{}^{cis}$ and $J_{H_tH_t}{}^{trans}$; vi) geminal coupling between terminal protons, $J_{H_tH_t}{}^{gem}$; vii) boron-boron coupling, J_{BB}; and viii) the bridge-proton—terminal-proton chemical shift, $\delta(H_t\text{-}H_b)$.

Verify that the following three parameters may be determined directly from the spectra:

$$\delta(H_t\text{-}H_b)=-4.50 \text{ ppm}, \quad J_{H_bH_t}=7.2 \text{ Hz}, \quad J_{BH_b}=46.2 \text{ Hz}.$$

The observed separation of 137.0 Hz which appears prominently in both boron and terminal proton spectra was assigned to the sum $J_{BH_t}+J'_{BH_t}$ by analogy with less-complex centrosymmetric systems (ref.10 and 13); this restriction was verified by the calculated spectra.

There were too few sharply resolved lines in the $^{11}B_2H_6$ spectrum for any iterative computer program to be completely successful; only trial-and-error methods for parameter adjustment were possible. A total of 32 ^{1}H and 32 ^{11}B spectra were calculated. The spectra which agreed most closely with the observed spectra are shown in Fig. 3. The parameters used to generate these spectra are given in Table I. Are all but one of the interactions listed above observed to make a definite contribution? Is the one exception compatible with the question asked towards the bottom of p. 276 herein and why?

CHAPTER XIII

OTHER TYPES OF SPECTROSCOPY
OTHER METHODS OF DETERMINING STRUCTURES
MISCELLANEOUS

150. "Report on Notation for the Spectra of Polyatomic Molecules" (prepared at the request of the Joint Commission for Spectroscopy of the International Astronomical Union and the International Union of Pure and Applied Physics) 23, 1997 (1955).

Though the identity of the author is not obvious, he(she?) acknowledges extensive consultation with G. Herzberg; is this surprising? The latter's book IRRS(referenced on p. 58 herein) is ref. 2 and is referred to repeatedly. Ref. 1 gives a standardized notation for diatomic molecules and their spectra.

You may wish to browse through this Report, or at least to be aware of its existence. Contained therein are some 42 recommendations and 5 suggestions; the distinction between the two categories being immediately obvious to parents. In this exercise let us touch upon a few points in the Report. One rather long exercise which will definitely not be suggested is to work through this reader from front to back picking out violations of these recommendations.

In the Introduction we see that, pending final decision by the Joint Commission, cm^{-1} and ν are both used for wavenumber (which itself is written as two words), rather than K(for ?) or σ, and in addition ν may serve as the frequency in Mc/sec(now denoted itself by ?). Table I, running to almost three full pages, is a summary and index of symbols and definitions some number of which we have used in these chapters.

Verify that the sentence is correct which immediately precedes REC. 2a, that neither normal vibrational modes nor orbitals(which are exactly what?) of species σ^- may exist. Generally speaking, the notation for linear polyatomics is to follow that established for diatomics as closely as possible. REC. 5a,b for choice of axes for C_{2V} and D_{2h} molecules, are designed to save confusion in discussing what type of spectral transitions in planar molecules?

What letter is recommended for the vibrational quantum number and what is your guess as to why? With what usage might the other common textbook choice possibly conflict? Rydberg transitions?

Are the designations of electronic configurations for molecules familiar to you? With near-Hartree-Fock wavefunctions now becoming available for an ever-widening class of molecules, and reasonable SCF wavefunctions for an even bigger class, is it likely that the parts of REC. 13 will be followed rather closely?

Many if not most of the recommendations should be second nature to anyone working in the field; do you agree?

151. "Zero-Field Electron Magnetic Resonance in Some Inorganic and Organic Radicals" T. Cole, T. Kushida, H.C. Heller 38, 2915 (1963).

The peroxylamine disulfonate ion, $(SO_3)_2NO^=$, has a single unpaired electron as you should verify by drawing a structure. Show that the possible values of the total angular momentum quantum number F(REC. 28!) are 3/2 and 1/2. Why is the vector relation $\underline{F}=\underline{I}+\underline{S}$ quite general for nonlinear molecules or molecular ions? What about the electronic angular momentum vectors $\underline{L}$ and Λ? Be as general as possible.

Given the spin Hamiltonian as

$$H = a\underline{I}\cdot\underline{S} \text{ , with a the (empirical)hyperfine parameter,} \quad (2)$$

show that both F^2 and F_z commute with it. The book "The Theory of Atomic Spectra" by E.U. Condon and G.H. Shortley (Cambridge University Press 1963) has information in Chapter 3 that will allow you to walk through this proof doing little more than changing letters. While in this book you may enjoy the short story in the first paragraph of the Preface to the 1963 Impression. At any rate having proved the commutation properties of F^2 and F_z, of what utility are the eigenfunctions of these operators? Show from the vector model that the energy levels in the absence of a magnetic field are

$$E(F) = \tfrac{1}{2}a[F(F+1)-S(S+1)-I(I+1)]. \quad (3)$$

Do you happen to know whose idea it was to replace X^2 by $X(X+1)$ for the magnitude of the square of the angular momentum vector $\underline{X}$? This would be prior to 1926 of course. Show that the difference between zero-field levels is as given in Fig. 2(a).

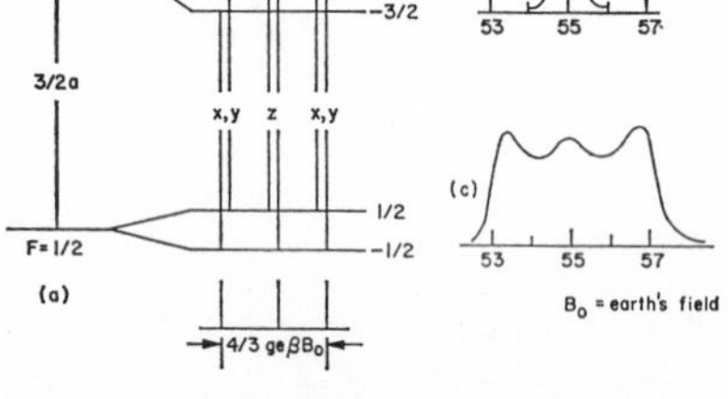

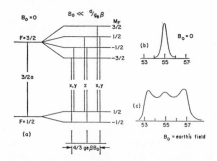

FIG. 2. (a) Energy levels and transitions for $(SO_3)_2NO^=$ radical, $I=1$, $S=\frac{1}{2}$ in both zero field and in a weak applied field. Polarization of the transitions is indicated by x, y, and z for the case where $\mathbf{B}_0 \parallel z$; (b) observed resonance line from $(SO_3)_2NO^=$ in zero field, and (c) in the earth's field ($\sim$0.5 G).

From studies on this ion in solution(as done here) in fields of 3-50 gauss, it was found that the value of a agreed very well from that determined from the zero-field hyperfine splitting in this work, 36.65 Mc/sec; cf ref. 1. Is this value of a indicated in Fig. 2?

What does the condition for a weak applied field given at the top of Fig. 2, $B_0 << a/(g_e\beta)$ with the g factor for a free electron and the Bohr magneton represented, say about the relative values of the magnetic and hyperfine energies and why? Is this requirement met for the earth's field? What field strength would be required for the two to be comparable? Demonstrate that the total weak-field splitting shown, together with the spectrum in (c), yield a value for the earth's field that is somewhat higher than that indicated in the caption. Could the free-electron g factor be that different from that in the ion? What might be the (really very minor)difficulty here? A clue may be found in the answer to the question: what method was used to cancel the earth's field so that spectrum (b) could be obtained?

152. "Structure and Rotational Isomerization of Free Hydrocarbon Chains" L.S. Bartell and D.A. Kohl 39, 3097 (1963).

Draw an end view of ethane, with a single substituent on each carbon, with the substituents in both the trans and gauche conformations. The substituents become alkyl groups in describing the rotational isomers of the gas-phase normal alkanes. Fig. 1 presents the reduced experimental data determined from electron diffraction. We have seen these curves before in exercise no. 70 for solid and liquid gallium. Note in every molecule the very strong peaks at about 1.0 and 1.5 Å; to what do these correspond? Is the area under a given peak proportional to the number of bonds of that length with the same proportionality constant for all pairs of nuclei? Why or why not? Is your answer borne out in Fig. 1? Can you identify two or three other characteristic-distance peaks from these radial distribution curves? The relative strength of the trans and gauche nonbonded peaks beyond 2.7 Å is a measure of the equilibrium distribution among the rotational isomers. The electron-diffraction analysis of this equilibrium was simplified by the consideration of just one parameter $\Delta G°$, representing an average free-energy difference between a given gauche and its corresponding trans conformation. Demonstrate that gauche twists may be either right- or left-handed. Is this possibility open for the trans conformation? It follows that there are more ways of realizing gauche conformations than there are of trans. These ideas are brought together into the equation

$$N_i/N_j = (m_i/m_j) \exp [-(n_i-n_j)\Delta G°/RT] \quad (1)$$

with N_i, m_i and n_i the number of molecules of isomer i in the sample, the multiplicity of that isomer, and the number of gauche conformations in that isomer. If the units of $\Delta G°$ are kcal or cal/mole, it is appropriate to ask per mole of what? Estimate a rough value of $\Delta G°$ solely from a knowledge of the geometries of gauche and trans butane and the barrier to internal rotation in ethane. Less than 1 kcal/mole? Why? Is the

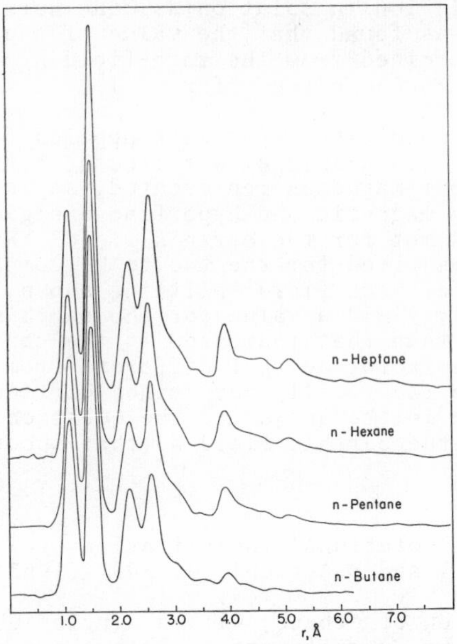

FIG. 1. Experimental radial distribution curves for various *n*-hydrocarbons, as determined by electron diffraction.

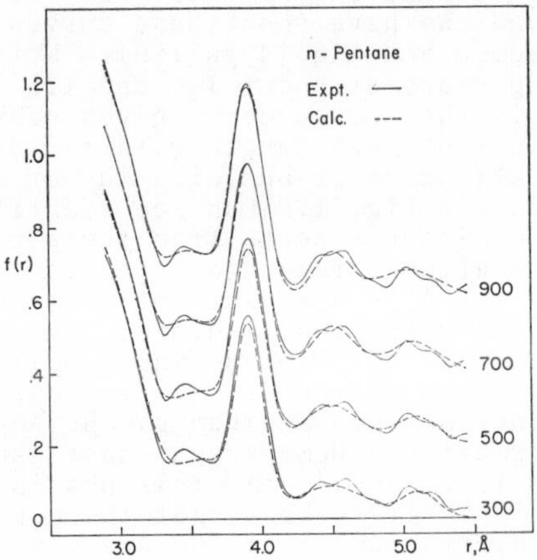

FIG. 2. Experimental and calculated radial distribution curves for the nonbonded skeletal distances in *n*-pentane. All peaks involving hydrogen have been subtracted except for those at 2.8 and 3.5 Å. Comparisons are made for assumed *trans–gauche* isomerization free energies ranging from 300 to 900 cal/mole.

above equation classical? What about the multiplicities?

Fig. 2 shows experimental and calculated curves(radial distribution functions) for the nonbonded carbon-carbon distances in n-pentane. Fig. 5 gives us the standard deviations as a function of assumed $\Delta G°$ for the four alkanes. How was the final value of 610 cal/mole obtained?

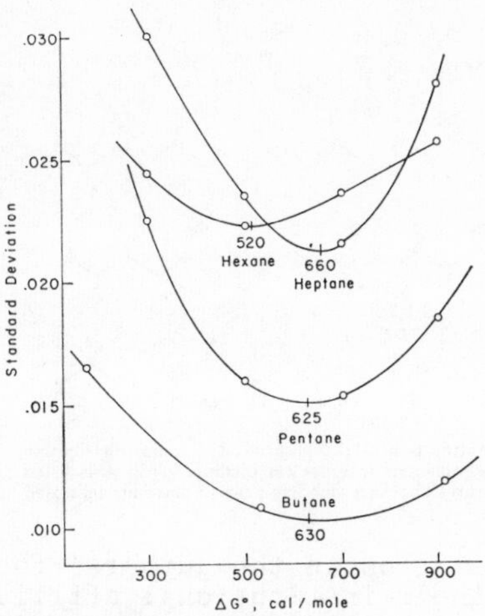

FIG. 5. Standard deviations between experimental and calculated radial distribution curves as a function of the assumed conformational free-energy difference.

Compute the abundances of the three rotational isomers of n-pentane from the values of n_i and m_i given in Table I. Compare with the results of the last column. Can you derive the multiplicities yourself?

TABLE I. Distribution of rotational isomers in gas phase.

Molecule	Isomer	m^a	% present[b]
n-butane	T	1	59.1
	G	2	40.9
n-pentane	TT	1	38.4
	TG	4	52.7
	GG	2	9.0
n-hexane	TTT	1	24.5
	TTG	4	33.6
	TGT	2	16.8
	TGG	4	11.6
	GTG	4	11.6
	GGG	2	2.0
n-heptane	TTTT	1	15.7
	TTTG	4	21.5
	TTGT	4	21.5
	TTGG	4	7.4
	GTTG	4	7.4
	TGTG	8	14.8
	TGGT	2	3.7
	TGGG	4	2.5
	GTGG	8	5.1
	GGGG	2	0.4

[a] Multiplicity [see Eq. (1)] takes into account both G and G' conformations. Conformations with G and G' adjacent are considered to be sterically unreasonable and are given zero weight.
[b] Calculated from Eq. (1) assuming $T=287°K$ and $\Delta G^0=610$/mole.

Fig. 6 identifies the observed C-C radial distribution peaks, some of which are not obvious on the scale of Fig. 1. See if you can compute one or two values for the trans isomers. Would the gauche values be more difficult? Read about the analog computer used to evaluate the slight shrinkage effects

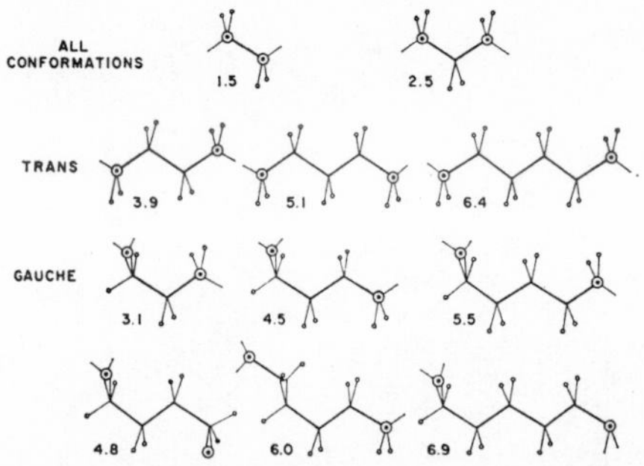

FIG. 6. Identification of the observed CC radial distribution peaks. The approximate internuclear distance r_{ij} in Å is listed under each figure. The corresponding atoms i and j are indicated by bull's eyes.

due to torsional oscillation in the authors' footnote 7. If you happened to miss their institutional affiliation under the title is it now obvious that it is neither an industrial nor a government laboratory?

A number of smaller effects are discussed. They state that the individual chains, when free to move independently(as in the vapor phase?), do indeed develop frequent gauche links along their length. Does the information of Table I bear this out? Other points of interest may be found under Discussion.

153. "Synthesis of Diatomic Radicals" H.E. Radford 40, 2732 (1964).

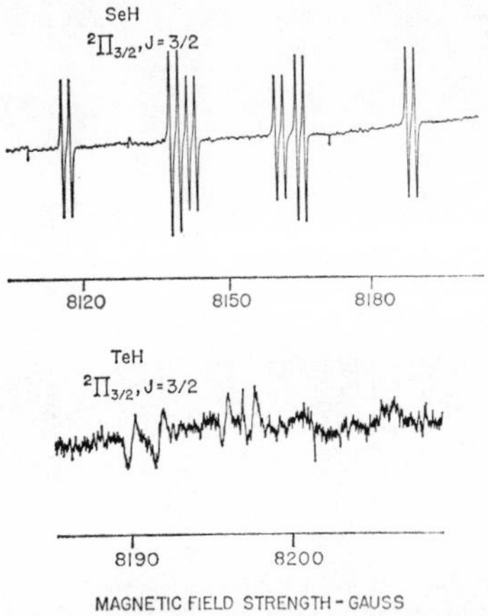

FIG. 1. Paramagnetic resonance spectra of free SeH and TeH radicals, recorded at 9215 Mc/sec with a superheterodyne reflection-type spectrometer. Only the four strongest lines of the TeH spectrum are shown.

The paramagnetic resonance spectra of Fig. 1 were observed when the products of an electrical discharge in hydrogen gas were conducted through a microwave absorption cell whose walls had been coated with elemental selenium in the one case and elemental tellurium in the other. In what basic, fundamental way do these spectra differ from those examined heretofore? These are first-derivative spectra; by drawing an ordinary absorption line and then plotting the first derivative of this line directly underneath convince yourself of this fact.

Table I gives the estimated partial pressures of these and other radicals obtained from intensities and dipole moments. How does the intensity of a line depend upon the dipole moment? Prove your statement from the data given. Is this the reason that the dipole moments were determined in magnitude only in the exercises of the chapter on rotational spectroscopy? Is there an experiment that will yield the sign of the dipole moment? What radical was used to calibrate the data of Table I?

TABLE I. Partial pressures of synthesized radicals, estimated from microwave absorption intensities. Total pressure in all cases is approximately 100 μ Hg.

Radical	Relative microwave absorption intensity	Permanent electric dipole moment (10^{-8} esu)	Partial pressure (μ Hg)
SO	1550	1.4[a]	50
SH	240	0.3[b]	10
SeH	60	0.2[b]	5
TeH	3	0.1[b]	1

[a] D. R. Lide, Jr. (private communication).

[b] Crude estimates based on electronegativities [L. Pauling, *The Nature of the Chemical Bond* (Cornell University Press, Ithaca, New York, 1960), 3rd ed. p. 98].

From measurements on the spectra, check the values of $\bar{g}_J$ and A_1, the molecular g factor and hyperfine coupling constants for both radicals, with the values of Table II. The former constant is determined from the energy to achieve resonance at the value of the field ignoring the fine(and of course the hyperfine) structure, or vice versa in the present case. Is this energy the separation between molecular Zeeman levels? Take the magnetic field at the "band" center, ca 8150 for SeH. Is this use of the g factor strictly analogous to our previous usages?

The hyperfine coupling constant may be determined from the smallest spacing in the spectrum. Which nucleus is causing this splitting for SeH? for TeH?

TABLE II. Molecular constants of the $^2\Pi_{\frac{3}{2}}$, $J=\frac{3}{2}$ ground states of SeH and TeH, derived from paramagnetic resonance spectra.

Molecular constant[a]	SeH	TeH
Molecular g factor, $\bar{g}_J$	0.80800±0.00010	0.80366±0.00020
Λ-type doubling frequency, ν_Λ	(14.4±0.1) Mc/sec	(6.6±0.2) Mc/sec
Hyperfine structure coupling constant, A_1	(1.9±0.1) Mc/sec	(1.8±0.2) Mc/sec
Rotational constant, B_0	(7.98±0.08) cm^{-1}	(5.56±0.15) cm^{-1}
Spin–orbit constant, A_0	−(1600±50) cm^{-1}	−(2250±200) cm^{-1}

[a] H. E. Radford, Phys. Rev. **122**, 114 (1961); **126**, 1035 (1962).

154. "Molecular Structure of Dicyclopentadienylberyllium $(C_5H_5)_2Be$" A. Almenningen, O. Bastiansen, and A. Haaland 40, 3434 (1964).

This compound was first reported(ref. 1) in 1959. It was unstable in air and reacted violently with water. The surprising physical property was its dipole moment in solution; in benzene it was 2.46±0.06 D and in cyclohexane it was 2.24±0.09 D. Yet a centrosymmetric molecular structure was indicated in the crystal(ref. 5). If the vapor-phase molecule is symmetric, then the dipole moment in solution could be due only to what? An electron diffraction investigation was undertaken in this work with the resultant distances shown in Table I. The distances

TABLE I. Molecular parameters of $(C_5H_5)_2Be$.

	r^a (Å)	u (Å)
C_1–H_1	1.070±0.005	0.084
C_1–C_2	1.424±0.002	0.047
C_1–C_3	2.304±0.004	0.055
C_1–H_2	2.230±0.010	0.120
C_1–H_3	3.340±0.010	0.090
Be–C	1.915±0.005	0.080
Be–C′	2.320±0.010	0.080
Be–H	2.720±0.020	0.120
Be–H′	3.020±0.020	0.120
C_1–C_1'	3.450±0.030	0.150
h(Cp–Cp′)^b	3.370±0.030	
h_1(Be–Cp)^b	1.485±0.005	
h_2(Be–Cp′)^b	1.980±0.010	
$h_1+h_2{}^b$	3.470±0.020	

[a] The indicated limits are limits of reproducibility.
[b] Calculated from observed parameters.

prefaced with a lower case h are measured perpendicular to the cyclopentadienyl planes. It is seen that the beryllium atom is not in the center but displaced up(or down) along the fivefold axis in one of two equivalent positions as shown in Fig. 4. To what point group does the C-H skeleton belong? Do you see why the introduction of the acentric Be atom lowers the symmetry to C_{5v}?

Show that the molecule would have a dipole moment of 2.5 D if completely ionic, $Be^{2+}(C_5H_5)_2^-$. Is there any reason, symmetry or otherwise, that would require the rings to carry the same net charge? It was pointed out to the authors by a very famous scientist, acknowledged in footnote 8, that if the ring closest to Be is neutral a small fractional charge on Be suffices to produce this same dipole moment. What is this charge? They suggest that since Be is small the ring-ring distance will be determined by repulsion between electrons in the ring π orbitals and that the metal ion then will be free to move within the resulting cavity, at least to some extent. With a simple electrostatic calculation show that the barrier to axial motion of the beryllium ion is 2-5 kcal/mole. Sketch the potential energy curve showing the double minimum. Approximate the well depths with a parabolic potential which goes smoothly (continuous first derivative) from one minimum to the other. See if you can determine the force constant to one significant figure. Is it reasonable to treat the ring framework as rigid with only the metal ion moving? Why or why not? Assuming that it is, compute the frequency of the axial motion of beryllium. Is it at all close to what you would expect? Might it be good to one and one-half significant figures? Check with the IR work on the compound from ref. 4 and see if this fundamental may be there. Of course this work on the frequency just barely deserves to be taken seriously; perhaps it is still a useful exercise.

Does the molecule seem to beg for a semiempirical quantum mechanical calculation? Look in some formula-abstracting scheme and see if this has been done.

Finally, how is the apparently centrosymmetric molecular structure in the crystalline phase explained (p. 3437)?

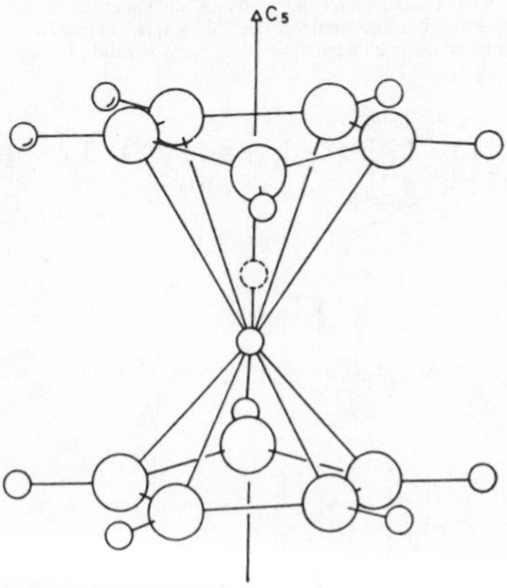

FIG. 4. The molecular structure of $(C_5H_5)_2Be$. Point group symmetry C_{5v}.

155. "Electron Paramagnetic Resonance of Vanadyl Ions Trapped in RNA and DNA" W. Snipes and W. Gordy 41, 3661 (1964).

Other metal ions have already been identified by EPR in RNA and DNA. In this work the spectra of Fig. 1 were obtained for three commercially obtained samples of RNA. The well-resolved eight components appeared to be due to the hyperfine structure arising from the ^{51}V nucleus. What is its nuclear spin? What about the isotopic abundance complications?

What reasons do they advance for assuming the metal ion is vanadium? Further, why do they suggest that it is in the form of the vanadyl ion VO^{2+}? What experiments did they perform to demonstrate that indeed the vanadyl ion is responsible for the EPR spectra? Note that these ions were not found in the commercial DNA samples, only the RNA.

Do the authors think it probable that vanadium would be introduced as an impurity during sample preparation? Do they speculate as to whether VO^{2+} has a biological origin or function?

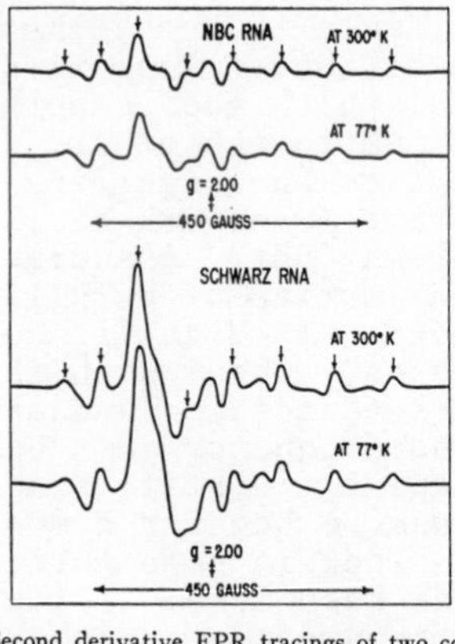

FIG. 1. Second derivative EPR tracings of two commercially obtained samples of dry, unirradiated RNA. The magnitude of the magnetic field increases from left to right, with the double arrow indicating the field position of the resonance of DPPH for which $g = 2.0036$. The eight arrows pointing downward indicate the position of the eight components of the ^{51}V hyperfine structure. Observations were made at a microwave frequency of 9200 Mc/sec.

156. "ESR Spectra for the t-Butyl Radical Trapped in a Solid" H. Shields and P. Hamrick 42, 443 (1965).

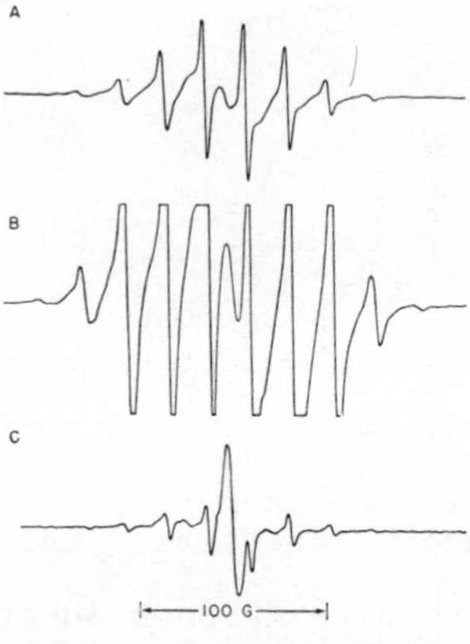

FIG. 1. Tracing of the ESR first-derivative absorption spectra for x-irradiated pivalic acid as displayed on a chart recorder. Spectrum *A* was observed at 130°K after x irradiation at 130°K. Spectrum *B* is the same as Spectrum *A* except displayed at higher gain to show the outside absorptions. Spectrum *C* was obtained by letting the sample warm to 250°K.

The ESR spectra for pivalic acid shown in Fig. 1 were obtained after exposure to $\sim 10^6$ R of x irradiation at 50 kV. The samples were irradiated at 130°K and observed at different temperatures. The spectra shown are a superposition of those from two different radicals. One contained ten equally spaced absorption peaks with intensity ratios of 1:9:36:84:126:126:84:36:9:1. Show that this is the pattern to be expected from the t-butyl radical $\dot{C}$-$(CH_3)_3$, with the unpaired electron interacting equally with the nine hydrogen nuclei. From the structure of pivalic acid, show how the t-butyl radical can be produced upon x irradiation. The authors say that the α-carbon bond to the acid group must be broken; do you agree? What is the most likely accompanying product or products of irradiation damage?

The second spectrum of the superimposed spectra was a centrally located resonance having a g value approximately equal to that of a free electron, 2.0023.

Verify their statement that the measured hyperfine coupling constant is 22 G. What is the value for the t-butyl radical in the liquid state(ref. 2)?

Upon warming the irradiated sample from 130°K to 250°K spectrum C was obtained. Is it clearly due to a different species? Why? Make one or two suggestions as to its nature.

157. "Crystal and Molecular Structure of Hydrogen Peroxide: A Neutron-Diffraction Study" W.R. Busing and H.A. Levy 42, 3054 (1965).

The molecule is now generally accepted to have C_2 symmetry. Show that its structure can be described by four parameters, two bond distances and two angles. One of the latter, the dihedral or azimuthal(demonstrate that this adjective is also appropriate) has remained the most elusive of the four. Their ref. 6 reported a careful analysis of the IR vibrational-rotational bands(symmetric or asymmetric top?) yielding 119.8±3° for the dihedral angle in the vapor phase. Ref. 7 made an x-ray study of the crystalline solid, finding a relatively short O···O distance, 2.78 Å, which was presumed to be that of a hydrogen bond. Would the x-ray method give the positions of the hydrogen atoms? Ref. 9? What about the neutron-diffraction method? Why? In the x-ray study the H atoms were assumed to lie exactly on the O···O lines, giving a value of 93.8° for the dihedral angle. What two other values of the dihedral angle were reported and in what compounds? "Apparently the azimuthal angle is rather sensitive to the surroundings of the molecule."

Read about the preparation of the single crystals. What was the name used for the instrument on which the diffraction measurements were made?

Fig. 1 shows the Fourier projections of (a) the hk0 zone and (b) the h0ℓ zone. What exactly does this mean? Note the lattice parameters as coordinate axes; does their placement correspond to the use of the Miller indices hkℓ as zones? To what crystal system does H_2O_2 belong? Do you see that there are four molecules per unit cell? If you still need practice in computing the density, a=4.06±0.02 and c=8.00±0.02 Å.

Fig. 2 shows the molecular geometry, hydrogen bonding, and thermal motion in the solid. From (b) do you see their value of 90.2±0.6° for the dihedral angle? Is the angle mentioned in the last sentence of the caption, and shown in (b), completely compatible with the assumed dihedral angle mentioned above from the x-ray work of ref. 7? The hydrogen bond O-H···H is not quite linear; show that the two O atoms subtend an angle of 167.6±0.4° about the H atom. Since the four atoms O-O-H···O are nearly coplanar, the authors attribute this nonlinearity almost entirely to the resistance of the O-O-H angle to bending. First show that the coplanarity and linearity under discussion are

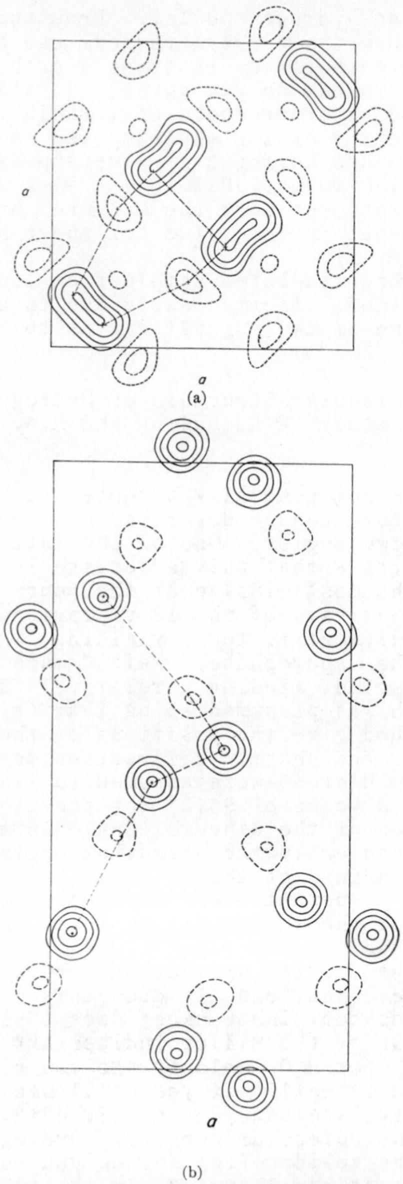

FIG. 1. Fourier projections of (a) the *hk*0 zone and (b) the *h*0*l* zone of H_2O_2. The solid curves represent the positive contours of oxygen and the dashed ones are the negative contours for hydrogen and for a small spurious negative region in (a). The contour intervals in (a) and (b) are 0.56 and 0.59×10^{-12} cm/Å^2, respectively. The zero contour has been omitted. The crosses represent the positions determined by the least-squares refinement, and the straight dashed lines indicate intramolecular and hydrogen bonds.

related yet partially independent. To be specific can we have

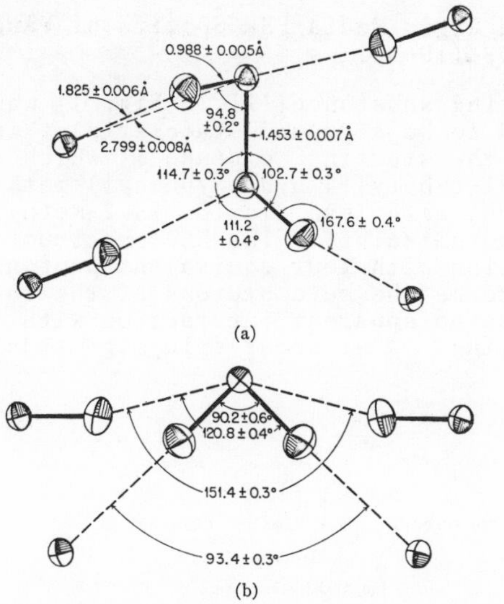

FIG. 2. Molecular geometry, hydrogen bonding, and thermal motion in H_2O_2. The oxygen and hydrogen atoms are represented by the smaller and larger ellipsoids, respectively. The principal axes of these ellipsoids are the rms principal thermal displacements of these atoms drawn to the same scale as the interatomic distances. (a) A general view of a molecule and its neighbors showing interatomic distances and angles. (b) A view down the O–O bond of a molecule showing dihedral angles measured about this bond. (The angle 93.4° is measured between the O–O···O planes.)

linearity without coplanarity? vice versa? Now discuss the authors' claim of resistance to bending.

Table III gives rms principal thermal displacements. Why are those for the smaller atom H greater? This is also shown in Fig. 2. For the H atom the surface of equally probable thermal displacement is nearly an oblate spheroid as shown in

TABLE III. Root-mean-square principal thermal displacements in angstrom units.

O	H
0.138±0.003	0.155±0.005
0.142±0.003	0.202±0.005
0.156±0.003	0.210±0.005

Table III, with a short axis which makes an angle of only 17±4° with the O-H bond direction. Hence expand upon their statement that the motion is consistent with the picture of an O-H bond which tends to resist stretching but which permits motion of the H atom by bending or torsional vibrations or by libration of the molecule as a whole. The distribution of the O atom is described by a slightly prolate spheroid, the long axis of which is essentially parallel(or collinear?) with the O-O bond; the angle is 3±9°.

Is this usage of the words prolate and oblate in accord with that of the chapter on infrared spectroscopy? Explain.

158. "Solution and Rigid-Media ESR Spectra of Yang's Biradical" R. Kreilick 43, 308 (1965).

This interesting substance(III in Fig. 1) was prepared in 1960 and predicted to be a ground state triplet in 1962. Initial oxidation of the starting compound I, which is 2,6-di-t-butyl-4-[bis(3,5-di-t-butyl-4-hydroxyphenyl) methylene]-2,5-cyclohexadien-1-one, with lead dioxide in 2-methyltetrahydrofuran gave the monoradical II. The ESR spectrum of this radical showed an interaction with four equivalent protons, which the authors presumed to be the meta protons of the conjugated rings. Discuss. There was no apparent interaction with the protons in the unconjugated ring. What about spin migration?

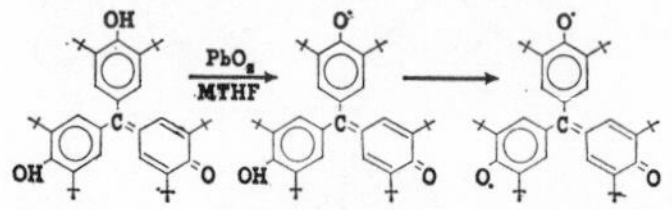

FIG. 1. Oxidation steps.

As oxidation proceeds, the spectrum of the biradical began to appear as they observed a superposition of the spectra of II and III. Finally the monoradical spectrum was completely replaced by the spectrum of the biradical, as in Fig. 2, which

FIG. 2. ESR of III in a rigid glass.

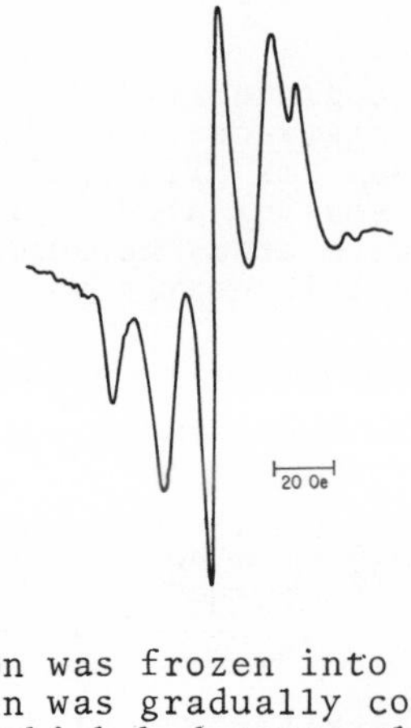

was obtained when the solution was frozen into a rigid glass at ∿-160°C. When the solution was gradually cooled the room-temperature spectrum of III, which had seven slightly overlapped lines(coupling to six equivalent protons? now what about spin migration?), began to broaden and the lines overlap; upon further cooling the lines collapsed into a single broad one which itself finally disappeared into the baseline at ∿-85°C; can you suggest a reason for the different behavior with temperature of the suddenly and gradually cooled solutions?

In Fig. 2 one observes the dipolar splitting. The spectrum consists of two pairs of lines disposed about g=2 and a line at g=2 from the monoradical. Show that the separation of the inner more intense lines is 38.4 Oersted. The author says that this indicates a separation of 9 Å between the two unpaired spins. Using the method outlined in ref. 3, show that this is so. The comment in exercise no. 13 about the units of the magnetic moment in that instance may be of use to you.

Whether or not you are able to check this distance by means of calculation, assume the geometry of the benzene ring applies to III and scale the distance between the unpaired spins as shown in Fig. 1. Is it close to 9 Å?

159. "Observation of the $2p^3 3s(^5S^\circ)$ Metastable State of Atomic Oxygen" G.O. Brink 46, 4531 (1967).

During a molecular-beam magnetic-resonance investigation of the products produced by a microwave discharge in O_2, a strong resonance was observed in atomic oxygen with a g factor of 2. By examining the levels for neutral oxygen (OI) in the volume by Moore, referenced on p. 70 herein, substantiate the author's claim that there are only two metastable S states of atomic oxygen, the $2p^4(^1S)$ state that lies 4.2 eV above the ground state and the state of the title lying at 9.1 eV. Why is he restricting the search to S states? Why to those S states that are metastable? Are there actually other metastable S states lying yet higher? Why were these not considered? What is the meaning of the right superscript ° in the state of the title? Is it used correctly here?

Of the two metastable S states that lie lowest, why is the first ruled out as the state participating in the Zeeman transitions? Fig. 1 shows the Zeeman spectrum; compute the g factor for comparison with the author's value of 1.97±0.05 for the ^5S state. What is the Landé g value? Also compute the g factor for the state identified as ^3P to compare with its Landé g value. What restrictions are placed on J for this last state(s) and why?

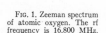

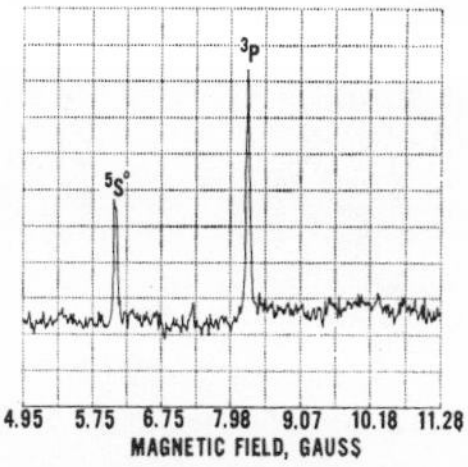

FIG. 1. Zeeman spectrum of atomic oxygen. The rf frequency is 16.800 MHz.

How was it known that these resonances were due to atomic oxygen?

A microwave discharge produced the atoms; the optical spectrum of the discharge was observed in emission and OI lines were observed at λ3947 and 7774. Show from Moore's volume that these two transitions originate on states from which the $^5S^\circ$ state may be reached by emission. What word or words should we use in place of state in the present circumstances? At any rate the optical spectrum provided further evidence for the presence of the $^5S^\circ$ state in the discharge. Why would its presence have important consequences for the chemical reactions occuring in systems containing oxygen with other gases? Interestingly enough this resonance is not always observed; would the optical lines mentioned above still indicate its presence? What might be the major mechanism for production of oxygen atoms in the $^5S°$ state?

160. "Studies of IBr, ICl, and I_2Cl_6 Crystal Properties by Means of the Mössbauer Effect in ^{129}I. I. Chemical Bonds" M. Pasternak and T. Sonnino 48, 1997 (1968).

Once again we have applications to chemistry of an effect predicted and then demonstrated by physicists.

In these experiments the gamma ray source was an excited nuclear state of the decay product of $^{129}Te(T_{\frac{1}{2}}=72$ min). For resonance between this gamma-emitter and the absorbing atom of the sample, what is the obvious identity of the decay product? By what mode does it decay? It in turn has $T_{\frac{1}{2}}=1.6\times10^7$ y.

The ^{129}I atoms of the absorber must of course be specially introduced. Starting with NaI, with 86% ^{129}I, write an equation for the formation of elemental iodine by the use of sulfuric acid and hydrogen peroxide. How was IBr then produced?

The gamma ray emitted had an energy of 27.8 keV. Show that its linewidth of 0.278×10^{-7} eV is equivalent to 6.7 Mc/sec and 0.029 cm/sec. This Doppler width is illustrated in the spectrum of the absorber IBr in Fig. 1. Notice that there are seven

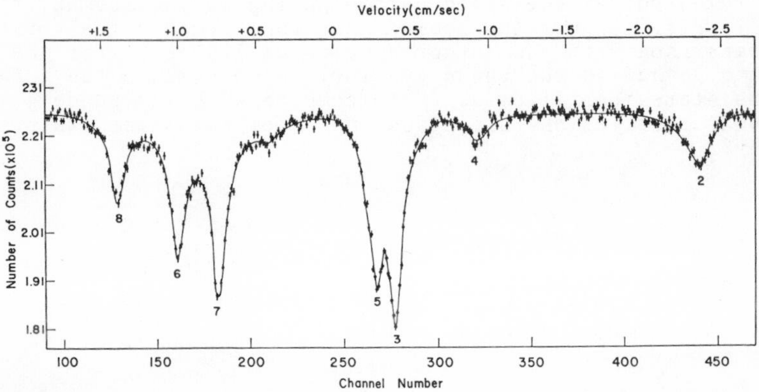

FIG. 1. Spectrum of IBr absorber at 80°K. The source is $^{66}Zn^{129}Te$. The numbers identify the transitions (see Table I).

well-resolved transitions. Table I gives line positions for the 5/2→7/2 transition(meaning exactly what?) split by quadrupole interaction. What is the nuclear spin of ^{129}I? What name is given to the combination e^2qQ? What is the meaning of the asymmetry parameter η? It is dimensionless as you see. δ is the isomer shift shown as I.S. in Table II.

Measuring line positions in Fig. 1(in what units? not channel number?) and using the results of Table I for line position, show that the values of Table II for IBr follow. The easy way of course is just the reverse of this, isn't it? There would seem to be a simple misprint in Table II; is it indeed that?

TABLE I. Positions of the eight lines resulting from the transition $5/2 \to 7/2$ split by quadrupole interaction in ^{129}I, for $\eta^2 \leq 0.2$. $|CG|^2$ are the Clebsh–Gordan coefficients. $A = e^2qQ/4$.

Line	Transition	$\|CG\|^2$	Δm	Position
1	$5/2 \to 3/2$	1	± 1	$A(1.654-0.392\eta^2)+\delta$
2	$5/2 \to 5/2$	6	0	$A(1.087-0.051\eta^2)+\delta$
3	$5/2 \to 7/2$	21	± 1	$A(0.230+0.035\eta^2)+\delta$
4	$3/2 \to 1/2$	3	± 1	$A(0.469+0.964\eta^2)+\delta$
5	$3/2 \to 3/2$	10	0	$A(0.178-0.091\eta^2)+\delta$
6	$3/2 \to 5/2$	15	± 1	$A(-0.389+0.250\eta^2)+\delta$
7	$1/2 \to 1/2$	18	$\pm\frac{1}{2} \to \pm\frac{1}{2}$	$A(-0.269+0.159\eta^2)+\delta$
8	$1/2 \to 3/2$	10	± 1	$A(-0.555-0.897\eta^2)+\delta$

TABLE II. Quadrupole-coupling η and isomer-shift values. The isomer shifts are with respect to the standard ZnTe source. Quadrupole coupling values obtained by the NQR method are given for comparison.

Compound	Mössbauer data e^2qQ (^{127}I) (Mc/sec)	η	I.S. (cm/sec)	NQR data e^2qQ (^{127}I) (Mc/sec)	η	Ref.
I_2	-2156 ± 10	0.16 ± 0.03	0.083 ± 0.001	2156 (4°K)	0.16	7
IBr	-2892 ± 10	0.06 ± 0.02	0.123 ± 0.002	...	...	
ICl	-3131 ± 20	0.06 ± 0.03	0.173 ± 0.005	3046 (77°K)	0.03	9
I_2Cl_6	$+3060\pm10$	0.06 ± 0.02	0.350 ± 0.002	3035 (77°K)	0.08	9
$I_2Cl_4Br_2$ I	$+3040\pm10$	0.06 ± 0.02	0.348 ± 0.002	...	...	
I′	$+2916\pm10$	0.06 ± 0.02	0.282 ± 0.002	...	...	

The derivation of the line position expressions of Table I is sketched in the article, as you would expect. Now one of the applications to chemistry is illustrated from their expression for the isomer shift δ in terms of h_p, the number of p holes in the outer configuration of the I^- ion $5s^2 5p^6$

$$\delta = 0.136h_p - 0.054 \text{ (cm/sec)} \qquad (7)$$

where the isomer shift is with respect to the standard source ZnTe, as before. Compute the value of h_p. From another method making use of the electric field gradient, the eq of e^2qQ, they find that h_p is 1.26. Do you see then the basis of their statement that in the IBr molecule 0.28 e^- is removed from the iodine shell to the covalent bond? They also find solely from the δ data that the iodine-halogen(Br and Cl) bonds in the molecular crystals are chiefly formed by p electrons without sp hybridization.

The very good agreement between the two methods of determining h_p show that the assumption of negligible intermolecular influence on the charge density and distribution at the iodine atom is valid. The crystal structure of IBr is shown in Fig. 5. Why were the compounds mixed with glass powder before transferring them to the liquid nitrogen cryostat?

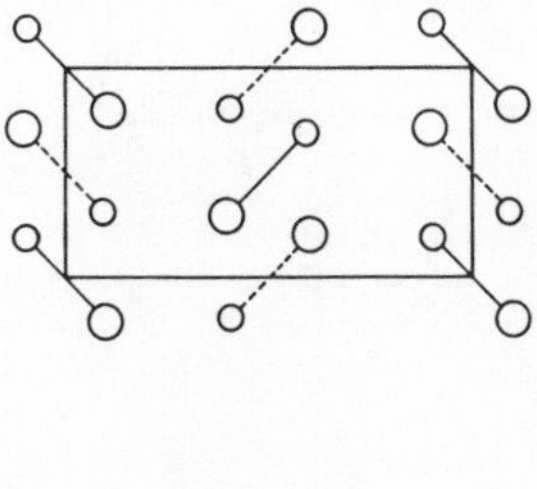

FIG. 5. Arrangement of IBr molecules in orthorhombic unit cell. The dashed lines indicate molecules above or below the plane of the paper.

161. "Interpretation of Spectra" J.W. Perram 49, 4245 (1968).

Since infrared and Raman spectra of hydrogen-bonded substances, in particular water, have been characterized by broad overlapping bands, several authors(ref. 1 and 2) had attempted to extricate the concealed structure using numerical or mechanical methods, by fitting several Gaussian components to the experimental contours.

The author gives a simple counter example to show that within the experimental errors and the resulting uncertainty in the actual component shapes, which are known to be only approximately Gaussian, that a Gaussian component may be further decomposed into two subcomponents.

Taking the single Gaussian function

$$I_0(x) = \exp(-x^2/2.05) \tag{1}$$

shown as Curve I in Fig. 1, consider also the function

$$I_1(x) = i(x) + j(x) \tag{2}$$

where

$$i(x) = a \exp[-(x+d)^2/2\sigma^2] \tag{3}$$

$$j(x) = a \exp[-(x-d)^2/2\sigma^2] \tag{4}$$

with $a=0.515359$, $d=0.244622$, and $\sigma=0.988937$.

$I_1(x)$ is also plotted in Fig. 1 as Curve I; $i(x)$ and $j(x)$ are plotted as Curves II and III respectively. The differences between $I_0(x)$ and $I_1(x)$, said to satisfy the inequality

$$|I_0(x) - I_1(x)| < 0.00038, \tag{5}$$

are also said to be somewhat smaller than one-third of the thickness of the lines drawn in Fig. 1.

If true, would these differences seem to lie within the errors of any conceivable experiment?

Show that $I_1(x)$ is an even function of x. Evaluate the difference shown in equation (5) for four or five evenly spaced values of x between 0 and 2, including the end points, to test the inequality shown. Given that then scale the line thickness as best you can to compare with the claim stated above. Note that the phrase "line width" has a completely different meaning!

Can a single counter example disprove a claim? If not, how many does it require? an infinite number? Invert the question;

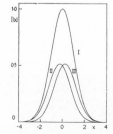

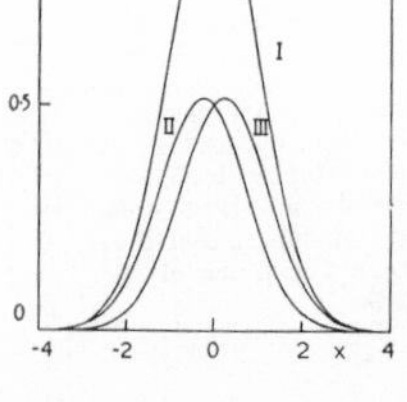

FIG. 1. Curves II and II are a Gaussian decomposition of Curve I, which is itself Gaussian.

can a single example prove a general claim? If the answers to these questions are not obvious, think a bit about the distinction between necessary and sufficient conditions.

"Thus the numerical decomposition of a structureless contour is not unique, so that the results of such a process provide no evidence as to the number or nature of the component contours present."

Notice the author's departmental affiliation; are you surprised?

162. "Investigation of the Magnetic Moments of S_2, Se_2, Te_2, Se_6, and Se_5 by the Stern-Gerlach Magnetic Deflection Method" D.J. Meschi and A.W. Searcy 51, 5134 (1969).

The elements of Group VI, oxygen through tellurium, all have triplet atomic ground states with $^3P_2$ terms. These atoms combine to form diatomic molecules which may also have low-lying multiplet electronic states that are therefore paramagnetic (why?). O_2 and S_2 for example have triplet ground states with $^3\Sigma_g^-$ terms. However for Se_2 ref. 3 had shown that a different coupling scheme for the angular momentum vectors is applicable, and what would otherwise be a $^3\Sigma_g^-$ ground state splits into two subsystems designated as 0_g^+ and 1_g, with the former being both nonparamagnetic and lower in energy by an amount equal to 2λ with λ the splitting constant of ref. 4. Given the five experimental points lying on a straight line in the semilog plot of λ(in cm^{-1}) vs molecular weight in Fig. 1, is it a reasonable

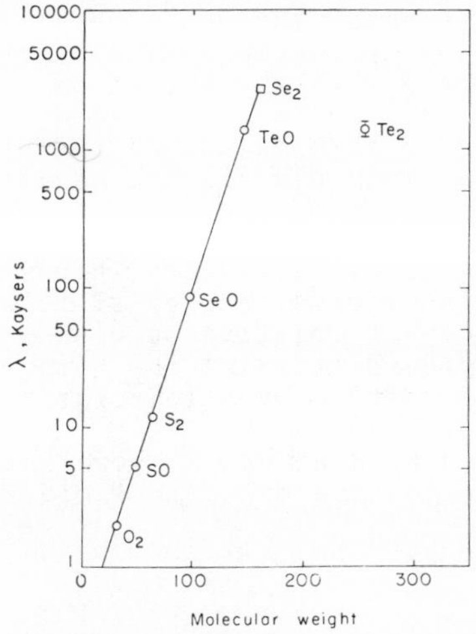

FIG. 1. Plot of splitting constant λ versus molecular weight for Group VI diatomic molecules, λ for Se_2 from extrapolation.

extrapolation to obtain the splitting constant of Se_2? The experimental value for Te_2 may make the extrapolated value questionable, yet it would seem to be considerably in excess of 100 K in accord with the estimate of ref. 3 and 5. Why would a better knowledge of λ be of interest to the thermodynamicist as well as to the spectroscopist? The object of the present set of experiments was to evaluate and compare the splitting constants for S_2, Se_2 and Te_2 for comparison with and extension of the spectroscopic values of Fig. 1.

The method used is that of the title, one of the very famous and early(1921 and 1922) demonstrations of the existence of quantized angular momentum components. The apparatus of Fig. 2 was housed in a vacuum tank. The quadrupole mass filter

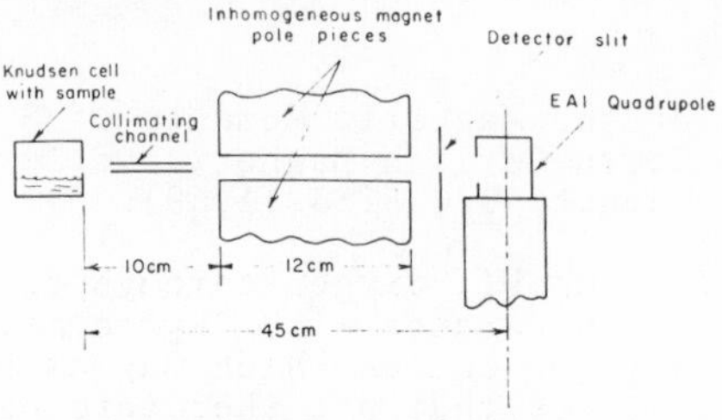

FIG. 2. Diagram of apparatus.

is shown on the right; what is the obvious advantage of a mass spectrometer type of detector? As outlined in the article, the technique involved several steps: i) with the detector slit fully open, the vapor pressure of the desired species was raised (how?) until an easily detectable signal was obtained from the mass filter; ii) the detector slit was closed to 0.62 mm; iii) with the magnet still off, the profile of the beam was obtained by moving the detector slit perpendicular to the nominal beam axis in very small increments, through a total distance of about 0.5 cm, and taking intensity readings as a function of position; iv) step iii) was repeated with the magnet on.

In Fig. 3 the beam profile of S_2 is seen to differ greatly when the magnet is off and then on; would this indicate that a large fraction of the S_2 molecules have a magnetic dipole moment and why? For which value of the magnetic field(on or off) is the beam more diffuse spatially and why? Show that the accepted spectroscopic value of λ, 5.276 K, gives 15.2°K as the temperature for which the thermal energy kT is equal to $2hc\lambda$; what does this say about the relative occupancy of the three components of the $^{3}\Sigma\bar{g}$ ground state at the temperature of this experiment? Would this last be oven temperature, beam temperature, or both? Discuss.

Fig. 4 for Se_2 generated at 980°K shows that the beam was not measureably affected by the magnetic field. From the close

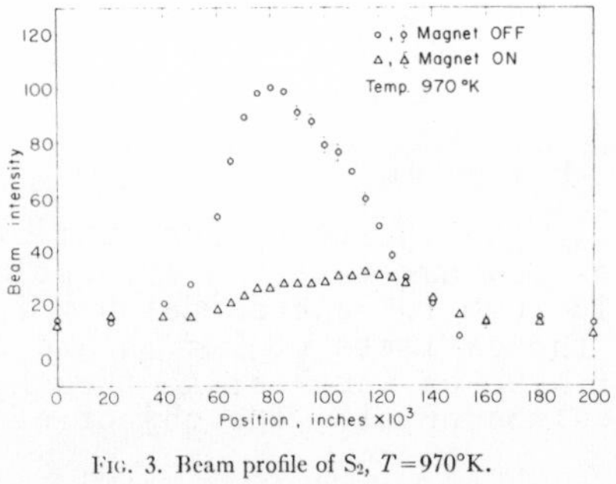

FIG. 3. Beam profile of S_2, $T = 970°K$.

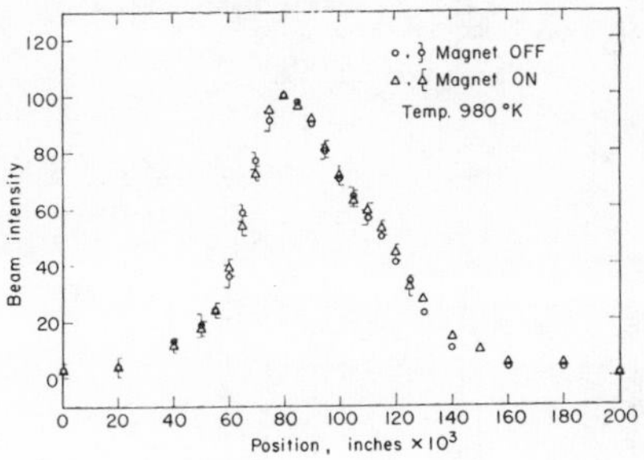

FIG. 4. Beam profile of Se_2 from CdSe, $T=980°K$.

coincidence of the beam profiles the authors estimate that less than one-tenth of the Se_2 molecules could have a magnetic moment of the order of a Bohr magneton. By assuming a Boltzmann distribution show that there is no singlet paramagnetic electronic state within 1500 K of the ground state. What does this computation imply about λ? If the two components of the 1_g state are degenerate or nearly so as expected from spectroscopy(ref. 3), then show that they must be at least 2000 K above the 0_g^+ ground state, and λ would exceed 1000 K. How does this compare with the extrapolation of Fig. 1? The argument was carried further to show support for the high value of 78.63 kcal/mole for the dissociation limit of Se_2; this from among the three possible spectroscopic choices of ref. 3. Is this another illustration of a method with much lower inherent "resolution" being able to assist in making a choice between two or more spectroscopic dissociation energies, each determined with very high accuracy? What was the "low resolution" method of exercise no. 4? What were the molecules involved there?

The case of Te_2 was quite analogous to that of Se_2.

The magnet used was capable of producing a field of 16 kG; should that be enough to "drive" the Stern-Gerlach experiment? Read about it in a textbook, if you are not already familiar with the theory, to see that an inhomogeneous field is required (why?), as shown in Fig. 2. The gradient obtainable in this set of experiments was 55 kG/cm.

163. "Inelastic Neutron Scattering Spectra and Raman Spectra of $CsHCl_2$ and $CsDCl_2$" G.C. Stirling, C.J. Ludman, and T.C. Waddington 52, 2730 (1970).

What are the selection rules for neutron(collisional) spectroscopy? Are there any? We will see that this fact is used to advantage in this paper to establish the geometry of the hydrogen dichloride ion(HCl_2^-) as something other than linear symmetric, $D_{\infty h}$; together with the Raman spectra this result leads to a bent structure, either C_{2v} or C_s.

What was the source of the neutrons? How was their energy reduced? In these experiments neutrons with initial wavelength of 5.4 Å were used; show that this corresponds to an energy of about 3 meV(m for mega- or milli-?). What energy in eV does the thermal energy kT correspond to at room temperature? What is the reason for the use of the adjective cold in the spectrum of Fig. 3? In cold-neutron scattering an energy-gain spectrum is then

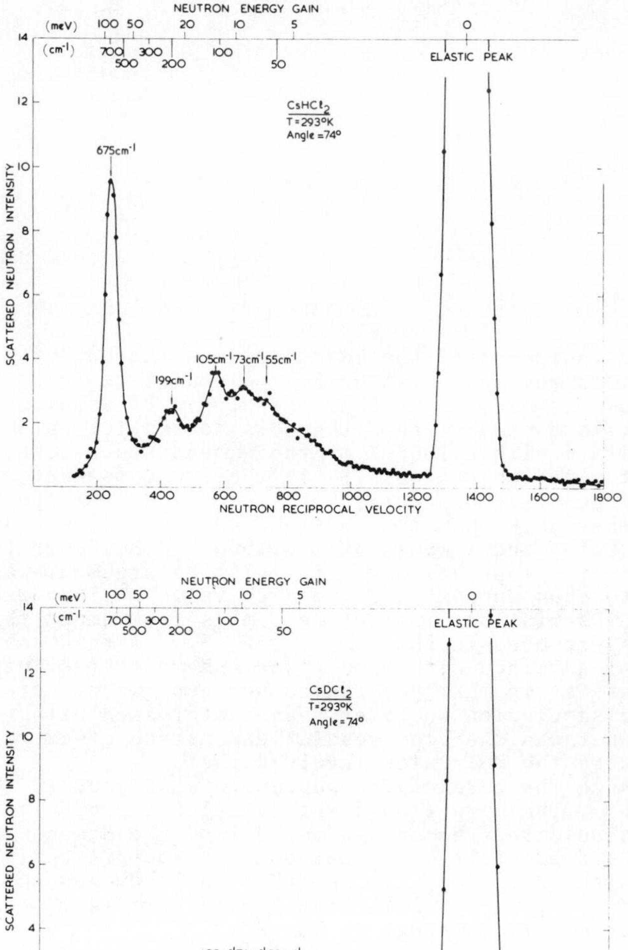

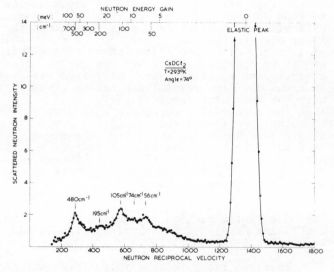

FIG. 3. The cold neutron spectra of $CsHCl_2$ and $CsDCl_2$ at the same sample thickness (8-mm tubes) and neutron flux. Velocity is in units of microsecond per meter.

observed; from where could the neutrons gain energy from the sample? From the excited vibrational levels, of course. What fractional population are they assuming as a limit by their statement that the available spectral range is limited in practice to the region below about 800 cm^{-1} for room-temperature specimens? Higher energy transfers can be observed in neutron energy loss experiments, where vibrational modes of the sample are excited in the scattering process. Discuss the correspondence to anti-Stokes and Stokes scattering in Raman spectroscopy.

In Fig. 3 the reciprocal velocity(time-of-flight?) scale is a measure of neutron energy; is the scale polarity correct?

The energy-gain scale gives the energy transferred to the neutron in the scattering process.

The neutron inelastic scattering method can be used to special advantage when studying hydrogen-containing compounds; this is because the neutron inelastic scattering cross section of hydrogen is an order of magnitude greater than for most other nuclei, so that the observed spectrum reflects primarily the motion of the hydrogen atoms. Explain carefully the reasoning used in reaching this last conclusion. This is a distinct advantage in making assignments as Fig. 3 shows for the 675 cm^{-1} bending vibration which shifts to 480 cm^{-1} upon deuteration. Is this ratio about what you would expect for a light atom bound to two massive ones?

By using the spectra of Fig. 3, the symmetry species and activities plus the possible proton motions of Table I, and the assignments of Table II construct an argument leading to the conclusion stated in the beginning of the exercise that the ion is bent.

TABLE I. Vibrational symmetry species, with infrared and Raman activities for possible Cl–H–Cl⁻ structures, together with whether the proton moves in the particular vibration.

Symmetry of the ion	$D_{\infty h}$	$C_{\infty v}$	C_{2v}	C_s
ν_1, symmetric stretch	Σ_g^+ (Raman, proton does not move)	Σ^+ (Raman and infrared, proton moves)	A_1 (Raman and infrared proton moves)	A' (Raman and infrared proton moves)
ν_2, bending mode	Π_u (Infrared, proton moves)	Π (Infrared, proton moves)	A_1 (Raman and infrared, proton moves)	A' (Raman and infrared, proton moves)
ν_3, asymmetric stretch	Σ_u^+ (Infrared, proton moves)	Σ^+ (Raman and infrared, proton moves)	B_2 (Raman and infrared, proton moves)	A' (Raman and infrared, proton moves)
$2\nu_2$	$\Sigma_g^+ + \Delta_g$ (Raman, proton moves)	$\Sigma^+ + \Delta$ (Infrared, Raman, proton moves)	A_1 (Raman and infrared, proton moves)	A' (Raman and infrared, proton moves)

TABLE II. Proposed assignments of vibrational spectra of $CsHCl_2$ and $CsDCl_2$.

	$CsHCl_2$				$CsDCl_2$			
			Neutron scattering				Neutron scattering	
Assignment	Infrared	Raman	Energy gain	Energy loss	Infrared	Raman	Energy gain	Energy loss
$L_1(T)$			55				56	
$L_2(T)$			73				74	
$L_3(T)$			105				105	
$\nu_1(A_1)$			199				195	
$\nu_2(A_1)(0^- \to 1^+)$	602.0 (vs)	612			429.7 (vs)	450		
$\nu_2(A_1)(0^+ \to 1^-)$	660.5 (s)		652	652	468.0 (s)		480	438
$2\nu_2(A_1)(0^- \to 2^+)$	1168.5 (vs)	1165		1198	851.5 (vs)	850		
	1208 (m)							
$2\nu_2(A_1)(0^+ \to 2^-)$	1294 (m)				925 (s)			
	1340 (w)							
$\nu_3(B_2)$	1670 (vs)	1630			1320 (sh)			
$\nu_3 + L_3(T)$	1790 (vvs)				1370 (vvs)			
$\nu_3 + \nu_1$	1875 (vvs)				1420 (sh)			

$L_1(T)$, $L_2(T)$, $L_3(T)$ are the three translational modes of the bichloride ion in the lattice

164. "High-Resolution He I and He II Photoelectron Spectra of Xenon Difluoride" C.R. Brundle, M.B. Robin, and G.R. Jones 52, 3383 (1970).

What two transitions in the helium atom and ion give emission photons at 21.22 and 40.82 eV?

Discuss Koopmans' theorem, relating one-electron SCF energies to the negative of the experimental ionization potential from that particular atomic or molecular orbital. What exactly is assumed about the wavefunctions as far as their similarity or equality before and after ionization is concerned? Is the theorem of Koopmans' then exact? From ref. 4 and 5 it had been shown that SCF calculations with Gaussian-type orbitals(cf exercise no. 25) could predict photoelectric ionization energies with some success if all Koopmans' theorem ionization potentials were reduced uniformly by the factor 0.92. The empirically adjusted theoretical values are listed in Table I, together with the adiabatic and vertical(difference?) I.P.'s observed experimentally. Which of these two experimental I.P.'s should correspond more closely with the corrected calculated value and why? One sees that the transitions in the photoelectron spectrum may thereby be assigned.

TABLE I. Observed and calculated ionization potentials (in electron volts) of XeF_2.

Adiabatic obs	Vertical obs	0.92 K.T. calc
12.35±0.01[a]	12.42±0.01 }	12.51 ($5\pi_u$)
12.89±0.01	12.89±0.01 }	
≈13.5	13.65±0.05	11.79 ($10\sigma_g$)
14.00±0.05	14.35±0.05	14.71 ($3\pi_g$)
15.25±0.05	15.60±0.05 }	15.92 ($4\pi_u$)
	16.00±0.05 }	
16.80±0.05	17.35±0.05	16.93 ($6\sigma_u$)
	≈22.5	25.24 ($9\sigma_g$)
		37.10($5\sigma_u$)
		37.20 ($8\sigma_g$)

[a] There is an indication of a weak shoulder centered at 12.30 eV which may correspond to the true adiabatic I.P. for this band. However, it seems more likely to us that this lowest band is a hot one, arising from the appreciable excitation of the low-frequency bending mode in the ground state of the neutral molecule.

The first two I.P.'s of XeF_2 are shown in the spectrum of Fig. 1. One of the bands(still another usage of this term?) is

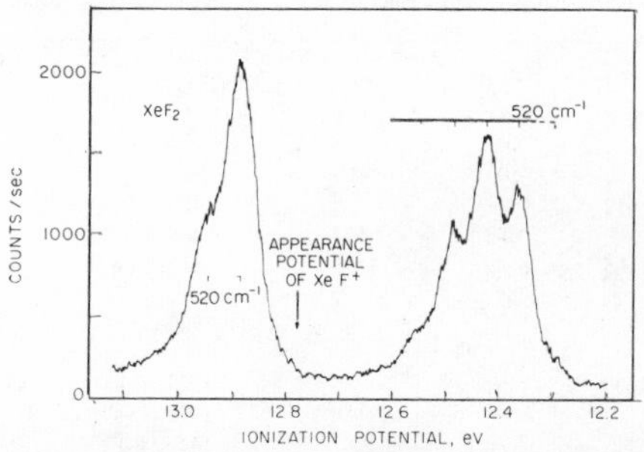

FIG. 1. Details of the first two bands in the photoelectron spectrum of XeF_2.

assigned to the transition to the $^2\Pi_{3/2}$ and the other to the $^2\Pi_{1/2}$ state as shown in Table II. It is believed that the

TABLE II. Term energies in Xe and XeF_2.

Xe[a]		XeF_2	
Upper state	Term (cm^{-1})	Upper state	Term (cm^{-1})
$5p(^2P_{3/2})6s$	30 400	$5\pi_u(^2\Pi_{3/2})6s$	30 865
$5p(^2P_{1/2})6s$	31 433	$5\pi_u(^2\Pi_{1/2})6s$	30 080
$5p(^2P_{3/2})6p$	19 322		
$5p(^2P_{1/2})6p$	19 317		
$5p(^2P_{3/2})5d$	16 628	$5\pi_u(^2\Pi_{3/2})5d$	17 860
$5p(^2P_{1/2})5d$	16 567	$5\pi_u(^2\Pi_{1/2})5d$	16 600
$5p(^2P_{3/2})7s$	12 551		
$5p(^2P_{1/2})7s$	12 590		

[a] C. Moore, Natl. Bur. Std. (U.S.), Circ. **467**, Vol. 3, 113 (1958).

assignment of the two bands to the correct transitions from the data of Table II is a nontrivial exercise. It may be best to first understand the atomic case, making good use of the reference of footnote a. There should be a superscript ^{5} right after the 5p under the upper state label, although the shorthand used would be obvious to anyone but the initiate. These states are excited states of the atom with the promoted electron shown being eventually lost to form the ion whose term symbol is given in parentheses. Notice that there are two different limiting states of the resultant ion, $^2P_{1/2,\,3/2}$. From what states are the term values shown measured? To be sure the scheme is understood, compute the first two and last two terms for the case of the atom. Now move to the molecular case to make the assignment of the bands of Fig. 1. For the atomic case, did you find that the present authors took the average-of-configuration energy in each case? For both the atomic and molecular ions, does the regular(or inverted) multiplet order hold for the energies of $J=1/2,\ 3/2$ and $\Omega=1/2,\ 3/2$? Discuss. What is the name given to this type of splitting, in the molecule as well as the ion? Are we surprised at its magnitude with a heavy atom like xenon?

Other features of interest remain to be examined in connection with Fig. 1. Why did the authors presume that the orbital which yielded the ionized electron from the neutral molecule was nonbonding? This would lead us to expect that the observed vibrational spacing would approximate one of the neutral molecule fundamentals; for $\nu''_{1,2,3}=513$, 213, and 557 cm^{-1} does this appear to be the case? Is what we see actually the excitation of vibration in the ion? Does the quite reasonable agreement in the frequencies then justify the assumption that a nonbonding electron was ionized? What molecular orbitals, in terms of bonding, antibonding, or nonbonding lie highest in the usual molecule when filled in the neutral's ground state? Is XeF_2 what you might call the usual molecule?

You might find the very first sentence of the article somewhat amusing; it is surely historically accurate. Might there be a touch of irony in the last phrase?

165. "X-Ray Photoelectron Spectroscopy of Carbon Monoxide" T. D. Thomas 53, 1744 (1970).

The exciting radiation was that given in the caption of Fig. 1. The binding energies of electrons in each molecular orbital were determined and shown in Table I. Was there ample

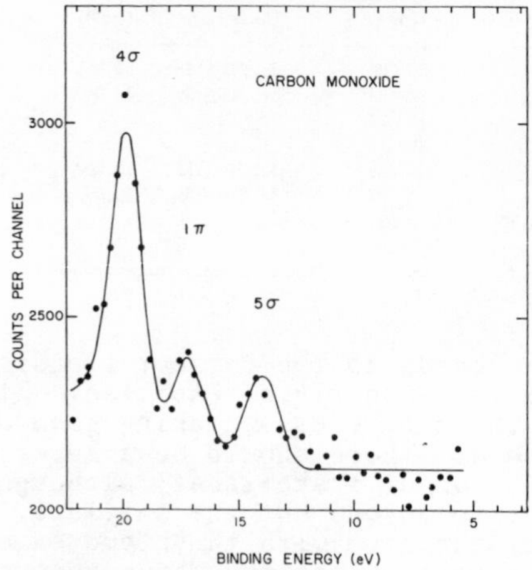

FIG. 1. Spectrum of outer electrons of carbon monoxide. The solid line represents a least-squares fit to the data. Exciting radiation is magnesium K_α x rays.

TABLE I. Electron binding energies in carbon monoxide and oxygen determined by x-ray photoelectron spectroscopy.

	Orbital	Binding energy (eV)
Carbon monoxide	5σ	14.0
	1π	17.2
	4σ	19.8
	3σ	38.9
(carbon $1s$)	2σ	296.2[a]
(oxygen $1s$)	1σ	542.3
Oxygen	$1s$	543.5[b]

[a] 5.4 eV greater than that of methane.
[b] This is a weighted average of the $^4\Sigma$ and $^2\Sigma$ states. The splitting between these is 1.13 eV, with the $^4\Sigma$ state having the lower energy.

energy in the magnesium K_α (meaning exactly what?) x rays? Now from ref. 19 and 31 check the factor of 0.92 of the previous exercise, where it was used to bring the negative of the one-electron orbital energies into closer accord with the photoelectron ionization potentials. From these SCF calculations you should subject the proposed correction to a critical test for the carbon and oxygen K shells.

As shown in Table I the carbon 1s binding energy in CO is 5.4 eV greater than that in CH_4. Is this shift in the expected direction? Why or why not? Why doesn't the comparison with methane allow us to assign a charge to the carbon of CO with any great confidence? Why wasn't a comparison made with measurements on either atomic carbon or C_2?

The oxygen 1s binding energy in CO is less than that in molecular oxygen by 1.2 eV, but greater than that in water by 2.9 eV and greater than that in a variety of compounds of oxygen with tetravalent carbon by 1.5-4 eV. Does this then indicate that the oxygen in CO is negatively charge wrt oxygen in O_2? not so negatively charged as oxygen in some other compounds?

The author discusses the possibility of obtaining population analyses(cf exercise no. 23) resulting from molecular orbital calculations by improvements in the measurement of photoelectron cross sections and the underlying analysis.

In this chapter we have seen several relatively new applications of physical techniques to chemistry. Perhaps you yourself will find new applications or else use those already known to determine properties of new materials or systems.

166. "Departmental Distribution Analysis" Anon circa 1970.

Quickly run through a recent issue and list departmental affiliations of the authors. This should give you an idea of where chemical physics research is done. There may be some areas that appear to defy classification; perhaps someday they will move out of the hybrid class into recognized disciplines just as chemical physics itself has, and then again perhaps not.

If you have not done so already, read the first paragraph of the inside front cover of any issue, even back to Volume 1.

AUTHOR INDEX

PRINTED AND BOUND IN GREAT BRITAIN
BY CHAPEL RIVER PRESS, ANDOVER, HANTS